Light microscopy in biology

a practical approach

D1215652

TITLES PUBLISHED IN
THE
PRACTICAL APPROACH
SERIES

Series editors:
Dr D Rickwood
Department of Biology, University of Essex
Wivenhoe Park, Colchester, Essex CO4 3SQ, UK
Dr B D Hames
Department of Biochemistry, University of Leeds
Leeds LS2 9JT, UK

Affinity chromatography

Animal cell culture

Antibodies

Biochemical toxicology

Biological membranes

Carbohydrate analysis

Centrifugation (2nd Edition)

DNA cloning

Drosophila

Electron microscopy
in molecular biology

Gel electrophoresis of nucleic acids

Gel electrophoresis of proteins

Genome analysis

HPLC of small molecules

HPLC of macromolecules

Human cytogenetics

Human genetic diseases

Immobilised cells and enzymes

Iodinated density gradient media

Light microscopy in biology

Lymphocytes

Lymphokines and interferons

Mammalian development

Microcomputers in biology

Microcomputers in physiology

Mitochondria

Mutagenicity testing

Neurochemistry

Nucleic acid and
protein sequence analysis

Nucleic acid hybridisation

Oligonucleotide synthesis

Photosynthesis:
energy transduction

Plant cell culture

Plant molecular biology

Plasmids

Prostaglandins
and related substances

Spectrophotometry
and spectrofluorimetry

Steroid hormones

Teratocarcinomas
and embryonic stem cells

Transcription and translation

Virology

Yeast

Light microscopy in biology

a practical approach

Edited by

Alan J Lacey

Department of Biology and Biochemistry, Brunel,
The University of West London,
Uxbridge, Middlesex UB8 3PH, UK

IRL PRESS
at
OXFORD UNIVERSITY PRESS
Oxford New York Tokyo

IRL Press
Eynsham
Oxford
England

British Library Cataloguing in Publication Data

Light microscopy in biology. (Practical approach series)
  1. Biology. Optical microscopy
  I. Lacey, A. J. II. Series
  578'. 4

Library of Congress Cataloging in Publication Data

Light microscopy in biology : a practical approach / edited by Alan J. Lacey.
  Includes bibliographies and index.
  1. Microscope and microscopy—Technique.   I. Lacey, Alan J.
  QH207.L49 1988   578'.4—dc19   88-23505

ISBN 0 19 963036 4 (hardbound)
ISBN 0 19 963037 2 (softbound)

Previously announced as:

ISBN 1 85221 083 4 (hardbound)
ISBN 1 85221 092 3 (softbound)

Typeset by Infotype and printed by Information Printing Ltd, Oxford, England.

Preface

The aim of this book is to give guidance at the practical level to people who wish to make more extensive use of their microscopes and also to offer encouragement to those attempting, perhaps for the first time, to visualize specimens and record their observations at the microscopic level.

One purpose of a microscope is to enable the observer to see or to record the fine detail of a specimen. It is also possible, however, to use the microscope to make measurements of the specimen. Another more recent development has been a technique for indicating the interaction of light and matter with the intention of determining such parameters as the refractive index, mass and chemistry of small objects. By repeated recordings or measurements over time, movements or other changes can be detected and studied.

To render fine detail visible requires the microscopist to achieve magnification, resolution and contrast. It is quite obvious that what the observer actually sees is a magnified replica of the specimen. In the art world, replicas are viewed with caution and it is hoped that sufficient detail is given here to provide an understanding of microscopic techniques and thereby encourage a critical, cautious approach in the microscopist towards his/her observations.

Light microscopes have been in use for more than 200 years but techniques change as demands alter. Technological advances in both hardware and software also stimulate new investigations; automative image recording processing and analyses are perhaps the latest of these developments.

Throughout microscopy the most significant step or skill is the production of a good image, that is an image in which there is optimum contrast and resolution as well as magnification. Any image analysis attempted depends for its results on the quality of the first image. Hence, image analysis, whether it is by visual methods or by the use of automative television systems, demands a good microscope. The cost of a versatile microscope is considerable (sometimes comparable to that of an electron microscope) but it is the tool by which the primary image is constructed and is therefore a valuable investment. The electron and light microscopes fulfil different functions but the light microscope has the advantage that it can make available information about living material in real time and as such is applicable to a wide range of studies in life science subjects.

Chapter 1 explains the need for, and how to achieve, even illumination of the specimen and Chapter 2 gives the general principles portrayed by some simple experimental procedures which determine the limits of what is achievable using light microscopy. Subsequent chapters cover the recording of images produced by the widely used illumination techniques. The use of stains in histochemistry and particularly the specificity achieved by immunofluorescence staining techniques are given in Chapters 4 and 6. A chapter on chromosome studies, Chapter 9, illustrates the strength of the combination of microscope techniques. Chapter 8 gives an account of the equipment required for, and offers instructions on, the production of improved images by the use of video microscopy. This chapter challenges, at least in part, some of the concepts on contrast, resolution and appropriate magnification that are set out in the first chapters but the general maxim that a major requirement in all microscopy is the preparation of as good a primary image as possible is greatly reinforced.

I personally would like to thank Professor A.S.G.Curtis for his contribution on interference reflection microscopy in Chapter 2. Dr Peter Eagles of King's College, London, has given me the opportunity of seeing video microscopy in action and help with the editing of the chapter on that subject. I am also grateful to many of the present authors who have worked with me in the past in teaching microscopy and those who more recently have showed me the applications of microscopical techniques to various areas of biology. I owe a great deal to the publishers for their patience and their assistance with the art work as well as with the preparation of the manuscripts.

Finally, I would like to acknowledge the help various members of the Biology Department at Brunel University have given me in making up for my deficiencies in word processing skills.

Alan J.Lacey

Contributors

S.J.Bradbury
Department of Human Anatomy, University of Oxford, South Parks Road, Oxford OX1 3QX, UK

P.J.Evennett
Department of Pure and Applied Zoology, University of Leeds, Leeds LS2 9JT, UK

R.W.Horobin
Department of Biomedical Science, University of Sheffield, Sheffield S10 2TN, UK

S.F.Imrie
Institute of Cancer Research, Royal Cancer Hospital, The Haddow Laboratories, Clifton Avenue, Sutton, Surrey SM2 5PX, UK

A.J.Lacey
Department of Biology and Biochemistry, Brunel, The University of West London, Uxbridge, Middlesex UB8 3PH, UK

W.Maile
Institut für Zoologie, Technische Universität München, Lichtenbergstrasse 4, D-8046 Garching, FRG

M.G.Ormerod
Institute of Cancer Research, Royal Cancer Hospital, The Haddow Laboratories, Clifton Avenue, Sutton, Surrey SM2 5PX, UK

J.S.Ploem
Department of Histochemistry and Cytochemistry, Sylvius Laboratories, University of Leiden, Leiden, The Netherlands

A.T.Sumner
MRC Human Genetics Unit, Western General Hospital, Crewe Road, Edinburgh EH4 2XU, UK

D.G.Weiss
Institut für Zoologie, Technische Universität München, Lichtenbergstrasse 4, D-8046 Garching, FRG

R.A.Wick
Photonic Microscopy Inc., 2625 Butterfield Road, 204 South Oak Brooks, IL 60521, USA

Contents

Abbreviations

2×SSC	0.3 M NaCl + 0.03 M tri-sodium citrate
ABC	avidin−biotinylated peroxidase−complex
AFS	acriflavin−Feulgen−SITS
AgNOR	silver staining for NORs
APAAP	alkaline phosphatase−anti-alkaline phosphatase
ASA	American Standards Association
ASG	acetic acid−saline−Giemsa
AVEC-DIC	Allen video-enhanced contrast using DIC
AVEC-POL	AVEC using polarized light techniques
BCECF	2,7-bis(carboxyethyl)-5(6)carboxyfluorescein
b.f.p.	back focal plane (usually of objective lens)
BG	blue glass filter
BP	band-pass
BrdU	bromodeoxyuridine
BSA	bovine serum albumin
B/W	black-and-white (photography and television)
C	contrast
CBS	chromatic beam splitter
CC	colour compensating
CCD	charge-coupled device
CCIR	Consultative Committee of International Recording
CEA	carcinoembryonic antigen
CI	colour index
(CMS)2S	spermidine bis-acridine
CSF	cerebrospinal fluid
DA	distamycin A
DAB	diaminobenzidine
DABCO	1,4-diazobicyclo-2,2,2-octane
DAPI	4′,6′-diamidino-2-phenylindole
dC	deoxycytidine
d.f.	depth of field
DIC	differential interference contrast
DM	dichroic mirror
DMA	direct memory access
DMF	dimethyl formamide
DNP	dinitrophenol
DTT	dithiothreitol
dU	deoxyuridine
ECD	equivalent circle diameter
EDTA	ethylenediaminetetraacetic acid
EIA	(EIAJ) Electronic Industries of America and Japan
EM	electron microscope
E-to-E	electronic-to-electronic
FAC	fluorescence analogue chemistry
FDA	fluorescein diacetate
FdU	fluorodeoxyuridine
FITC	fluorescein isothiocyanate

FPG	fluorescent plus Giemsa (e.g. Hoechst 33258 + Giemsa)
f.p.s.	frames per sec
FRAP	fluorescence recovery after photobleaching
f-stop	relative aperture of the diaphragm (camera)
FVN	field of view number
GAG	glycosaminoglycan
HbA	haemoglobin-A
HbS	haemoglobin-S
HBO	high pressure mercury burner
HRB	high-resolution banding
HRP	horseradish peroxidase
Ig	immunoglobulin
ISIT	intensified silicon intensifier target
ISO	International Standards Organisation
IVEC	Inoué video-enhanced contrast
λ	wavelength
LED	light emitting diode
LEYTAS	Leydon Television Analysis System
LP	long wavelength pass filters
LSM	Optical Laser-Scanning Microscopy
MCP	microchannel plate
MDCK cell	Madin Darby Canine Kidney cell
MP	total magnifying power
n	refractive index
NA	numerical aperture (usually of objective lens)
NOR	nucleolar organizer (cytology)
NTSC	National Television System Committee (USA)
o.p.d.	optical path difference
PAL	Phase Alteration Line (FRG)
PAP	peroxidase−anti-peroxidase
PBS	phosphate-buffered saline
PHA	phytohaemagglutinin
PVA	polyvinyl alcohol
Q	quantum efficiency
RGB	red, green and blue
RSE	relative standard error
SDPD	N-succinimidyl 3-(2-pyridyldithio)propionate
SECAM	Séquentielle à Mémoire (France)
SEM	scanning electron microscope
SIT	silicon intensifier target
SITS	4-acetamido-4′-isothiocyanostilbene-2,2′-disulphonic acid
SLR	single-lens reflex (camera)
S/N	signal-to-noise ratio
SP	short wavelength pass filter
TBC	time-based corrector
TBS	Tris−HCl-buffered saline
TEM	transmission electron microscopy
TRITC	tetramethyl rhodamine isothiocyanate
TRSP	time-resolved spatial photometry
TSRLM	tandem scanning reflected light microscopy

TTL	through-the-lens
VEC	video-enhanced contrast
VIM	video-intensified microscopy
VTR	video tape recorder

CHAPTER 1

The principles and aims of light microscopy

A.J.LACEY

1. INTRODUCTION

Köhler illumination is a basic requirement of all modern light microscopy and the first part of this chapter covers the procedure by which it is established. The next subject to be covered is the nature of the interactions between light and matter. Some of these interactions are illustrated with practical exercises which should enable the reader to conclude that what is actually seen in the microscope is really a magnified replica of the specimen and not the specimen itself. The replication process is achieved by bringing about the interference of several trains of light waves issuing ultimately from the back of the objective lens of the microscope.

At some stage you are going to ask what is the potential of the microscope you, yourself, have available. You may like to start by following *Table 1* but most of its contents will be more fully appreciated after you have read, and carried out, the rest of this chapter's contents.

1.1 Köhler illumination

All modern microscopy, which seeks to understand the relationship of the image obtained in the microscope to the physico-chemical nature of the specimen, begins with Köhler illumination.

The principle and purpose of the system is firstly to obtain an evenly lit field of view against which the detail of the specimen will be plainly recognizable, and secondly to light the specimen with as wide a cone of radiation as is possible in order to achieve maximum resolution of fine detail.

1.1.1 *Principles of the method*

(i)   A lens in front of the source of light places an image of that light source at a position which is *not* in the plane containing the specimen.

(ii)   A second lens (that of the condenser) puts an image of the surface of the first lens onto the specimen to be examined and does so with as short a focal length as possible to achieve as wide a cone of light as possible illuminating the specimen.

1.1.2 *Preliminary checks of the equipment*

Microscope manufacturers issue detailed instructions for each type of microscope purchased but often in laboratories the manuals have become separated from the

1

Table 1. Conjugate planes in Köhler illumination.

Series A
  Lamp filament (source)
  Condenser iris (aperture iris)
  Back focal plane of objective
  Ramsden disc
Series B
  Field iris
  Specimen
  Primary image
  Retina of eye

instruments or familiarity has dispensed with them! (Refer to *Figure 1* for the names and symbols given in the procedure.)

(i)   Check that your microscope has a lamp collector lens, that is a lens between the lamp filament and the front of the condenser. It may be built into the base of the microscope but will be there to project an image of the light source onto the condenser iris. When the latter is closed, and with the assistance of a mirror, it will be possible to see whether the lamp lens actually does this. In many cases however, the lamp filament is not seen because a piece of ground glass breaks up its image. In such cases the illumination seen on the condenser iris should be of an area large enough to cover the iris adequately. There may be no adjustments possible in several makes of microscope but the principle remains the same. There may be centering screws for the lamp filament with respect to the lamp collector lens. These can be used now or at stage (vii) in the method (Section 1.2). There will be an iris associated with the lamp collector lens and this is known as the *illuminated field iris* (Section 1.2).

(ii)   Check the presence of the microscope condenser and its diaphragm. In a well-built system the diaphragm will be in the front focal plane of the condenser lens. This iris is known as the *aperture iris*.

    The condenser may have supplementary adjustments. A flip-top condenser will have the ability to have its top lens either in or out of the light path—note the control lever and its setting. An alternative make might have a lens capable of being swung into and out of the light path underneath the main condenser. The two alternatives have opposite effects. When the flip-top lens is in place the condenser has a short focal length and, in the second case, the extra lens underneath has the effect of giving the condenser a longer focal length. These settings are related to the size of the image of the field iris in the plane of the specimen (see Section 1.2). Note the possible centering facilities of the condenser and even those of its iris.

(iii)   The eyepiece can be taken out and replaced by a phase telescope which itself can be focused onto the back of the objective or, more accurately, onto the back focal plane of the objective. Some microscopes have an eyepiece changer which rotates in a supplementary lens to turn the eyepiece itself into a phase telescope. There will be a focusing capacity on this accessory also. Check the situation in your instrument.

1.2 **Procedure for setting up Köhler illumination**

(i)   Switch on the light and centre it, following the manufacturer's instructions.

(ii)   Using a fairly contrasty specimen, that is one with some colour in it, focus on the specimen using a ×10 objective.

(iii)   Looking through the eyepiece partially close the field iris until it is seen in the plane of the specimen. It may or may not be clearly focused and may or may not be in the centre of the field of view.

(iv)   Focus the image of the field iris sharply by focusing the microscope condenser nearer or further away from the specimen. Note in passing that the focused edge will have either a blue or yellow margin. If it has both colours unevenly around its margin this may be due to poor alignment, and the alignment instructions should be reworked.

(v)   Centre the image of the field iris, in most cases by using the centering screws of the condenser. This can be checked finally by opening the field iris until its margin is just inside the field of view.

(vi)   Open the field iris until its boundary is just outside the field of view. It is at this stage that in really low-power objectives the image of the field iris is insufficiently large to fill the field of view, in which case the supplementary lens is required or the top lens should be taken out of the optic axis by the flip-top lever. The condenser may then have to be refocused as described above.

(vii)   Take out the eyepiece, insert the phase telescope (or swing in the eyepiece changer lens) and focus it onto the back focal plane (b.f.p.) of the objective. This can be recognized by seeing the *condenser* iris in view. Test that this is the case by opening and closing the condenser iris. There may be capacity for centering it in respect of the b.f.p. If so then carry this out (be careful not to meddle with the centering screws of the condenser which you have already set).

Adjust the image of the *aperture iris* (condenser iris) so that it is about 70% of the b.f.p. of the objective. You will notice that the glare disappears from the side of the body tube of the microscope as you make this adjustment.

You may also notice an image of the filament in the centre of this bright illuminating aperture. This is correct. If there is a facility for centering the aperture iris, centre it with respect to the objective and then, by using the lamp filament centration screws, centre its image in turn. The presence of ground glass will not allow this image to be seen clearly so the maximum brightness should be centered.

(viii)   Replace the eyepiece and assume this to be the best situation for this particular combination of objective and condenser settings. Excessive brightness in the image *should not* be controlled by reducing the size of the illuminating aperture. Refer to stage (x) for details of controlling the brightness.

(ix)   For a change in objective check stages (iii)−(vi). You will find that some adjustment is needed in stage (vi) but little in the other stages. Then carry out stage (vii) where there will be considerable change in the size of the illuminating aperture diaphragm in order to match the numerical aperture (NA) of the new objective.

(x)   In visual microscopy the intensity of the image brightness can be controlled by

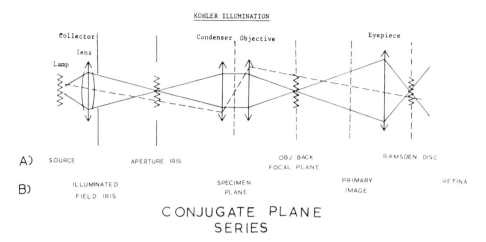

Figure 1. A ray diagram of Köhler illumination showing the position of the two sets of conjugate planes. Solid lines represent rays moving from left to right arising from the centre of the lamp filament to the near margin of the lens. The dotted line is a ray passing through the centre of the lamp collector lens. Vertical solid lines with arrowheads are lenses, while vertical solid lines without arrows are diaphragms. Vertical dotted lines are plane in which specimen, b.f.p., objective (obj.) and image are situated. Zigzag represent the lamp filament and its subsequent images.

altering the voltage setting on the lamp. However, a reduction will increase the warmth of the tones because of the extra red in the light or the loss of blue. This method is not acceptable for photographic purposes and you should refer to Section 8.3.3 in Chapter 3 for further discussion on this point. In both visual microscopy and photomicrography the intensity of the image can be reduced by inserting neutral density filters in the path of the light before it enters the condenser.

1.3 Conjugate planes in Köhler illumination

During the procedure for Köhler illumination you will have noticed a series of planes containing images of the lamp filament or the specimen itself. One further plane can be found by placing a piece of lens tissue at right angles to the eyepiece, a few mm above the eye lens. A small (~3 mm) disc of light will be seen on the paper. This is the Ramsden disc. It is at this point that the actual pupil of the eye is placed so that all light passing through the Ramsden disc passes into the eye. Withdraw the tissue and it can then be examined with a magnifying glass and be seen to contain images of the lamp filament, the aperture iris and b.f.p. of the objective.

To manipulate the microscope successfully in virtually all techniques the concept of *conjugate planes* must be understood. There are two sets of planes in Köhler illumination which are at right angles to the optic axis of the microscope and they are named according to which step in the illumination they represent (see *Figure 1*). They are described as conjugate if the features in an earlier plane are repeated in a subsequent one; thus the plane containing the lamp filament, or source of light, is conjugate with the plane containing the condenser iris, which itself has the function of controlling the aperture of the condenser. It again is conjugate with the plane containing the back focus of the objective and is finally conjugate with the plane containing the Ramsden disc. Therefore, whatever is contained in irregularities of intensity, for example in the lamp filament, will be present and will be seen in the aperture iris plane of the condenser. Both of these will also be seen in the b.f.p. of the objective and finally in the Ramsden disc. You will recall from the procedure that the lamp filament image was deliberately focused onto the condenser iris. Focusing of the condenser onto the specimen necessarily brings the front aperture of the condenser conjugate with the b.f.p. of the objective.

Similarly there is a set of planes amongst which is that of the specimen. Beginning as close to the light source as possible there is the plane containing the field iris. It is focused by the condenser onto the specimen so it is then conjugate with the plane that contains the specimen. Both are seen in focus therefore in the plane containing the primary image and, by implication, in the retina of the eye. Therefore these are also conjugate. In photomicrography the film emulsion is put into the plane which corresponds to the same conjugate series. The two conjugate series are named A and B in *Table 1* and *Figure 1*.

The practical significance of these two separate series is that irregularities in the light source are not present in the image plane and therefore will not be perceived by the eye or recorded on the film. If you are not familiar with them then it is well worth while memorizing them and understanding their whereabouts in the microscope as they can be used as reference points in a shorthand version of protocols and interpretative statements. Study *Figure 1* which incidentally you can relatively easily draw if you remember the following points.

(i)   Light passing through the centre of focus on one side of a lens passes out on the other side parallel to the lens axis.

(ii)   A ray of light entering a lens in a direction parallel to the lens axis passes through the centre of focus on the other side of the lens.

(iii)   A ray of each sort leaving, for example the specimen, will cross on the other side of the lens. Where they cross will be the image point.

Recall that measuring graticules are normally placed in the primary image plane but could be placed in the field iris plane. Their use will be discussed in Chapter 7.

2. LIGHT−MATTER INTERACTIONS AND THEIR SIGNIFICANCE IN LIGHT MICROSCOPY

All light−matter interactions may be grouped into seven categories: refraction, reflection, absorption, transmission, fluorescence, polarization and diffraction.

2.1 **Refraction**

Refraction is the change in the velocity of light when it enters a medium. Thus light moves more slowly in glass than in air and the ratio of the two velocities or, more exactly, the ratio of the velocity of the light in a vacuum to that in the glass is a number which is called the refractive index (n) of the glass.

All lens systems in microscopes utilize the refractive power of glass for the purposes of focusing the light and for correcting aberrations in the lenses, as well as presenting

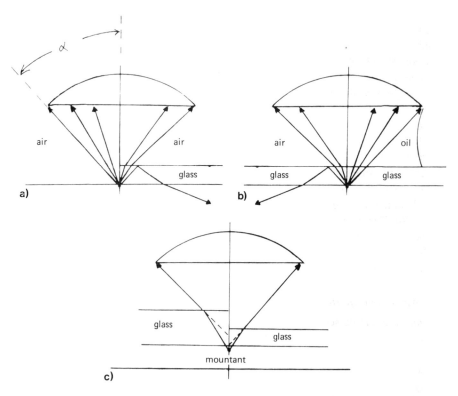

Figure 2. Coverglass, refraction and NA. (**a**) Left side without cover. Marginal ray at single (α) NA $= \sin\alpha$. Right side with coverglass, rays refracted at glass − air interface. (**b**) Left side with glass − air for dry objective. Right side with glass − oil for homogeneous immersion giving effective larger NA than left side. (**c**) Dotted lines show apparent sources of marginal rays from specimen under two different coverglass thicknesses. Note also specimen immersed in mountant may effect condenser cone effectiveness.

the specimen as a magnified image to the observer's eye. The refractive index is related to the wavelength of the light (λ), being greater for shorter wavelengths (blue light) and smaller for longer wavelenths (red light). Some significant and important roles are played by this variation with wavelength. White light passing through a lens will be brought to a series of foci depending on its component wavelengths, blue being nearer to the lens (shorter focal length) than red (longer focal length). The physical difference between these foci is a measure of the chromatic aberration of the lens. The spread of the refractive index over the wavelength is termed the dispersive power and is a property of the material. Glass of differing dispersive powers is used to correct aberrations within lens systems. This same phenomenon can be used to identify features within a specimen—for example the recognition of an asbestos fibre.

The degree to which light bends as it enters the medium is markedly affected by the angle at which it strikes or enters the medium. Thus a ray of light entering a glass surface at right angles to the surface passes on unchanged while one entering at an angle will be bent away from the line at right angles to the surface (termed the normal). In the case of a lens such a ray will exist passing through the centre of focus on the other side of the lens. The difference of foci between rays entering close to the axial direction of a lens and those entering near to the margin of the lens is a measure of the spherical aberration of that lens. An objective lens should be chosen for a minimum of chromatic and spherical aberration and it should be used in such a way as to maximize its capabilities.

The angle at which the light strikes the glass is also important in the exploitation of the critical angle concept. If light strikes at too great an angle it will not enter the glass at all and again, when light is leaving glass to enter air, it may well be internally reflected back (*Figure 2a*) and fail to leave the glass. Both these situations are possible occurrences in the collection of light from a specimen into an objective or in the illumination of that specimen by a condenser lens. The full cones of light in some cases can only be utilized if the glass−air interfaces are reduced in their impact by placing oil of about the same refractive index as the glass between one glass surface and another (*Figure 2b*). For example, there may be a need for oil between the top lens of a condenser and the bottom surface of a slide, as well as the more frequently occurring situation of oil-immersion objective lenses. Some lenses are designed to take other immersion media, for example water, with some fluorescence or interference objectives.

2.1.1 *Numerical aperture*

In *Figure 2* it can be seen that there is only a limited angle of acceptance of light that a lens is capable of taking from a specimen. The half angle of this total angle is given here the symbol α and the value of the angle can be seen to be related to the distance between the lens and the specimen and to the size of the lens. These relationships are usually combined and can be represented as $\sin\alpha$.

On the right hand side of *Figure 2a* the space between the lens and the specimen is occupied by air ($n = 1$). Therefore, $\sin\alpha$ represents the NA of the lens. For dry lenses the top value of NA is limited by the critical angle of the light ray in the glass coverslip covering the specimen. A top working value of 0.95 is usually available for such a dry lens. If, however, oil is placed between the objective lens and the specimen to lose any glass−air interfaces and provide a homogeneity in which the light rays are travelling (*Figure 2b*) then a much wider angle of light can be accepted by the lens. The value of this light gathering power is called the NA as before but is now the product of the value of the refractive index of the immersion liquid and $\sin\alpha$. A top working value of NA = 1.3 is nowadays generally available. Condensers also have a NA and it serves to provide as wide a cone of rays as possible when light is passing out of the top lens into the specimen. A high NA condenser will produce a wide angled cone of light.

Numerical aperture, lack of spherical and chromatic aberration and flatness of field are the aims of the lens manufacturer and form the basis of the cost of production. High NA dry objective lenses are very finely corrected for aberrations and they need to be used with a very precise coverslip thickness over the specimen. The basis of the correction is given in *Figure 2c*. Too thick a coverslip will not be sufficiently corrected for, while the lens will over-correct an excessively thin one. Some lenses have a correction collar for adjustment to coverslip thickness but others state the precise thickness required alongside the NA and magnification. See Section 5.2.3 in Chapter 3 for further details.

Table 2. Examination of the microscope potential.

Examine and make a note of the types of objectives, eyepieces and condensers that are available.

Objectives

Look for and interpret the symbols engraved on the sides of the lenses noting particularly:

NA	e.g.	0.65	or 1.3
Magnification		×40	×100
Infinity corrected		40×	100×
Coverslip (thickness)		0.17	−
Immersion (otherwise dry)		Usually oil but sometimes water	
Lens type		Plan, Phaco, Apo etc.	
Manufacturer and code number			

Eyepieces

Magnification	×10	×15
Field number	18	14
Type (field)	Wide field (WF)	
(corrections)	periplan, compensating	
(high eyepoint)	spectacles symbol	

Note whether there is a magnification changer in the tube of the microscope and note the presence of a phase telescope setting and its focusing device.

Condensers

Note carriage of condenser and find centration screws.

Note on universal condensers the settings for various illumination techniques. Note any centration capacity for, e.g. phase rings or condenser irises. Inverting the condenser and rotating the settings will show the various features when they are in place.

Engravings may include the NA with and without the top lens.

Note the methods of removing the top lens either by unscrewing or using a lever to control a flip-top. Corrected condensers may have symbols like achr. apl. Special condensers will be marked as such.

It is important of course to match the lenses in use at any one time. Thus the very highest corrected objective should be used with as highly a corrected condenser as is available, as well as with the requisite eyepieces.

To examine the potential of your microscope, carry out the examination suggested in *Table 2* for the lenses available both on your immediate instrument and in your laboratory as a whole.

2.2 **Reflection, absorption and transmission**

Reflection from surfaces of materials, the converse of absorption, is extensively used in microscopy and is again related to wavelength. If all wavelengths are absorbed equally, and therefore not reflected by a feature, then there is a tonal difference of that feature from the reference area which would be differentiated as varying shades of grey. If absorbance is different for different wavelengths then the features become coloured, the resulting colour being that produced from white light after removal of the wavelengths which have been absorbed. The same can be said for the reflection image. If all wavelengths are equally and totally reflected then the surface feature will be indistinguishable from bright light. If one wavelength is reflected and the other component wavelengths of white light are absorbed then the feature will be coloured according to the reflected wavelength.

Absorption with its converse of transmission and reflection is utilized in microscopy by the use of dyes, stains and metal-deposition-type preparations. The various absorbing materials are selectively placed onto the chemical or physico-chemical features of the specimen to be examined.

Table 3. A test object for fluorescence.

Requirement: a defunct colour TV tube which has been opened by a *skilled glass-blower*.

1.   Wearing gloves remove the phosphors from the inside of the broken tube by placing a piece of sticky tape against them. Peel the tape off.
2.   Place the sticky side of the tape against a coverglass and invert onto a slide on which some mountant has been placed.
3.   Allow the mountant to spread and anchor the coverslip onto the slide.
    Notes: The phosphors are highly toxic and should not be allowed to come into contact with the skin. There will be in some cases an aluminium backing to the phosphors but using the procedure above the phosphors are just beneath the surface of the coverslip and are very satisfactory for incident light excitation.
4.   Excite the phosphors with a range of wavelengths and note that with UV three colours are seen brilliantly while with other wavelengths of excitation one or other of the three primary colours is missing.

Considerable effort is made both by manufacturers and operatives of microscopes to reduce unwanted reflections. Two examples are the production of coated lenses and the matt black finish inside the body of microscopes and cameras. By strict control of the illuminating aperture in Köhler illumination, the operator can prevent stray light reflections off the edges of lens mountings and walls of the microscope tube spoiling the quality of the image.

2.3 **Fluorescence/phosphorescence**

Some materials absorb light of one wavelength and re-emit the energy at a longer wavelength. If there is a time delay between the absorption and re-emission then the material is said to be phosphorescent, but if the re-emission stops as soon as the exciting radiation ceases then the material is fluorescent. A test object for fluorescence is given in *Table 3*.

The wavelength of the re-emitted radiation is usually longer than that of the exciter and therefore of a greener or redder colour. This phenomenon is of course the basis of fluorescence microscopy, with excitation by UV, blue or green wavelengths, and will be described in detail in Chapter 6.

2.4 **Polarization**

Polarization of light by matter is familiar to everyone (although normally in a rather limited way). Light rays are pulses of energy whose amplitude of energy changes is measured at right angles to the direction of the ray. Statistically, normal light is said to vibrate in all possible directions at right angles to the ray direction. Some matter will transmit (or reflect) light differently depending on the relation of the vibration direction to some feature of the matter. Such a substance may then transmit a ray with its vibration only in one direction, for example in a plane at right angles to it, and the ray emerging would be of plane polarized light. By introducing such a material into the microscope between the light source and the specimen the observer is able to determine how the specimen affects plane polarized light. Rotating either the specimen or the polarizer, to see whether the effect is different in different relative directions,

Table 4. To find the permitted vibration direction of an unknown piece of polaroid (see also *Figure 3*).

Principle: When an unpolarized beam of light is reflected off a transparent surface the reflected beam is plane polarized and its vibration direction is parallel to that of the surface of the reflecting material and at right angles to the ray direction.

Method

1. Place a piece of clean fairly thick glass on a dark bench some distance in front of a lamp such as is used for illuminating a microscope of the older type.
2. Place your unknown piece of polaroid in front of your eyes in such a position as to observe through it the reflected beam of light coming from the surface of the glass.
3. Rotate your polaroid while at the same time noting the change of intensity of the perceived light beam.
4. Note the orientation of the polaroid with respect to the plane of the glass surface when the intensity of the beam is at a minimum.
5. Conclude from 4 that the permitted vibration direction of the polaroid is now at right angles to the glass surface.
6. Check by observing the orientation of the polaroid when the perceived beam is at its maximum intensity.
7. Conclude from 6 that the permitted vibration direction is now parallel to the glass surface.
8. Mark your polaroid in one corner with an appropriate symbol.

There are several other methods involving the comparison of the unknown with known polaroid but the one given above is derived from basic physics.

will determine whether the specimen demonstrates the phenomenon of pleochroism. Other material may show different refractive indices depending on the direction of polarization of the light ray passing through it. Such material will then have two principle refractive indices and will be known as birefringent. By putting the specimen between two pieces of polaroid at right angles to each other (crossed polars) on the microscope the observer can, by rotating the specimen, show the presence of birefringence and produce information of a very precise kind. With suitable safeguards, it is possible to identify the material to a remarkable degree.

A newly acquired piece of polaroid should be marked with its permitted vibration direction; *Table 4* gives a practical way of determining that direction from first principles. See also *Figure 3*.

2.5 **Diffraction**

Diffraction is perhaps not so universally understood as the other relationships of light and matter that have been discussed so far. It does however have fundamental significance in the microscope. Some diffraction effects occur in normal vision situations but usually light is considered to travel in straight lines and shadows are thought to have clean edges. Diffracted light at the edges is considerably weaker than the direct or undiffracted light and thus is overlooked in normal vision.

A few simple practical illustrations (*Tables 5 – 7*) will be given here to demonstrate the importance of diffraction in light microscopy. They seek to bring to the attention of the intending microscopist that what is seen in the microscope is a replica of the specimen and not the specimen itself. It is a common experience to be sceptical about the degree of accuracy of replication, for example in art or sculpture, but it seems that this same scepticism must be learned by the person using a microscope. In all replication techniques the quality of the copy produced is related to the quantity of information transferred and is thus a measure of the constraints of the apparatus or technique used.

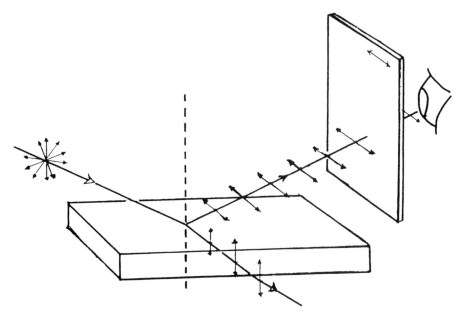

Figure 3. Method for determining permitted vibration direction of a piece of polaroid. See *Table 3*.

Table 5. Demonstrating the scattering of light by the specimen.

1.	Take any specimen containing fairly well stained and detailed structure. One example would be a Masson trichome-stained dermis section.
2.	Examine it through the microscope using the condenser iris in a nearly closed manner. Note the spurious details on the edges of all the smaller features of the specimen. These are diffraction effects.
3.	Take the eyepiece out of the microscope (or make the eyepiece into a telescope such as would be used for phase-contrast systems). Note the bright central disc of light coloured with the principal colour of the specimen but surrounded with a fuzz of light. Move the specimen in and out of the line of the objective and note that the fuzz of light is present only when the specimen is present. The fuzz of light will also be approximately of the intensity to compensate for the loss of intensity in the central disc when the specimen is present compared with when it is not. The fuzz of light is the scattered light and that scattering is largely the result of diffraction by the specimen.
	Again it is important to note that the objective lens has collected both the central disc of light and also some of the scattered light. This will be more obvious if the condenser iris is more closely shut and also will be clearer if there is greater periodicity in the specimen. The next exercise uses a deliberately designed specimen to illustrate the importance of diffraction in image formation.

These illustrations are preferably carried out with a correctly set up Köhler system of illumination. Some approximation to them can be achieved, however, by using the field iris as the illuminating iris if the condenser is not capable of giving a sufficiently narrow beam of light (see *Table 5*).

2.5.1 *Abbe test plate*

Experiments with Abbe diffraction plates or other diffracting specimens are intrinsically very interesting. They basically compare what is seen in the image with what is seen and is passing through the b.f.p. of the objective.

Table 6. Experiments with Abbe diffraction plate (see *Figure 4*).

A. *Specimen, diffraction and objective*

1. The Abbe plate will have a position in which a series of opaque lines alternate with transparent ones (effectively a diffraction grating). Observe it through the microscope set with the condenser iris closed to a small hole (ai).

2. Take out the eyepiece and examine (with a phase telescope) the back of the objective. The back lens will show a diffraction pattern of coloured spots (spectra) on either side of the bright central disc (bi).

3. Rotate the specimen, watching as you do so the change of the direction of the diffraction pattern.

4. Note the relative intensities of the diffraction spots and also note that the blue of each spectrum is nearer to the centre than the red.

5. Make a series of inferences from the above on the angle and direction of diffraction by the specimen and the collecting ability of the objective as seen by what is in the b.f.p. See the text for discussion and equations.

B. *To determine the value in image terms of the various components of the diffracted light in image formation*

1. A series of screens can be placed in the back of the objective. Two are considered here but others can be made out of round coverslips of appropriate size or small discs of opaque material.

   (a) Place a small washer-like screen in the back of the objective used in the experimental set up in protocol **A**. Make the aperture in the washer sufficiently small to stop the diffracted light from passing through the objective.

   (b) Observe the effect on the image by replacing the eyepiece and see no detail in the image of the lines. These may only be represented by a smudge of lower intensity.

   (c) Infer that without the diffracted light the central bright disc cannot produce an image of the specimen and only if the diffracted light is present with it will an image be produced. Without the contribution of the first order to image formation the microscope fails to resolve the lines.

   (d) Obtain a three slit screen of suitable size and insert it into the b.f.p. to modify the appearance of the b.f.p. from (bi) to (bii).

   (e) On returning to the image note the new image is now as shown in (cii), that is a series of lines which are apparently much closer together than they were in the original image (ci).

2. Move the Abbe plate to cross grating position (aii). Observe the back of the objective as before. Note the diffraction spectra are arranged in at least three directions about the central bright disc (biii).

3. Place a slit like screen in the back of the objective in such an orientation as to allow through only one set of diffraction spectra (biv). Replace the eyepiece and observe the new image (civ). You will see the the image has changed dramatically from the original wire netting-like one to a series of lines running at right angles to the direction of the spectra that you have allowed to pass through the objective.

   Infer from 1 and 3 that what passes through the back of the objective controls the details of the image.

Detailed investigation of the significance of diffraction to microscopy is not possible in the space available; an extended version of what is given here will be found in ref. 2.

Abbe was the person who first elucidated the theory of image formation as demonstrated by the experiments listed in *Table 6*. The Abbe test plate was produced by Zeiss, but unfortunately has recently gone out of production. If your laboratory has one then you should experiment with it for yourself but if you do not have one available you can achieve the same effects by using specimens such as diffraction gratings obtainable from Graticules Ltd. Biological specimens, such as the wing scales of butterflies for straight gratings and the multiple eye structures of insects as a substitute for cross gratings, can be used. A similar hexagonal pattern to the latter is available

Table 7. Alternatives to the Abbe plate.

Any specimen having fine and repeating detail will provide an opportunity to carry out an approximation to the Abbe experiments. One commonly available combination is given here.

1. Set up a microscope to observe the diatom *P.angulatum*.
2. Observe the diatom with an oil-immersion high power objective say with NA 1.3 and which has a diaphragm mounted in its b.f.p.
3. With the b.f.p. diaphragm of the objective open observe the hexagonal pattern of the diatom. Close the diaphragm down and note the pattern disappears.
4. Re-open the b.f.p. iris to return to the pattern in the image. Close down the condenser iris noting that the pattern remains visible.
5. Take out the eyepiece and examine the b.f.p., noting the diffraction pattern as a central bright white spot (the zero order) and diffraction spectra in three directions but close to the boundary of the b.f.p. (*Figure 6a*). Observe the effect of opening and closing the b.f.p. diaphragm and finally close it sufficiently to prevent the first order spectra from being seen. (*Figure 6b*). Replace the eyepiece and see that the pattern in the image has gone (*Figure 6d*). Re-open the b.f.p. iris and the detail will reappear (*Figure 6c*).
6. Infer from these adjustments of the b.f.p. iris that the first order light must pass through the b.f.p. of the objective and thereby contribute to the image for any hexagonal pattern to appear in the image. The microscope fails to resolve the pattern if it fails to collect the first order diffracted light.
7. This exercise should be repeated with the dark ground condenser and will show the effect of the zero order light on image formation—see Chapter 2, Section 4.2. Meanwhile, refer back to the text summary for the discussion of the relationships between the specimen, NA, b.f.p. and the image detail and quality.

in the frustule of the diatom *Pleurosigma angulatum* but at a much smaller dimension (*Table 7*). These substitute specimens should be observed with appropriate objective lenses—for example, the wing scale and the diatom need a NA of about 1.3 while the insect's eye is visible with a low power objective of about NA 0.17.

The understanding of the way in which the image is produced in the microscope is the beginning of an appreciation of how far the image is an acceptable replica of the specimen being observed. The exercises given in *Tables 6−7* show that in the final analysis the resolving power of the objective, discussed in practical terms in Section 2.5.3, is determined by the ability of the objective to collect the diffracted light.

The stages in image formation in the microscope can be deduced from the steps set out in *Table 6A*.

(i) Step 1 infers that there is a relationship between the image and the specimen and that the image is a magnified version of the specimen, for example a series of lines running north−south.

(ii) Steps 2 and 3 infer that the specimen scatters light by diffraction in a direction at right angles to the lines but parallel to the direction of periodicity (repeat) of the opacity and transparency. The objective collects some of that diffracted light and allows it to pass through. What is collected by the objective is seen as a series of diffraction spectra in the b.f.p.

(iii) Note that the blue light is diffracted at a smaller angle (i.e. nearer to the central undiffracted light) than the red light. The position and pattern of the diffraction maxima in the b.f.p. form what is termed an optical transform of the specimen.

These inferences can be drawn together in mathematical form by relating the equation

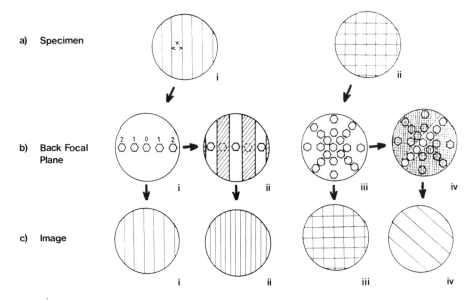

Figure 4. Experiments with the Abbe test plate or other test gratings. (**ai**) line specimen of spacing (*x*); (**aii**) cross line specimen of spacing (*x*). (**b**) Back focal plane without and with slit screens. (**c**) Images (of the specimens) without and with the screens present in the b.f.p. Arrows indicate modifications in the b.f.p. and relationships of the images to the b.f.p.

for the diffraction angle θ to that of the quantitative statement of the collecting power of the objective, that is its NA.

For each order of diffraction the angle θ is given by the equation:

$$\sin\theta_i = \frac{i\lambda}{x} \qquad \text{Equation 1}$$

where θ = the angle of diffraction, λ = wavelength of light and x is the spacing (distance) between the lines. The letter i is the order number and must be an integral number. Thus for third-order diffracted light the angle ($\sin\theta_3$) will be three times larger than that for the first order ($\sin\theta_1$).

If NA, which is equal to $n \times \sin\alpha$, is greater than $\sin\theta_1$, then the diffracted light at angle θ_1 will be collected by the objective and will be seen to be so by the presence of a spectrum in the b.f.p. of the objective.

This series of experiments can be repeated with other diffraction gratings if they are available. The greater the ruling number then the smaller will be the value of x in Equation 1 and the further out the spectra will be from the centre.

When a cross grating is observed the b.f.p. of the objective will have the appearance of *Figure 4b(iii)* it can be seen that the diffraction spots on the diagonal have a slightly different position from those on the vertical and horizontal if the grating lines are at right angles to each other.

The appearance of the b.f.p. is sometimes referred to as a Fourier transform of the specimen produced by trains of light waves.

The next step in appreciating how the image is built up is to discover what part each order of diffraction plays in forming the image. A practical exercise is set out in *Table 6B*,

IMAGE FORMATION IN THE MICROSCOPE

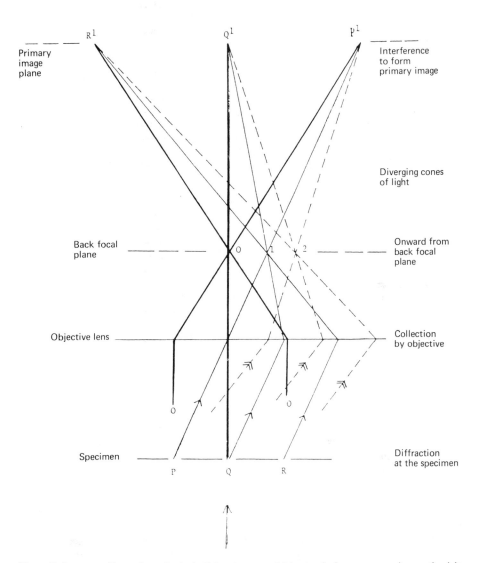

Figure 5. Summary of image formation in the light microscope. Main steps in the process are given on the right.

in which it is assumed (see *Figure 5*) that each spectrum in the b.f.p. is the apex of a cone of diverging light leaving the back lens of the objective. By suitably blocking the passage of each order of diffraction in turn by an opaque patch, the part played by each can be determined.

Table 6B deals with the first order and higher numbered orders of diffracted light. The significance of the zero order is more suitably discussed under factors affecting contrast and an experiment to evaluate it is given in Chapter 2, Table 8.

15

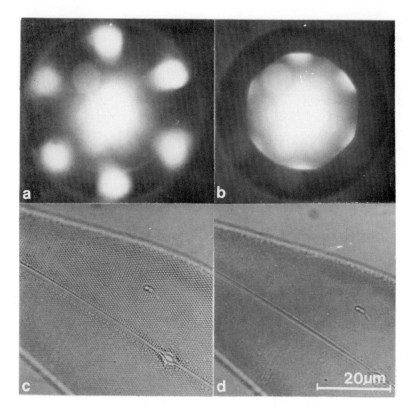

Figure 6. The relationship of image to what passes through the back focal plane (b.f.p.) of the objective. The objective is a Zeiss Planapo NA 1.3−0.8, ×100, oil-immersion lens. The specimen is the diatom *P.angulatum*. (Scale bar = 20 μm). (**a**) The b.f.p. of the objective with fully opened objective iris. The figure shows the zero order in the centre and the first order diffraction spectra arising from the diatom and being collected and allowed to pass through the objective. (**b**) The same situation is in (**a**) except for the closed down objective iris occluding (almost completely) the first order diffraction spectra. (**c**) The image from the situation in (**a**), that is where the various diffraction orders contribute to image construction. (**d**) The image from situation (**b**) where the first order diffracted light is prevented from contributing to image construction. Note that in (**c**) the hexagonal pattern is resolved while in (**d**) it is not.

As the Abbe test plate is no longer commercially available, a series of alternative specimens which show the same phenomena can be found in most biological laboratories. *Table 7* describes the use of one such alternative and *Figure 6* illustrates the steps.

Tables 6 and *7* describe only a few experiments which can be performed with simple accessories and are given here to illustrate the basic details of image formation in the microscope. They indicate that the constraints in the use of the microscope are very largely determined by the ability of the objective to collect the light issuing from the specimen. *Figure 5* summarizes image formation and the following verbal summary is effectively the description of that figure.

2.5.2 Primary image formation

The specimen is illuminated by a train of light waves and scatters some of this light by various means including diffraction. The direction and degree, or angle, of diffraction is determined by the direction and fineness of the periodicity in the specimen. A number of orders of diffracted light are recognizable including the zero order (bright central disc) and the first order spectra. The angle of the diffracted light ray is given by the formula $\sin\theta = \lambda/x$ where x is the spacing in the specimen.

The objective collects the light scattered by the specimen. The amount or numbers of orders of diffracted light collected are determined by the angle of acceptance of the objective ($\sin\alpha$) or more accurately by the NA of the objective, which takes into account the refractive index of the medium between the specimen and the objective lens. The NA must be numerically larger than the $\sin\theta$ for diffracted light of that order to be collected.

If the eyepiece of the microscope is removed the b.f.p. of the objective can be seen (with the aid of a telescope). It contains multiple images of the illumination, the number being related to the numbers of orders of diffracted light collected by that objective. The b.f.p. is effectively an optical transform of the specimen.

Light issuing from the b.f.p. spreads out in a series of cones whose apices are in fact the diffraction spectra seen in the b.f.p. The rays from each of these cones, in so far as they originate from the same light source, are capable of interference. An interference pattern is produced in the primary image plane by this light issuing from the zero order spot and the first order spectra. This interference pattern is the primary image and is in form a magnified replica of the specimen.

It is possible to show that if the zero order light is prevented from passing through the b.f.p. by placing a central opaque patch over it, thereby preventing it from contributing to image formation, the image is one of reversed contrast. If the first order is prevented from contributing and the image is made from the zero order and the second order then a spurious doubling of the information in the image is produced. A one-to-one relationship between the image detail and the specimen detail only occurs if the first order light helps to make the image. Any additional orders contributing to the image improve the quality (i.e. give the features sharper boundaries) and there is a better discrimination between intensity differences in the image. The primary image is itself a Fourier transform of the b.f.p. of the objective and, as such, improves with the increasing number of terms contributing to that transform. If the first order diffracted light fails to contribute to the formation of the image, the microscope fails to resolve the detail of the specimen giving rise to that diffraction.

The b.f.p. of the objective will be discussed repeatedly in Chapter 2 on the setting up of the phase-contrast microscope.

2.5.3 *Resolution*

The experiments with the Abbe test plate and the diatom have described the way in which images are built up in the microscope from what passes through the b.f.p. of the objective. If the detail is in the image then the microscope system has resolved the detail and made it visible in the image. We have seen that it is capable of resolving such detail only if the objective collects the first order diffraction sufficiently well for an interference pattern to develop in the primary image plane. That ability is determined by the NA.

The NA has no units of length being the product of two ratios. We have seen that blue light is more easily collected than red light for any value of the NA. Thus the resolving power is proportional to the NA and inversely proportional to the wavelength of the light used to illuminate the specimen. The resolving power can be expressed as the ability to resolve a dimension (between two points) and this dimension is called the resolution. It is given by the formula

$$\text{Resolution} = 0.61 \times \lambda/\text{NA} \qquad\qquad \text{Equation 2}$$

Substituting in this formula the wavelength of green light, for example 500 nm and an NA of 1.3, the resolution of the objective is about 250 nm. Thus in theory the microscope should be able to allow the observer to see two points 250 nm apart. It will not be able to display them as two separate points if they are less than 250 nm apart.

2.5.4 *Significance of the condenser in resolution*

Equation 2 uses the objective NA but in actual practice the condenser plays a considerable part in achieving resolution. All the experiments that have been discussed so far have used very narrow illumination beams to portray the angle of diffraction. However, the condenser throws a wide angled cone on to the specimen. In Köhler illumination this cone is set at about 66% of the acceptance cone of the objective. Thus a 1.3 NA objective will be illuminated by about 0.9 NA of condenser. This will mean that some illuminating beams will be leaving the specimen at an angle of about 70° to the vertical. They will however have first and higher orders of diffraction on the objective side which will enter the objective, while those on the slide side will be lost. If there is no NA in the condenser then there will be no light entering the specimen: thus the formula for the limits of resolution must also take into account the condenser. In practice this is as follows.

$$\text{Resolution} = 0.61 \times \lambda/(\text{NA}_{\text{obj}} + \text{NA}_{\text{cond}})/2 \qquad\qquad \text{Equation 3}$$

The eye itself has a resolution limit which will be of the order of 2 min of arc in weak-eyed, or 1 min of arc in sharp-eyed people. Hence, the resolution of the microscope must be made to subtend an angle of this value at the eye and this is where the magnifications of the microscope lenses are of significance. This will be discussed in the next section.

2.5.5 *Magnification*

The geometric optics diagram of the light microscope (*Figure 7*) indicates that magnification is a two stage process.

First, a real magnified image (*H*) of the specimen (*h*) is produced in the primary image plane by the objective lens. Mathematically, the magnification is the ratio of the image size to the specimen size (when placed at the same distance from the eye or screen). This is the primary magnification and its value is usually engraved on the objective. The image produced by the objective will usually be found inside the eyepiece of the microscope and will be inverted as far as the specimen is concerned. A small disc of ground glass can be placed in the plane which normally holds measuring graticules and will readily display this primary image particularly if the specimen is a contrasty one.

Secondly, light issuing from this primary image plane enters the eye lens of the microscope which then modifies its path and angle so that the eye receives the light

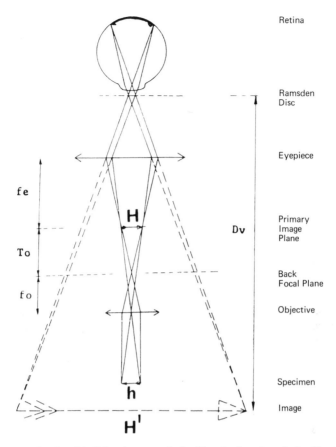

Figure 7. Geometrical optics of the light microscope. *fe*, focal length of eyepiece; *fo*, focal length of objective; *To*, tube length; *h*, specimen dimension; *H*, intermediate (primary) image dimension; *H'*, final (virtual) image dimension; *Dv*, least distance of distinct vision (250 mm).

as though it came from a much wider angle. This is the secondary magnification and its value is usually engraved on the eyepiece. The final image with this angle would have a dimension of *H'*. The total magnification product (MP) of the microscope is the product of the primary magnification of the objective and the secondary magnification of the eyepiece.

Values of objective magnification are usually within fairly narrow bands associated with the NA of the lens itself: thus an NA objective of 0.65 will have an MP of ×40 and an NA lens of 1.3 magnifies about ×100. Eyepieces will again be of about ×10 or ×6.3 but other values can be found and some microscopes have magnification changers for their eyepiece systems. There is however a significant relationship between the total MP of the microscope and the NA of the objective used, as given in Equation 4.

$$Dr = (2 \times Dv \times NA)/MP \qquad \text{Equation 4}$$

where *Dv* (the least distance of distinct vision) has a value of 250 mm. For practical reasons the diameter of the Ramsden disc (*Dr*) must be a value between 1 and 3 mm. This limits the relationship of MP to NA to the extremes of ×170 NA and ×1000 NA.

The practical use mentioned above, incidentally, is related to the pupil size of the observer's eye. The observer's pupil is the entrance pupil of his eye and the Ramsden disc is the exit pupil of the microscope. They work together best when of matching size as well as when correctly positioned. (See *Table 8* for a method of finding the Ramsden disc and its size.)

2.5.6 *Magnifying power and resolution*

A further constraint on magnifying power is in relation to the resolution limit of the objective lens (see Section 2.5.3). The purpose of magnification is to make the detail of the image subtend sufficient angle at the eye for the eye to resolve that detail. The angle was stated to be between 1 and 2 min of arc. A dimension of 0.1 mm is required at the standard distance 250 mm from the eye to produce this angle. Thus the resolution of the objective must be magnified up to that level. For example, the minimum MP of a microscope using a 0.65 NA objective would theoretically have to be ×200 in order to cause the resolution of the objective of 500 nm to be resolved by the eye. In practice a small extra MP is added. These values can be ascertained by substituting in the various formulae set out above. It can be seen that for an MP of ×400 and objective of 0.65 NA with a primary magnification of ×40 there would be a requirement for at least a ×5 eyepiece (see *Table 9*).

Any excess magnification over the minimum needed is described as empty magnification. It merely coarsens the image and no further information is added or obtained.

2.5.7 *Summary of the signficance of diffraction in light microscopy*

Sufficient has been said in Sections 2.5.1−2.5.6 to clarify individual points and it is only necessary here to bring those points together. Diffraction is only one of the ways

Table 8. A method to find the Ramsden disc and its size.

1.	Using ground glass or a piece of lens tissue observe the Ramsden disc by holding the glass at right angles and ~3 mm from the top of the eyepiece when the microscope is set for viewing a specimen. The Ramsden disc will be seen as a minimum disc of light at that point.
2.	Note its diameter and the distance it is above the surface of the eye lens.
3.	Eyepieces used by spectacle wearers will have that distance large enough for them to have their spectacle lens between their eye and the surface of the eyepiece. The size of the Ramsden disc is proportional to the NA of the objective and inversely proportional to the total magnification (MP) of the microscope as shown in Equation 4.

Table 9. To demonstrate empty magnification.

1.	Observe a diatom such as *Navicula lyra* with an objective of ×40 and NA 0.65 together with an eyepiece of ×6 giving a total magnification of ×240. Note the detail is present in the image as the objective is capable of resolving spacings of ~500 nm.
2.	Repeat with an objective of ×10 and NA 0.25 together with an eyepiece of ×25 which will give about the same magnification. The general outline of the diatom is present but its surface contains no detail.
3.	Infer that there is excess magnification in the second case. The objective is capable of resolving spacings of only 1200 nm and the pattern in the diatom is much smaller. Therefore there is surplus or *empty magnification* present.

in which the specimen scatters light but it has been shown that a quantitative relationship occurs between it and the lateral resolving power of the microscope. The resolving power of the objective is determined by its ability to collect the diffracted light, the relevant parameter of the objective being its NA. The resolution limits set by the objective are then brought to the resolution limits of the observer's eyes by the magnification, the degree of which is determined by the information available. Any excess provides no further information although it may in fact be beneficial in avoiding eye strain.

The lateral resolution we have been discussing so far is that in the x and y directions in the specimen. The resolution in the z direction is yet to be considered. It too is dependent on the NA of the objective and is discussed in the next section.

3. DEPTH OF FIELD

The optics of the phenomenon of depth of field (d.f.) is outside the scope of this book but the limits of depth are a very practical issue in microscopy. Large d.f. is a most attractive feature of scanning electron microscopy but scanning optical microscopy has not yet been developed or used to the same extent. The very shallow d.f. of high power objectives however is used to provide information in one plane at a time in the specimen in the z direction. (The use of stored images of very shallow depths in the specimen which are produced by scanning lasers as illuminating systems in microscopes is now being developed.) Improvement to the contrast in images is achieved by removing the out-of-focus light from planes above and below the immediate one and the techniques are mentioned in Section 3.1.

The d.f. is the result of three factors in visual microscopy: the axial resolving power of the objective in the z direction, the geometrical d.f. and the accommodation range of the eye (1). These combine to produce a formula (Equation 5) which indicates proportionality with the wavelength of light used, inverse proportionality with not only the NA of the objective but with the NA squared, and again inversely with the magnifying power. James (1) has calculated the instrumental d.f., that is the d.f. minus the eye factor. Some of his values are given in *Table 10*.

$$\text{d.f.} = n \left[\lambda/2 \times (NA)^2\right] + \left[0.34/(MP \times NA)\right] \qquad \text{Equation 5}$$

The values in *Table 10* are only given as a guideline but it can be seen that they are very shallow for objectives of high NA, and if d.f. is of paramount importance it might, in some situations, be better to use a low NA objective with low primary magnification together with a high magnification eyepiece. Such a situation would of course make

Table 10. Total instrumental d.f. (μm) for a variety of objectives and total magnification ($n = 1.5$). After James (1).

| Objective (N/A) | Microscope total magnification | | | | |
	$\times 10$	$\times 50$	$\times 100$	$\times 500$	$\times 1000$
0.05	1187	371			
0.3			22	8	
0.65				3	2
0.85				2	1
1.3				1	0.6

a sacrifice in lateral resolution as portrayed in Section 2.5. The measurement of depth is described in Chapter 7, Section 2.3.

During visual use of the microscope the experienced observer is constantly focusing the machine up and down in order to build up an impression of an image in depth. Future developments will enable matching of this memory by the use of video recording and then displaying the combined images as already mentioned. The single photographic record of a field of view shows very clearly the limits of the d.f. in focus at any one time. Shallow d.f. can be an advantage if optical dissection in the z direction is required.

3.1 Tandem or confocal scanning microscopy

Developments exploiting the very shallow d.f. in light microscopy have been made recently whereby the specimen is scanned with a series of very tiny fields of view, the images of these tiny fields are recovered individually and, after storage, the complete image is built up as a mosaic of the combined parts. In one such microscope a rotating Nipkow disc of 100 mm diameter containing 48 000 holes illuminates the specimen. In this reflected light microscope, the same disc allows the image light to pass back through it. The scanning rate is 12 000 lines across the image and gives 20 full images/sec made up of 400 partial images.

The system has been called the Tandem Scanning Reflected Light Microscope (TSRLM). Petran *et al.* (3) have used the technique to obtain detail of bone in large objects, such as skulls, or from thick highly translucent three-dimensional specimens such as animal or plant tissues.

Several laser scanning microscopes have been developed in order to improve the quality of the image visibility by reducing the impact of the out-of-focus images from planes above and below the one of interest. The Laser Scan produced by Zeiss and the SOM and Laser-sharp of Bio-Rad are three examples. The SOM builds up a point by point image from a scanning focused beam of laser light. The processing of the images by electronics produces an image somewhat reminiscent of a scanning electron microscope (SEM). The instrument can be tuned to remove the out-of-focus images and therefore greatly improve detail visibility in the image. To achieve the removal a very narrow aperture is placed in the primary image plane, directly in front of the detector and therefore conjugate with the field illumination pin-hole. N.S.White has claimed (personal communication) that such a system improves the resolution of the microscope by a factor of 1.4.

A device for fitting to the trinocular head of a wide range of existing microscopes is the MRC Laser-sharp of Bio-Rad developed jointly by that company with W.B.Amos and J.G.White of MRC Cambridge. Applications of the confocal scanning microscope are expected to be particularly successful in removing the fuzzy glare that surrounds images produced by epi-fluorescence methods. Examples of its use include the characterization of the spindle in dividing cells, the layers of the brain and canals in developing chick embryos, and detail of nematode embryos in egg masses. These features can be seen clearly through a mass of surroundings which would otherwise produce a confused scatter of light. Because the amount of excitation radiation is so small the condition of the specimen is not seriously disturbed. Again as the images in the framestore can be processed onto film later there is less chance of photobleaching the fluorescent material in the specimen (4).

4. FIELD OF VIEW

The labelling of objectives and eyepieces for magnification needs perhaps a word of explanation. Where a lens does not produce a real image as, for example, in the case of the microscope eyepiece, the angle of the rays of light are altered by it. This is known as angular magnification and is conventionally designated by the ×10 symbol which is found on eyepieces.

Many objectives, but not all, produce real images. The size of the image compared with the size of a comparable feature in the object is a lateral magnification and is correctly stated as a ratio, such as 100:1. Some objectives are so marked, others are marked with a short-hand version of this ratio, termed the magnification number, which in this case would be simply 100. Infinity corrected objectives do not produce a real image. Their magnification is angular and so is correctly labelled as ×100. Old objectives, however, may have this form of labelling yet still produce a real image.

A beginner may well be impressed with the enlargement of the features he is studying when looking into a microscope but may not realize that the field of view which is being observed may be very small. If, for example, the total magnification of the microscope in use is ×1000, then what is being seen as a sample is equivalent in length terms to 1 mm in 1 m and, in area terms, the area of the image seen is only one millionth of the specimen area. This is certainly worth remembering when making predictions about the whole on the basis of a single sample.

It is worth taking a moment or two to determine for yourself the field of view dimensions seen with various combinations of eyepieces and objectives. Using a low power lens it is possible to observe directly a normal desk-type transparent ruler, seeing the millimetre rulings from one edge of the field to the other. With higher powers use a stage micrometer.

Many manufacturers offer wide field eyepieces and these will be engraved in a appropriate way. Compare for yourself their performance against normal ones. On many eyepieces there will be engraved, as well as the magnification, another number which is the field of view number (FVN). This is, in effect, the diameter in millimetres of the stop which limits the field of view but it is useful in the following way. The field of view of the specimens seen when using a particular eyepiece will be in mm, the equivalent of the FVN divided by the objective magnification. The relationship is modified by a tube factor for accurate purposes but can be used as an estimate. Thus for example a wide field eyepiece with an FVN = 18 when used with a 40:1 objective will present a view of about 0.4 mm linear dimension. If you wish to determine your tube factor, measure the actual field of view by the method given above for a number of objectives and substitute in the equation below:

$$\text{Field of view (mm)} = \frac{\text{FVN}}{\text{Objective magnification} \times \text{Tube factor}} \qquad \text{Equation 6}$$

5. SUMMARY

It has been shown that the specimen, on being illuminated by a train of light waves, scatters them to a varying degree depending on its structure. The objective lens, limited by its NA, collects a portion of that scattered light. This light together with the remaining unchanged light is caused, by the action of the lens, to form an interference pattern which is itself the primary image of the microscope.

The observer believes this to be the structure of the specimen but its authenticity as a magnified replica of the specimen is limited by a number of factors. There are aberrations in the lenses, the objective itself can fail to resolve the detail and the total magnification of the microscope may be either insufficient to portray this resolution to the eye or too great, resulting in coarseness in the image.

Stray light is a serious cause of poor image quality. The operator can, by using the method of Köhler illumination, achieve considerable control of stray light. In conventional microscopy of thick specimens unwanted light, coming from out-of-focus images of planes above and below the plane in question, considerably degrades the image. Some modern developments specifically utilize the very shallow d.f. obtainable with high power objectives. By scanning very tiny fields of view and storing their images and by rejecting out-of-focus images they can later build, using electronics, composite, good-quality images which would otherwise be unobtainable.

To be visible the detail in the specimen must be resolved in the image. The image must be sufficiently magnified to match the resolving power of the eye. There must also be sufficient contrast in the image for the eye to perceive the detail. The methods for achieving contrast by exploiting the seven basic interactions of light and matter will be given in Chapter 2.

The aims of light microscopy are therefore (i) to render fine detail visible, and (ii) to establish to what degree of accuracy those images portray the specimen.

6. REFERENCES AND FURTHER READING

1. James,J. (1974) *Light Microscopy Techniques in Biology and Medicine*. Martinus Nijhoff, The Netherlands.
2. Michel,K. (1964) *Die Grundzuge der Theorie des Mikroscops*. Wissenschaftliche Verlagsgesellschaft MBH, Stuttgart. (Full of diagrams and drawings which are easily understood by the non-German speaking reader. Contains many extensions of the Abbe experiments and an explanation of phase-contrast is included).
3. Petran,M., Hadravsky,M., Benes,J., Kucera,R. and Boyde,A. (1985) *Proc. R. Microsc. Soc.*, **20**, 125−129.
4. Amos,W.B. and White,J.G. (1987) *Proc. R. Microsc. Soc.*, **22**, 525.

CHAPTER 2

Rendering transparent specimens visible by inducing contrast

A.J.LACEY

1. INTRODUCTION

This chapter considers a range of widely available techniques which produce contrast either between the image of the specimen and its background or between features within the image. The section on interference reflection contrast has been written by Professor A.S.G.Curtis. Fluorescence is only briefly mentioned as it is the subject of Chapter 6. Each technique is considered in terms of its principles, procedure and applications, and a comment is given on what the enhanced contrast also implies about the nature of the specimen. There is an emphasis throughout on the production of contrast within living cells but not exclusively so. The suggested specimens are readily available and are chosen to illustrate specifically the technique. The final section attempts to apply all the techniques to a single specimen, firstly to achieve a summary, but also to show that each image can be interpreted in terms of the physico-chemistry of the specimen, whether or not it shows good contrast.

Highly versatile microscopes carrying both transmitted and incident modes of illumination are now available in many biology laboratories and these allow the observation of an undisturbed specimen by a range of contrast techniques taken sequentially. Most frequently the transmitted light phase is used for searching specimens and then, by an easily achieved switching, epi-fluorescence or other incident light method can be applied to observe cell detail. Rarely will they both be used together as, in general, reflections tend to confuse transmission images and vice versa. Their combined use is however described in ref. 23.

Bright-field microscopy is still the most widely used technique and is the assumed method in the majority of chapters in this book. It can be considerably strengthened by the use of filters but its full potential can only be realized if the illuminating light is carefully controlled by the method of Köhler given in Chapter 1, Section 1.1.

1.1 Definition of contrast

Contrast is the ratio of the difference between the intensity of the background (I_b) and that of the specimen (I_{sp}), to that of the background:

$$\text{Contrast} = (I_b - I_{sp})/I_b \qquad \text{Equation 1}$$

Thus, a brilliant white specimen on an equally brilliant white background will provide no contrast and so will not be visible. Nor, however, will a (small) opaque specimen

Table 1. Bright-field microscopy: contrast assessment.

A. Set up Köhler illumination as per Chapter 1, Section 1.2.
B. Low-contrast situation.
    1. Replace the specimen you used for the setting up of Köhler with a preparation of cheek cells (prepared as follows: touch the inside of the cheek with a clean finger or wooden spatula, transfer the cells to a clean microscope slide, use saliva as a mountant and cover with a coverslip) or onion epidermal cells as described in *Table 17*.
    2. Adhering strictly to the Köhler method observe the specimens. Note that virtually nothing is visible, i.e. no contrast is obtained by this method on these unstained biological materials. There may be some visibility of the vertical walls of the onion cells by virtue of their absorbing capacity and resultant reduction of intensity.
    3. Note that the image is determined by the degree of absorption.
    4. Try closing down the aperture iris and note the improvement of contrast, but see the criticism of this technique in the summary to this section.
C. High-contrast situation.
    1. Now replace the specimen with a stained section such as a dermis stained with a Masson trichome technique (ref. 1). Remember to reopen the aperture iris to its correct setting.
    2. Observe at once the very contrasty coloured layers of the specimen against a white background.
    3. Recall that this contrast is achieved by differential absorption of light achieved by the use of stains (and sectioning), all of which has resulted in the loss of life processes in the specimen.
    4. Observe the effects on contrast of the different coloured layers by the use of filters of the same and complementary colours, outlined in more detail in Section 7.2 of this chapter and referred to again in Chapter 3, Section 8.2.1.

---

be seen if the background is excessively brilliant. If, however, the specimen is bright while the background is black then the contrast will be very good indeed. Methods of achieving contrast are based on the relations of light and matter which were listed and discussed in Chapter 1, Section 2.

2. BRIGHT-FIELD MICROSCOPY

Instructions for setting up and appreciating the contrast potential of this method are given in *Table 1*. As you see in step B4, improved contrast can be achieved by closing down the aperture iris. There is a temptation to do this in all situations where contrast is low but resolution of fine detail is lost when using narrow illuminating apertures. If too small an aperture is used, spurious boundaries develop around small features as a result of diffraction effects. The method then is not appropriate for accurate observation of fine detail.

Thus under bright-field microscopy, living single cells or monolayers of cells are not usually capable of altering the light by absorption to allow contrast to be achieved. Supplemented by stains, bright-field microscopy is however a powerful, commonly used technique and can be further enhanced by additions of filters to the microscope.

3. PHASE-CONTRAST

Principle: In the back focal plane (b.f.p.) of the objective a plate is inserted which has two effects, (i) to alter the phase of the zero order diffracted light in relation to the other orders (see Chapter 1, Section 2.5 for the meanings of these orders), and (ii) to reduce the intensity of the zero order to bring it more into balance with the other orders. In modern microscopes the plate carries a circular annulus which is different

Table 2. Protocol for phase-contrast.

1.	Check that you have phase-contrast objectives and appropriate condenser and preferably a telescope to assist in viewing the b.f.p. of the objective.
2.	Observe a specimen of cheek epithelial cells (see *Table 1*), using Köhler illumination. Note the very poor contrast obtained with the normal bright-field condenser setting. Some improvement, and this might be necessary even to find the cells, can be obtained by closing down the aperture iris of the condenser.
3.	Having found the cells, reopen the aperture iris and then swing in the appropriate illuminating annulus of the condenser—it will be coded by colour banding or number to match the objective in use.
4.	Observe the b.f.p. with the focused telescope and align the image of the illuminating annulus with the phase ring seen as rather darker than the rest of the phase plate. If the image of the light ring is too big or markedly too small or is well out of focus while the phase ring is in focus suspect poorly set up Köhler or that the condenser has slipped in focus. Adjust the focus of the condenser accordingly. If of course the bright ring is grossly wrong in size then assume that a wrong match has been attempted and re-examine the setting on the condenser. You have now set up phase-contrast conditions except for the use of a green filter. Not every one uses such a filter but it is appropriate to do so for evaluating differences of wavelength of 360/4°, say for green light, while such a proportion is not a meaningful concept for the mixture of wavelengths of which white light is made.
5.	Replace the telescope with eyepiece and observe. Some increase in the voltage in the lamp might be necessary but be careful not to over-run the lamp.
	You may prefer to set up the microscope with a contrasty specimen first and then replace with the transparent layer of cells once you have established the plane of focus, assuming slide thicknesses are the same.
6.	Change to a higher power objective and repeat steps 1−4.
	Most microscopes are capable of holding their centration for a considerable number of changes of objectives and condenser matching annuli but it is worth checking the alignment from time to time.

in thickness to the remainder and which also is partially silvered to achieve absorbance of some of the direct (zero) order light.

In order to have a illumination train of light comparable to the phase ring, an annulus is mounted in the front focal plane of the condenser which is conjugate with the b.f.p. of the objective.

The two features are matched in size and in alignment by the aid of a telescope, in place of the eyepiece, which is capable of being focused on the b.f.p. of the objective. The method is given in *Table 2*.

Check the microscope and its accessories for these points and preferably also have available a green filter. Check the use of the alignment screws and separate them clearly in function from such things as condenser alignment. Note that when the microscope is set for Köhler illumination, the condenser annulus will be in a conjugate plane with the phase annulus in the b.f.p. and if the maker's code of matching is followed then they will overlap perfectly in size in the b.f.p.

In all microscopy the specimen scatters light and the light which has passed through the specimen can be seen to have lost intensity compared with the light that has not passed through. (Compare, for example the intensity of the zero order when a specimen is present and not present.) The scattering is achieved by processes including diffraction and refraction as well as reflection and absorption. In the case of a transparent specimen the scattered light comes out of the specimen with a different phase relation to the zero order than if the features of the specimen had been opaque. The phase difference between

Table 3. To show the separation of diffracted from undiffracted light in phase-contrast microscopy.

1.	Set up as in *Table 2*, steps 1−4 but replace the specimen of cells with a diffraction grating or the Abbe test plate mentioned in Chapter 1, Section 2.5.1.
2.	Examine the b.f.p. and note the appearance such as shown in *Figure 1*.
3.	Note that the zero order passes through the phase retarding ring as stated above and that the diffracted rays (seen as multiple rings) do not do so for by far the greater part of their circumference. Only in small areas do they themselves also pass through the phase ring. Note also that their relative intensities to the zero order is markedly changed if the zero order is out of alignment with the phase ring.
4.	Alter the centration, change the condenser annuli and observe the effect in both the b.f.p. and the image contrast. It is possible to use a much larger illuminating annulus than is appropriate for e.g. a ×10 objective, and the image is then seen in totally reversed contrast known as dark-field which will be discussed further in Section 4.

them is about one-quarter of a wavelength for monolayer cells instead of about one-half. Wave trains having this lower level of phase difference will not interfere sufficiently strongly to give amplitude change of sufficient size for the eye to determine. The phase-contrast microscope brings those phase relationships to a level where they can produce a perceivable interference pattern.

3.1 Interpretation of phase-contrast image

The image will consist of variations in intensity of green or grey to white depending on the presence, or otherwise, of the green filter. In cheek cells the cytoplasm granules and the nucleus are, in positive phase-contrast systems, seen as darker in intensity than the cytoplasm itself and the background. The positive phase machines render the greater optical path darker than the lesser up to a certain value. The optical path is the product of the distance the light has to travel (in the z direction) towards the eye (t), multiplied by the refractive index (n) of that portion of the specimen through which it is coming. Both the linear path and the refractive index will be different in other areas of both the specimen and the background.

$$\text{Optical path } (\Delta) = t \times (n_o - n_m) \qquad \text{Equation 2}$$

In Equation 2, n_o is the refractive index of the specimen and n_m that of the medium in which the specimen is mounted.

Negative systems are also available in which case the situation discussed above is of reversed contrast, the nuclei appearing bright against a shorter path in the surrounds. Ambiguity does arise when the path lengths are greater than 360/4 and the position of focus is also critical. Here, however, we are only concerned with contrast and we have achieved a contrast, sufficient to be perceived by the eye or recorded on a photographic emulsion, of features which would otherwise have been invisible or nearly so.

3.2 Diffracted light in phase-contrast

An exercise to show the separation of diffracted light in the phase-contrast microscope is shown in *Table 3* and *Figure 1*.

Specimen

a)

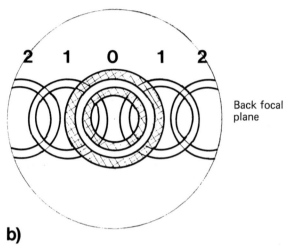

Back focal
plane

b)

Figure 1. Diffraction in the phase-contrast microscope. (**a**) Specimen of grating (phase) with line spacing ~ 10 μm. (**b**) B.f.p. of phase objective NA 0.25 showing zero order diffracted light (0) superimposed on darker phase plate and other orders of diffracted light (1 and 2) passing partly through the phase plate but mostly inside and outside it.

3.3 Use of phase-contrast to measure parameters of nuclei in cells (immersion refractometry)

The ability of the microscope to make optical path differences in the specimen visible as differences in intensity contrast allows the opportunity of measuring, for example nuclei inside cells for both their linear dimension in the z direction and also their refractive index.

For an extended discussion on this topic see Ross (2). *Table 4* gives the protocol for some sample measurements. These are merely illustrative examples and were originally published by Barer and Joseph in a series of papers (3 — 5). They have been used extensively by the Royal Microscopical Society in their teaching courses.

Phase-contrast microscopy has made a major improvement to the techniques available for observing living cells and is now routinely used in a wide range of work. It does have remarkable sensitivity, capable of distinguishing 7/360 of a wavelength. It has however a considerable disadvantage in that the image of each feature is surrounded

Table 4. Immersion refractometry of liver cells.

A. *To show changes in refractive index and to measure refractive index of living Amoeba.*

Requirements: sugar refractometer, bovine plasma albumin fraction V powder, phase microscope with green filter.

1. Prepare various strengths of the albumin dissolved in water. A stock solution may be very viscous but can be kept in a refrigerator for ~1 week.
2. Measure by the use of a sugar refractometer the refractive index of each solution by finding the equivalent % sugar and looking at the conversion tables supplied with it. Remember that the value is temperature-sensitive for any one strength. Take care to wash thoroughly the split prism after each measurement. A plot of the results will show that the refractive index changes linearly with concentration.
3. Prepare a single *Amoeba* in water on a slide. Cover.
4. Irrigate the edge of the coverslip with the albumin solution and draw through the solution with a filter paper placed on the opposite side of the coverslip. Repeat the irrigation until you are satisfied that the *Amoeba* is bathed in the new strength solution.
5. Observe the changes in the intensity of the various features of the *Amoeba* noting that as the albumin increases in concentration the *Amoeba* lightens in tone, disappears and then finally reappears in reversed contrast as the concentrations increase.
6. Reverse the irrigation series and obtain the reverse set of changes.
7. The point at which the *Amoeba* is invisible will be the concentration of albumin which has the same refractive index as the general features in the *Amoeba*.
8. Considerable change will occur in the visibility of the pseudopodia as they move. This is related to the changes of concentration of the cytoplasm within them.

B. *Blood cells.*

1. Make up a series of solutions of albumin in 0.7% NaCl solution and obtain the refractive index for each solution as described in the *Amoeba* exercise above.
2. Immerse a drop of blood in each of the media keeping the dilution error to a minimum.
3. Take five random fields of view and count the cells which are dark, bright and indeterminate. Do this for each solution concentration.
4. Tabulate against refractive index of medium.
5. Plot on graph paper and obtain the refractive index of the cells.

Table 5. Simple patch stop low power dark-field (and Rheinberg illumination).

1. Set up Köhler illumination and observe a slide on which a number of discretely separated features are visible, as, e.g. in a diatom preparation.
2. Remember that you adjusted the aperture iris to give 66% illumination of the b.f.p. of your objective. Examine the physical condenser iris and note the size of the opening.
3. Prepare a disc of transparent material of the size of the filter housing under your condenser. On the centre of this disc attach an opaque circular area of the same or slightly larger size than the aperture in your condenser iris.
4. Place this composite disc in the filter housing as near the condenser iris as possible.
5. Re-examining the specimen you will probably discover that nearly everything is black! Adjust the focus of your condenser slightly upwards and open up the condenser iris a small amount. This should enable the features to become brightly lit on a background of minimum intensity. They will be of reversed contrast.

by a halo of bright light. This makes linear measurement much more difficult, especially for very small objects.

3.4 **Meaning of phase-contrast image**

Phase-contrast microscope images are built up by the interference of wave trains of light whose relative phase has been altered by the machine used to collect the light. The image is effectively a map of the optical path differences between the features and between the specimen and its background. In so far as the refractive indices of protein solutions are related to their concentrations then the phase microscope image is also a map of the mass or concentration of proteins and other components of the protoplast. Note that for the microscope to achieve its production of contrast, the diffracted light must be sufficiently separated from the zero order light to pass along a different path, that is, one which excludes the phase ring in the b.f.p. of the objective.

4. DARK-FIELD MICROSCOPY

In setting up phase-contrast systems it is sometimes possible to obtain sufficient mismatch of the illuminating annulus to the objective that a dark-field results and the specimen shows in reversed contrast. You can try this using a ×10 objective, for example, with an annulus appropriate for a ×100 lens. Slight adjustment of the focus of the condenser will help as only the very marginal rays of light are coming through the condenser and their focal point is very close to the top of the condenser.

The principle of dark-field microscopy is that the zero order diffracted light does not contribute to the image formation. As Figure 5, Chapter 1, shows this can be demonstrated relatively easily by occluding the zero order in the b.f.p. with a central opaque stop. Alternatively a low-power lens can be used with a small patch mounted on a round coverglass put into the back of the objective. The illuminating aperture will be suitably reduced to match the central stop. In the author's laboratory this is achieved using the cornea of the butterfly as a specimen and the McCrone dispersion staining objective as the lens. The condenser of the microscope is usually best removed and the illuminating aperture is achieved by the original field iris which can be closed right down to produce a very narrow cone of light. This is then easily occluded by the patch stop in the b.f.p. of the objective. In bright-field conditions (without patch stop) an opaque, wire-netting-like image is produced by absorption but when the patch stop is in place, the wire netting becomes bright on a dark background. The new image is built solely of scattered light and lacks the zero order component.

In general use however, dark-field conditions are obtained by preventing the direct light from entering the objective at all. This is better than the use of a patch stop in the b.f.p. of the objective as described above. *Table 5* describes a method for constructing a condenser patch stop to achieve low-power dark-field.

It is worth having a set of such condenser patch stops (and they are cheaply made) in your accessory box. They provide a rapid and effective dark-field method for several low-power objectives. For higher power work special condensers are required and immersion conditions may be needed, see Section 4.2

Table 6. Protocol for high-power dark-field (see *Figure 2*).

1.	Set up the specimen (e.g. a single diatom) as though for bright-field conditions and locate it as far as you can noting the stage settings or other features enabling you to relocate it.
2.	Place a drop of oil on the underside of the slide and also onto the top of the condenser and place the slide carefully back onto the stage in such a way as to allow the two menisci of oil drops to push away any air bubbles.
3.	Recover the location of the specimen as far as you are able.
4.	In very high-power lenses there is a diaphragm in the b.f.p. of the objective which should be set at its maximum size, which is engraved on the side. This will enable bright-field conditions to be retained and will assist in locating the specimen further. Such lenses will be appropriately labelled and will probably also be used under oil-immersion conditions.
5.	Once you have located the specimen with the new objective carefully close down the diaphragm mentioned above and note that the background will lose its intensity to near zero and the image of the specimen will change to one of reversed contrast—that of dark-field microscopy.

Table 7. Condenser light cones in turbid cubes.

1.	Place a drop of oil onto the polished side of a turbid cube or one made from fluorescent perspex. At the same time place a drop of oil onto the top lens of the condenser. Place the cube over the stage and bring into oil contact with the condenser.
	NB This method ensures that any air bubbles are pushed to the side by the two merging curved surfaces of the oil drops.
2.	Using the bright-field setting of the condenser observe from the side the cone of light in the cube, noting its angle which is controlled by the iris under the condenser and the height of the apex of the cone above the top of the condenser—the working distance of the condenser. The apex of the cone, when used in Köhler illumination, is set into the specimen itself. Remember that in normal use there would be the slide thickness to consider.
3.	In the universal condenser the annuli can be rotated in for the phase settings. Observe the cones of light of each of these in turn and notice that they are hollow but are still sufficiently acute to enter the objective when in position.
4.	Rotate in the dark-field setting and observe the very obtuse angled cone which is again hollow but whose angle is so large as to be likely to go outside the objective.

Note that without the presence of oil between the top of the condenser and the base of the cube these rays of light would fail to enter the cube because they would be at an angle greater than the critical angle for the material and would be reflected from the surface.

4.1 Rheinberg illumination

The simple patch stop of the dark-field method that is described in *Table 5* can be made of coloured transparent material. Complementary colours are often used for the central patch in comparison with the outer zone. A blue patch on a yellow margin or green patch in the centre of a red margin are often used. It can be predicted from first principles what the colour of the specimens will be on the particular colour of background for the two combinations mentioned. The specimen takes the colour of the light which has been scattered by it while the background is that of the central patch.

Unstained, living material will be readily visible by the use of either dark-field or Rheinberg illumination methods to give an interesting contrast.

4.2 High-power dark-field

Universal condensers have a setting for high-power dark-field and there are also, in

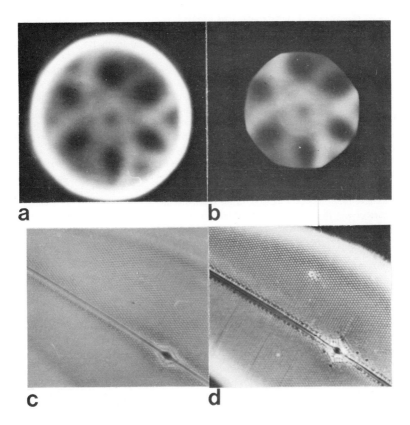

Figure 2. Significance of zero order light in image formation: dark-field image of *P.angulatum*. (**a**) B.f.p. of objective NA 1.3 with diaphragm open to allow bright peripheral zero order to contribute to image formation. (**b**) Same objective as for (**a**) but with diaphragm closed to occlude zero order. Note first order diffraction as series of intersecting arcs. (**c**) Image with setting (**a**). Detail is resolved but hardly visible. (**d**) Image with setting (**b**). Detail again resolved but much more visible than in (**c**). Note that detail is resolved in both (**a**) and (**b**) but much more visible in (**b**). (**c**), (**d**) ×900.

most laboratories, dark-field condensers of varying degrees of complexity.

Some dry working dark-field condensers are extremely easy to use, for example, that of Leitz for the Ortholux series of microscopes. After setting up Köhler illumination for bright-field, the condenser is wracked down to allow it to be slid out and replaced by the dark-field one which is then merely slid into place and focused up to stage level. Other makes have dark-field condensers which again can be slid in to replace bright-field ones but the majority of dark-field condensers used for high-power work need to be oiled to the base of the slide. A method for high-power dark-field is given in *Table 6*.

4.2.1 *Turbid or fluorescent cube demonstration*

If you have a universal condenser or a series of independent condensers which can be oiled then the exercise in *Table 7* is useful to set up and show what each type is doing.

33

Table 8. Image formation in dark-field microscopy.

1.	In the situation which has been achieved in *Table 6* (step 5) examine the b.f.p. of the objective with the phase telescope accessory.
2.	Open the iris in the objective and observe the zero order light as a brilliant ring around the outside edge of the b.f.p. (*Figure 2b*).
3.	Close the diaphragm down again and note that the edge of the diaphragm occludes the zero order light, preventing it from passing on to help construct the image (*Figure 2d*).
4.	Compare what is happening in the b.f.p. on various settings of the iris with what is happening to the image contrast. There is a reverse of contrast when the zero order is prevented from contributing to image formation. (Compare *Figure 2c* with *Figure 2a*.)
5.	If a specimen shows periodicity, e.g. as in the diatom *P. angulatum*, the first order diffracted light will be seen to be a series of crescents of coloured light intersecting in the centre of the b.f.p. The degree of symmetry of the intersection and also the degree of evenness of illumination in the bright outer zero order will be a measure of the centration and accurate alignment of optic axes of the condenser with that of the objective. If there is assymmetry in either of these aspects take steps to adjust by using the centering screws of the condenser.

4.2.2 *Significance of zero order light in image formation*

The significance of zero order light in image formation is shown by the exercise given in *Table 8*, for a specimen such as the diatom *Pleurosigma angulatum*. The b.f.p. of such a specimen and objective combination is, in dark-field microscopy, a very impressive sight (*Figure 2*) and should be sought for its own sake as well as for the understanding of the methodology.

4.2.3 *Constraints of high-power dark-field condensers*

There are sometimes difficulties in making high-power dark-field condensers work on previously prepared slides. The cause will already be plain to the reader in that the distance of the apex of the illuminating cone of light is too short to allow it to fall into a specimen if that specimen is mounted on a slide over a certain limited thickness. Nothing can be done about such a situation except to seek out a rather rare type of focusing dark-field condenser. The alternative would be to re-mount the specimen onto a thin slide!

A further constraint on the use of dark-field microscopy is that if too many features are present in the field of view, or if air bubbles or particles are present, then their presence will prevent a good image being obtained. The method is best used, therefore, with single small discrete structures well separated from each other.

4.3 **Dark-field image**

The dark-field image is an image produced by scattered light and so it is in effect a map of the discontinuities of refractive index and other light scattering features of the specimen.

5. POLARIZED LIGHT

This section introduces methods to produce contrast based on polarized light systems. It does not attempt to provide an introduction to crystallography.

Light waves have vibration directions in all planes at right angles to the direction

Table 9. Pleochroism.

1.	Obtain plane polarized light by placing a piece of polaroid (the polarizer) in the light before it enters the specimen.
2.	Looking at most specimens with such an arrangement and comparing with normally lit ones shows little difference beyond perhaps a slight reduction in intensity. This is because the majority of specimens, particularly biological ones, allow the light to pass through them equally whatever the vibration direction of the light. Such materials are described as isotropic.
3.	Rotate the specimen if you can or rotate the polarizer. If the specimen is isotropic then no changes in colour or intensity will occur on relative rotation.
4.	If possible examine a slice of mica sufficiently thin to be transparent, (a geological section of quartz tourmaline is very satisfactory). Rotate as above and take careful note of the changes of colour in relation to some linear feature of the specimen. See text (Section 5.1) for explanation.

of the ray. Plane-polarized light is light whose rays have vibrations only in one plane at right angles to their direction. Such light can of course be produced by simply filtering out all but the permitted vibration. Polaroid will produce this effect and to a high enough quality to be suitable for most of the purposes envisaged in this introduction. Chapter 1, Table 4 gives the method for determining the permitted vibration direction of a piece of polaroid.

A few simple exercises in *Tables 9−12* give a practical approach to the understanding of the way in which polarized light and matter interact and which therefore might be useful to achieve contrast in light microscopy.

5.1 Use of a single polar

The technique for producing contrast with plane-polarized light is given in *Table 9* under the heading of pleochroism.

From *Table 9* you will see that at one orientation the mica will be slate grey in colour and at right angles it will be straw yellow. The absorbance of the light is different depending on the relative orientation of the light vibration direction and some feature of the matter through which the light has passed. This is a phenomenon named *pleochroism*. The molecules of the dyes congo red and methylene blue are brick-shaped (*Figure 3a* and *b*). They are pleochroic in that colour is seen when the light passes through with its vibration direction parallel to the long dimension of the molecule. Such dyes can be useful in locating features in the specimen which have a directional bias such as, for example cellulose fibrils. The dye molecules could become arranged between and parallel to the microfibrils and would divulge this orientation when examined for pleochroism. If the molecule is colourless when the light vibration is parallel to the short dimension then it exhibits a class of pleochroism known as *dichroism*. (Not to be confused with dichroic in mirrors etc. used in fluorescence microscopy and other techniques.) Iodine and gold also show similar properties.

5.2 Use of crossed polars

If another polaroid is put across the light issuing from the polarizer it can either allow the light to continue by virtue of its permitted direction being parallel to that of the light vibration, in which case the two pieces of polaroid are termed to be parallel polars, or it can stop the light, the permitted vibration being at right angles to the light vibration

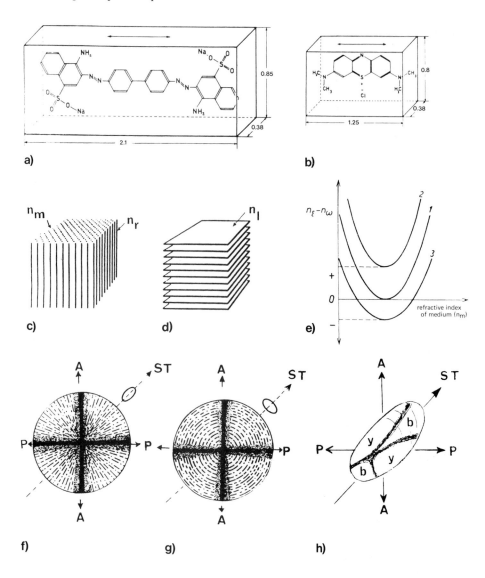

Figure 3. Polarized light for contrast in microscopy. Single polar and pleochroism: (**a**) congo red, and (**b**) methylene blue molecules showing direction of vibration (arrows) of light for strong absorption. Dimensions are in nm. Crossed polars and birefringence: (**c**) rodlet (n_r), and (**d**) lamina (n_l) forms lying in an inclusion medium (n_m); (**e**) change in value of birefringence ($n_\varepsilon - n_\omega$) with change of refractive index of inclusion medium (n_m) in such as (**c**) and (**d**). Lines: 1, positive form birefringence; 2, positive form and positive crystal birefringence; 3, positive form and negative crystal birefringence. (**f**) Negative spherite cross; (**g**) positive spherite cross with appropriate index elipses. (**h**) Potato starch grain between crossed polars and sensitive tint plate (ST = slow axis) showing addition colour blue (b) parallel to ST and subtraction colour yellow (y) at right angles to ST. Note also the isogyres are not parallel to analyser (AA) or polarizer (PP) directions. (**a**) and (**b**) after ref. 6 with permission of Harvard University Press and (**c**−**g**) after ref. 7 with permission of Unwin Hyman.

Table 10. Test for crossed polars.

1.	Set up the microscope with Köhler illumination in the usual way.
2.	Put in the polarizer.
3.	Take away the specimen.
4.	Place the second piece of polaroid over the eyepiece.
5.	Rotate the second polarizer which is now called an analyser until minimum intensity is achieved. This relative orientation is the crossed polars position. Note in some way the relative orientations and if possible set them in such a way as to have the polarizer with its vibration direction east−west.

Table 11. Examination of specimen between crossed polars for contrast between birefringent and non-birefringent areas.

1.	Set up for Köhler illumination.
2.	Set the polarizer to a known orientation.
3.	Immerse a few grains of potato starch in water, cover with a coverslip and examine. Very little will be seen other than perhaps the outline of the grains. Some contrast can be obtained by closing down the condenser iris, when perhaps the hilum and even annular structure might be visible. Recall however the inappropriate use of narrow beam illumination.
4.	Put in the analyser and note at once the suppression of the background to near zero intensity. The starch grains are however startlingly bright and there is a maltese-cross-like dark zone in each of them. This appearance is common to a number of crystalline structures and the image portrays the presence of a spherulitic structure (see *Figure 3f−h*).
5.	Rotate the stage with the specimen and observe that the isogyres (the dark lines in the grains) remain in the same orientation relative to the cross wires although the physical morphology of the grain changes in its relation to the cross wires.

direction and the polarizer. In such a situation they are described as *crossed polars*. How to obtain crossed polars and to examine a specimen between crossed polars are outlined in *Tables 10* and *11*.

Some polarized light microscopes have the following features which make for accuracy but which are not absolutely essential to achieve the level of use for contrast that is required here.

(i) The condenser and particularly the objectives are strain-free.

(ii) The polarizer is generally rotatable in a mount which will show the direction set.

(iii) The stage is rotatable and there are slots in the tube of the microscope to take various plates, one of which (that known as the Sensitive Red 1 Plate) will be considered in Section 5.2.2.

(iv) The second polarizer, the analyser is on an in/out slide carried in the tube of the microscope, closer to the eyepiece.

(v) Cross wires are usually in the plane of the primary image, that is inside the eyepiece. They are a most useful reference for making little sketches of orientation.

From the point of view presented in this chapter, sketches which include the cross wires as reference are essential in understanding the information likely to be made available in the image by the use of polarized light. From this point on, a simple polarized light microscope containing the above should be used for the exercises.

5.2.1 *Forms of birefringence*

The simple setting of a specimen between crossed polars as given in *Table 11* can be used for finding crystalline areas in tissue sections by their spectacular contrast from amorphous regions. The technique is used extensively for studies in bone, hair and teeth.

The contrast between the starch grains and the background, or between the crystals and the non-crystalline areas in a tissue, is due to the properties and orientation of those crystals and their effect on the light. A detailed interpretation will not be given here. Suffice to say that the bright areas are the result of their having two principle refractive indices. Light behaves differently depending on its vibration direction in relation to those refractive indices. The substance is said to be *birefringent* and *anisotropic*. Thus the placing of the specimen between crossed polars will give very satisfactory contrast between birefringent and non-birefringent features. It is sometimes helpful just to rotate the polars very slightly relative to each other, to make the non-birefringent areas visible by absorption.

Materials such as cellulose, starch, chloroplasts or mitotic spindle fibres have a complex structure which can give rise to at least two forms of birefringence. *Figure 3c* and *d* show structures made up of rodlets (or laminae) of refractive index n_r and n_l in a medium of a different refractive index n_m. Placed between crossed polars and rotated, these structures will exhibit birefringence which will be more or less determined by their form. If the medium is then replaced by one having a refractive index equal to n_r or n_l the form birefringence will disappear leaving any residual birefringence present as that related to the intrinsic birefringence of the rodlets or laminae. By measuring the change in birefringence $(n_e - n_\omega)$ against a changing medium refractive index, a series of graphs can be obtained (as in *Figure 3e*) and used to separate the various forms of birefringence.

5.2.2 *Direction of birefringence*

The direction of birefringence can be determined by the use of a comparator crystal such as that known as the Sensitive Tint Plate; a protocol is given in *Table 12*. Its use is also very rewarding in achieving contrast between areas of differing orientation or between birefringent areas against amorphous structures.

The Red Plate, when set at 45° to the crossed polars, will give a reddish pink to the background. What has happened is, in effect, that the plane-polarized white light entering the plate was split into two beams, one passing relative to one refractive index and the other relative to the second refractive index of this birefringent material. Because the refractive index of the plate in the direction of the marked slow axis (ST in *Figure 3h*, $n_\parallel$) is higher than the one at right angles to it $(n_\perp)$, the beam vibrating in the high direction has been slowed down relative to the other. On entering the analyser both beams were again resolved into two components and the components, passing through the analyser from each, interfered with one another. The difference in the optical path (o.p.d.) of the two beams resulted in the amplitude of the green light being reduced to zero. White light minus green is seen as the Red 1 colour. The practical point is that the arrow on the sensitive plate shows the direction of the high refractive index

Table 12. Use of Sensitive Tint Plate.

1.	Follow steps 1–5 in *Table 11* and then insert the Red 1 or Sensitive Tint Plate in the slot in the back of the objective or body of the microscope.
2.	Observe the colours in the starch grain and note them in respect of the cross wires' directions. Do this with different orientations of the starch grain or other specimen you are using.
3.	Note the colour of the background is the red of the first order of Newton's colour scale.
4.	Follow the interpretation of the perceived colours in the main text.

direction. The difference in the two refractive indices ($n_\parallel - n_\perp$), which is the degree of birefringence, is of course multiplied by each unit of thickness of the plate. Thus the o.p.d. is the product of thickness (t) and the birefringence ($n_\parallel - n_\perp$).

$$\text{o.p.d.} = (n_\parallel - n_\perp) \times t \qquad \text{Equation 3}$$

The brightness and colour of materials seen between cross polars when appropriately orientated provides information about the degree of crystallinity and the orientation of, or within, a structure. The Red 1 plate has an o.p.d. of 550 nm and this effectively destroys the green component of white light. Other colours can then be interpreted as o.p.d. values—see Newton colour charts. If the thickness of the specimen is known then the value of the birefringence can be calculated from Equation 3.

Consider the colours of the starch grains in *Table 12*, step 2 and *Figure 3h*. A remarkable colour change will have been observed from that seen in (*Table 11*, step 4). The grains now show blue and yellow quadrants with the isogyres Red 1 instead of black. The colours remain in the same quadrants relative to the cross wires although the starch grains rotate. The background remains the colour Red 1.

The result is that new contrast has appeared and it has value in this respect alone. As a result, the grains can be more easily photographed. The integrated exposure meter can handle the light measurement of the whole field (see Chapter 3). The individual grains require a spot photometer in the crossed polars situation if a satisfactory picture is to be obtained.

The second point is however perhaps more important in that further information is revealed about the specimen. The colours are either up the Newton scale (blue = an addition colour) or down the Newton scale (yellow = a subtraction colour). To produce an addition colour the total o.p.d. must be greater so any effect of that area of the specimen must have its birefringence in the same orientation as the Red 1 plate. In other words the high refractive index must be parallel to the slow axis of the Red 1 plate given by the arrow.

In the case of the starch grain, there are areas (blue) where the high refractive index is parallel and yellow areas where the low refractive index is parallel to the slow axis of Red 1. These areas actually radiate out from the hilum of the grain. By reference to *Figure 3f* and *g*, the spherulitic structure can be classified as positive or negative.

On rotating the grains the colours remain in the same orientation to the cross wires but the quadrants of the grains change colour according to their relative orientation to the cross wires. It was recommended earlier that sketches of the fields of view should

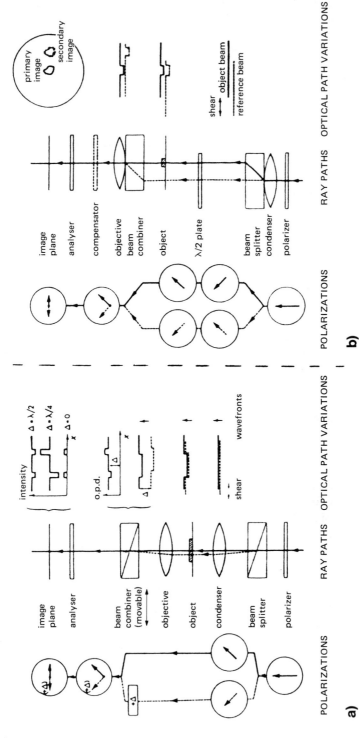

Figure 4. Ray paths interference microscope systems. (**a**) interference contrast (DIC), (**b**) Jamin-Lebedeff interferometer microscope. Arrows in polarization figures are indicating directions of travel of rays when seen end on. Other arrows indicate direction of travel of rays. (Δ = o.p.d.) Both drawings are after Spencer (10) with permission from Cambridge University Press. Note in (**a**) the intensity values across the feature and compare with the nucleus edges in *Figure 7a* and *b*. Note in (**b**) the lateral shear distances are for instance in the author's Smith version, 330, 160 and 27 μm for the $\times$10, $\times$40 and $\times$100 objectives, respectively.

be made in order to assist the memory and help the interpretation. For further studies on starch and polarized light see Frey-Wyssling (6).

5.3 Use of polarized light in biological microscopy

No attempt is made here to review the many uses of polarized light but two points might be included over and above the immediate study of pleochroism or birefringence given in Sections 5.1 and 5.2.

(i) On many occasions plane-polarized light can be useful to improve contrast by cutting down glare, (e.g. see Section 6.3). When light is reflected from a surface it will change its vibration direction and so it is possible to cut reflections out by using a polarizer to eliminate those vibration directions. One use recently mentioned is in examination of immunogold—silver stained specimens in immunocytochemistry. One investigator uses plane polarized light to examine his specimens stained in this way and the gold differentiated areas are visible as a green pleochroic colour (McKenzie, personal communication).

(ii) Polarized light is used in the majority of existing interference microscopes both of the differential interference contrast (DIC) type and also in some interferometric microscopes (see Section 6.1 etc.). In these microscopes the light—specimen interactions are a different issue and the polarized light is merely used to enable beam splitting and recombining to be achieved by the microscope itself.

Further reading and development of the use of polarized light to study biological materials is to be found in references such as Ruthman (7), Schmidt and Keil (8), Frey-Wyssling (6) and Roelofsen (9) which are mainly concerned with polymeric carbohydrates such as starch, cellulose and glycogen.

5.4 Summary of the contrast images produced by polarized light methods

The use of polarized light can therefore give excellent contrast by suppressing the intensity of the background to virtual darkness in crossed polars and displaying brilliantly, and by colours, the image of the specimen. Furthermore, the image can be interpreted in terms of refractive indices, optical path and orientation. The image between crossed polars, after allowing for orientation, represents a map of birefringent materials within the field of view. Using plane-polarized light without the analyser will, on rotating the specimen, give a map of the pleochroic features of the specimen. The presence of birefringence indicates the presence of some form of ordered arrangement in the specimen.

6. INTERFERENCE METHODS FOR ACHIEVING CONTRAST

6.1 Differential interference contrast (DIC)

A most popular method of achieving contrast, probably utilized more now than phase-contrast, is that called Normarski or Interference Contrast. It has long been popular in materials science where incident illumination is frequently used but the method is now rapidly overtaking phase in biological laboratories.

The principle (*Figure 4a*) is that a beam of plane-polarized light is split by a beam

Table 13. Protocol for DIC.

1.	The instruction manual for the microscope will indicate the coding of the condenser Wollaston prisms to the appropriate objectives.
2.	Prepare the specimen of cheek epithelial cells mounted in saliva or isotonic medium (see *Table 1*). Cover with a 0.17 mm thick coverslip.
3.	With Köhler illumination set the interference condenser to normal illumination setting. (You will perhaps notice that the direction of shear is shown on the condenser housing as you sort it from the accessory box!)
4.	Insert the polarizer.
5.	Locate the cells perhaps by reference to the air bubbles in the saliva and again by closing the aperture iris temporarily to improve absorption contrast.
6.	Rotate in the appropriate condenser setting (Wollaston prism) for the objective in use.
7.	Remove eyepiece to examine the b.f.p. of the objective.
8.	Cross the polars by inserting the analyser and by tuning the upper Wollaston prism obtain the minimum intensity band in the b.f.p. Close down the aperture iris to a little larger than the dark band.
9.	Replace eyepiece and note the appearance of the image. The cells will show a three-dimensional quality and the direction of apparent lighting should be noted.
10.	Adjust the second Wollaston prism while you watch the image and note that you can reverse the apparent direction of lighting and perhaps experience at the same time the change in perception of hills and valleys.
11.	Infer, and it is important to do so, that the image you see is a result of the microscope setting.
12.	The setting of the microscope in this way is displaying as a contrast image the o.p.d. of the different features and their surroundings. Each change of optical path is represented by an edge effect giving the three-dimensional impression.
13.	Colour can be obtained by referring back to step 8 and tuning the prism to obtain a coloured band on either side of the zero order grey.

splitter (Wollaston prism) into two beams. Both beams pass through the specimen very close together but sufficiently far apart for the eye to perceive an edge, or three-dimensional effect, by differences of intensity in the final image. The two beams are recombined in a second Wollaston prism mounted in the b.f.p. of the objective and then an analyser completes the image formation. The two beams are sheared laterally, relative to each other and this direction of shear is perceived as for example in *Figure 7a*, an apparent lighting of the image from one side. See *Figure 4a* for the light path and note that the second Wollaston prism can be moved relative to the other which gives a method for tuning the instrument. The o.p.d. of the two beams is usually set to give what is called the zero order grey as a background intensity and then the specimen is outlined by an edge of either increased or decreased intensity relative to it. However, by tuning the instrument the direction of shear can be altered by 180° and by so changing the direction of shear an interesting effect in perception by the observer becomes apparent. What at first looked like a fried egg on one setting can be made to look like a plate with a depression cavity in it on the new setting (see *Figure 7a* and *b*). Again this can be important in the presentation of micrographs taken with this method. If the picture is inadvertently turned round the same result may appear.

A major change of the settings of the Wollaston prism can allow the observer to have a coloured image on a very pleasing background colour. The colours are again on the Newton interference white light minus scale. Incident light microscopes have only one

Wollaston prism as the illuminating one is one and the same as the objective prism. A sample protocol is given in *Table 13* for first time users of DIC.

Note that the DIC microscope has a crossed polars situation in its light train and thus any birefringent material in the specimen will show up bright at certain orientations. Again note that the direction of shear may coincide with a linear feature in the image, orientated in the same direction, and thus this feature will not show the edge effect in the normal way. (Note this point is mentioned in practice in Chapter 9, Section 6.1.)

6.1.1 *Use of differential interference contrast*

There will be several references to DIC in subsequent chapters as it is used extensively for examining unstained specimens. It is an excellent technique for making o.p.d. in the specimen visible and does so largely by the shear effect at the boundaries of those areas of o.p.d. It is not appropriate to pursue this technique to make quantitative measurements of optical path and the Jamin-Lebedeff or other true interferometer type of microscope should be used (Section 6.2).

6.2 **Jamin-Lebedeff interference microscope**

A very valuable method which is, however, disappearing from many laboratories is the use of the principles of interference of two light trains originating from the same source. These can be made to convey information about the specimen by putting one beam through the specimen and the other around it and comparing the two. The two beams can be tuned together to give a background of interference fringes which can be interpreted as a series of contour lines of optical path length or by broadening one of these fringes produce a flat background. The specimen then distorts the fringes or shows up as different intensity in monochromatic light or a different colour from that of the background in a white light illumination system. The shift of fringes or changes in relative intensity or colour can be related to optical path (= thickness × relative refractive index) and thus be used to calculate such parameters as film thickness or mass of a nucleus even while it is within a living cell. This method has considerable advantages over that of phase contrast in immersion refractometry (see *Table 4*) in that there are no haloes around the image features and gradients of optical path changes are rendered visible. The instruments are however considerably more expensive than either the phase or the DIC systems.

A sample protocol is given in *Table 14* for this technique as many laboratories could well have these machines stored somewhere and they are certainly worth a try. Ross (2) describes their use in some detail and their potential can readily be shown by studying the cheek epithelial cells mentioned earlier. The machines vary in their method of separation of the two beams. Some shear the reference beam to one side (*Figure 4b*). The separation of the observed image and the reference image is a limiting distance making discrete objects more easily interpreted. Other machines are double focus systems and do not have this limitation. As *Figure 4* indicates these microscopes use polarized light and polars are also used for beam splitting and recombination.

Table 14. Protocol for using a Jamin-Lebedeff-type interference microscope.

A. *Setting up the interference*

1. Mount cheek epithelial cells in saliva on a 1 mm thick slide covered with a 0.17 mm thick coverslip.
2. Set up Köhler illumination.
3. Using an appropriate condenser for the ×10 objective and white light insert the polarizer.
4. Cells will be more easily found if reference is made to the air bubbles in the saliva. The liquid between the bubbles will contain the cells.
5. With the goniometer set on the working part of its scale and the quarter wave plate inserted, tune the condenser tilting screws to achieve an even intensity in the background. This is best achieved by replacing the eyepiece with a telescope and using the b.f.p. of the objective but you will have to place a piece of ground glass in the light path below the condenser to break up the image of the lamp filament which will be present if Köhler is strictly adhered to.

    When first examined the b.f.p. may have a series of coloured fringes across it; these can be centred by finding the white lying between two Red 1 colours. The fringes can be broadened or narrowed by turning the tilt screws in opposite directions to each other; rotating the goniometer will move the whole fringe pattern.

    Once you have found the central white band broaden one of the Red 1 bands to fill the b.f.p. or as nearly as possible. The aperture iris can be reduced a little if required. By broadening one of the fringes the two beams have been tuned to give an o.p.d. between them of known value for the whole field of view. Where there were multiple fringes showed effectively a contoured optical path increasing on either side of the zero white fringe.
6. Remove the ground glass, replace the eyepiece and observe the cells with their nuclei. They will, if using white light, show contrast by colour differences from the background and within their cytoplasm. If using monochromatic light (green) then the optical path differences will be manifest as intensity differences.
7. It may be preferable to examine one cell by a ×40 objective so change the objective and condenser pair and carry through the appropriate tuning (as in steps 5 and 6).

B. *Taking measurement with the interference microscope*

1. Make a tabulation of the colours of the background, the cytoplasm and the nuclei against the angular setting of the goniometer. By finding the angle adjustment needed for transferring the colour of, e.g., the nucleus to the background or the cytoplasm to background or vice versa it is possible to say that such an angle is equivalent to the o.p.d. between the background and the feature concerned. A calibration of the goniometer angle will then give this o.p.d. in units of, e.g. wavelength of green light (see further under next section).
2. If you illuminate with monochromatic light, e.g., wavelength 550 nm, then changes of the goniometer will give changes of intensity as already mentioned. Again tabulate but this time merely state that the feature is either getting lighter or darker because it is not easy to say that at a certain setting the feature is at minimum or maximum intensity. If you now plot these changes against goniometer angle a graph of the changes for the background can be compared with that for the feature of interest and the separation of the slopes will be the equivalent value of the goniometer angle to the o.p.d. between them. Calibration for the goniometer angle will again be needed for the particular wavelength in use.
3. Equation 2 relates the o.p.d. to the product of the thickness in the *z* direction and the difference in refractive index of the specimen to that of the immersing medium. So far only the o.p.d. has been measured, and the thickness and one or other of the refractive indices needs to be found in order to calculate the other. Do this by measuring the nucleus e.g. in the *x* and *y* directions and assume that its dimension (t) in the *z* direction is the same as one or other of these. Measure the refractive index of the medium by a refractometer as before.
4. Step 2 can be repeated using another known wavelength, e.g. that of red light. A comparison of the values of o.p.d. green and o.p.d. red gives a pair of quadratic equations from which to calculate two of the unknowns, while again measuring, e.g. the refractive index of the medium.

The protocol in *Table 14* is for the examination of cheek epithelial cells or other discretely separated specimens using a shearing system. The first section (A) demonstrates use of the microscope for obtaining contrast and the second part illustrates its use for quantitative purposes. The individual manufacturer's instruments will of course require their own instruction manuals but the principles will be much the same throughout.

Interferometric microscopes have been successfully used in determining, for example, the refractive indices of nuclei or sperm heads and, by knowing the refractive increment with changing concentration, it is possible to calculate the concentration or even the mass of such tiny features. The interference microscope can of course be used in the same manner as the phase-contrast in the technique of immersion refractometry and it is indeed preferable for the reasons already given. Significant contributions have been made by these techniques, for example to the study of muscle by Huxley and Niedergerke and to nucleic acids in nuclei by H.G.Davies in the 1950s. Further details can be found in Ross (2).

The usefulness of the interference technique for quantitative cytochemistry has now probably been overtaken by the alternative of photometric measurement of fluorescence from labelled highly specific antibodies.

At least two interference microscopes—the Interphako of C.Zeiss, Jena and the Pluta device from P.Z.O.—are, however, still available commercially.

6.3 Interference reflection microscopy

6.3.1 *Introduction*

Interference reflection microscopy (IRM) is a technique for investigating the thickness of a gap between two optically transparent structures. If it can be assumed that one of the surfaces is planar, the topography of the other can be appreciated or even measured. Thus, when used quantitatively the technique gives contours of gap thickness, that is, isopachytes. In effect, for reasons discussed below, it is limited to the study of substrate—cell separations in the range $10-30$ nm. When used qualitatively it provides an appreciation of such things as the closeness of approach of a cell to a surface to which it is adhering, provided that the surfaces approach within 30 nm. The technique was introduced to biology in 1964 (11).

6.3.2 *Basic physics*

The basic physics of the technique can be appreciated from *Figure 5*. Light is reflected at each interface of refractive index discontinuity it encounters, for example at the interface between a glass dish and the watery medium it contains and at the interface between a cell and the medium it is growing on. Thus a beam at normal or near normal incidence passing through a glass—water interface and then a water—plasmalemmal interface gives rise to two reflected waves which will be, unless very special conditions are met, out of phase. These two reflected waves will give rise to interference if they are out of phase. The basic optical physics of this situation has been described fully in ref. 12.

The intensity of the resulting reflected light is determined by the refractive indices, or the three different media, and by the gap dimension. The quantitative relationships

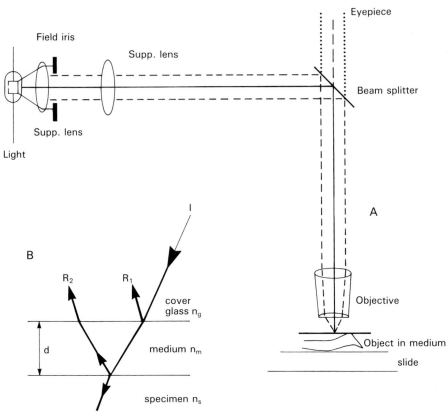

Figure 5. Interference reflection microscopy. (**A**) Ray diagram. Broken lines represent incident light, dotted lines represent reflected light beams. The ray paths for incident and reflected beams are not quite identical but the difference is too small to portray on this figure. (**B**) Detail at specimen: I = incident ray; R_1 and R_2 = reflected rays from the boundaries of cover−glass and medium−specimen respectively; d = distance from cell surface to coverglass; n = refractive index of glass, medium, specimen etc. Note phase difference (Δ) between R_1 and R_2 is proportional to distance (d). R_1 loses $\lambda/2$ on reflection at boundary and thus partial destruction by interference takes place between R_1 and R_2 with a total phase difference of ($\Delta + \lambda/2$).

are given for the simple three layer system or a narrow layer between two quasi-infinite structures by:

$$p = \frac{\left\{ n_1^2(n - n_0)^2\cos^2\left(\frac{\pi}{\lambda}\, 2n_1 d\cos\theta\right) + (n_1^2 - n_0 n)^2\sin^2\left(\frac{\pi}{\lambda}2n_1 d\cos\theta\right) \right\}}{\left\{ n_1^2(n + n_0)^2\cos^2\left(\frac{\pi}{\lambda}\, 2n_1 d\cos\theta\right) + (n_1^2 - n_0 n)^2\sin^2\left(\frac{\pi}{\lambda}2n_1 d\cos\theta\right) \right\}}$$

where p is the relative intensity of reflected light, θ is the angle of incidence, d is the thickness of the layer, n_0 is the refractive index of the glass, n is the refractive index of the cell and n_1 is the refractive index of the gap between cell and substrate. This treatment is based on normal incidence. In practice, of course, the microscope objective illuminates the specimen with a cone of light whose angle is determined by the numerical

Table 15. Basic instructions for IRM.

1. Use a microscope with an incident illuminator (often known as Ploem illumination) using a mercury (100 W minimum) or xenon arc lamp as a light source. The beam splitter must not be wavelength selective and thus any dichroic beam splitter must be replaced with a neutral pellicle, half-silvered coverslip or appropriate beam splitter. Prism-type beam splitters may lead to the appearance of subsidiary shifted images. A field iris should be fitted in the optical train and a slot for filters is almost essential so that monochromatic illumination can be used and/or the overall intensity reduced. If a mercury arc lamp is used bear in mind that there is very little red light in most such lamps so that any interference which should produce red light, or requires red light illumination, is likely to be relatively suppressed.

2. (i) Using incident light inspect a piece of clear muscovite mica (ruby mica) using a ×10 objective.
   (ii) Focus on the edge of the mica. Ensure that the mica has been slightly strained near the edge so that wedges of air have entered some of the cleavage planes. These form the sources of interference. The mica should be mounted in air. This gives a very high-contrast specimen.
   (iii) Check that the light is centred and that the image is relatively free from glare. If glare is present, close the field iris and if necessary paint any of the insides of the microscope tubes with optical matt black to reduce reflections from bright metal edges etc.

3. Observe zero order and higher order fringes by their colour in white light. Then observe under single colour illumination. Interpret the images in terms of the spacings between the cleavage layers that are being revealed.

4. Repeat steps 2 and 3 under ×40 and ×50 objectives.

5. Biological application.
   (i) Obtain cells grown on or adhering to a glass or other fairly high refractive index coverslip. Do not use polystyrene coverslips.
   (ii) Ensure that the coverslips are thin enough for use of a ×50 or ×100 objective.
   (iii) Mount the coverslips on a glass slide with appropriate measures to prevent dehydration during observation and ensure that the cells are on the side of the coverslip facing the glass.

6. Using transmitted phase-contrast illumination focus on the regions of the culture which you wish to examine by IRM. Use a ×50 or ×100 objective of NA of at least 0.8 and oil on the objective if this is appropriate to the design of the objective, having very carefully wiped off any moisture on the top surface of the coverslip.

7. Turn off the phase-contrast illumination and use the incident illumination. Very carefully focus the specimen. It will be in focus over a much smaller range than for a phase-contrast specimen. The focus will lie near the point at which the edge of the field iris is also in focus (provided that it is closed enough for its edge to be visible in the field of view) so start with the field iris fairly stopped down. Focus on the edge of a cell so that an object of contrast will be visible.

8. Adjust the field iris, filter set and any supplementary lenses for optimal contrast.

9. Note that the NA of ×10, ×20 and most ×40 objectives is too low for an adequate image to be formed in a biological system.

10. Use white light illumination to estimate which order the fringes belong to. Note the caution about red fringes given in step 1 above. Interpret the image in terms of the closeness of approach of the cell to the coverslip.

aperture (NA) of the objective (unless an aperture iris is fitted and used). It should also be noted that in practice very few objectives have the NA claimed for them.

The question of the effect of the non-normal incidence of the light has been very thoroughly investigated by Gingell and Todd (12,13). They find that for appreciable interferences, that is, those beyond the zero order, the effect of an appreciable illuminating NA is to introduce substantial changes from those expected from the normal incidence situation. However, the zero order interference arises from cell surface separations of less than 30−40 nm. At these separations, as Gingell and Todd (12) point out, the expected interference of the normal incidence situation approaches very closely that of the cone illumination situation. Since the gap separations of $0-30$ nm are those of most interest it is clear that using the normal incidence treatment is permissible.

There has been some discussion of the exact intensity relationship to be expected in a microscope system (refs $12-15$), in part because of uncertainties about the values of refractive index to be adopted for the various layers, but in subjective terms, for the refractive indices commonly found in biological systems of glass ($n = 1.52$), medium ($n = 1.34$) and plasmalemma ($n \sim 1.45$), maximal interference will be expected for minimal gap thicknesses and the first peak at thicknesses of about 50 nm. However, if the cell is thin, interference may arise by reflection from the opposite side of the cell (14). Since the light intensity to be expected at 0 nm separation differs very slightly from that to be expected at less than 10 nm separation, IRM is not really capable of distinguishing molecular contact of the plasmalemma and substrate from adhesion with an appreciable separation of the cell from the substrate.

Attention has also been given to the interferences arising in more complex systems which may, or may not, approximate still more closely to the real biological structures (see refs 12 and 13).

A set of basic instructions for examining a test specimen (mica) and such specimens as living chick fibroblasts grown on glass coverslips is given in *Table 15*.

6.3.3 *Interpretation of results*

(i) *Qualitative.* For most combinations of refractive index likely to found in biological systems, in which glass coverslips are used as the substrate for the cells, the closest approaches between plasmalemma and glass will appear black or very dark grey. If the glass is replaced by a polystyrene coverslip, the refractive index combination is such that the image has almost no contrast. Thus adhesions of fibroblasts and a range of other cell types are characterized by focal contacts which are small areas of relatively, perhaps very close, contact between substrate and cell. These appear as very dark discrete patches (16) while the grey areas are indicative of larger separations and have been termed close contacts (17). Interesting observations have been made on the effects of various types of protein coat on the substrate and the appearance of the contacts. It is important, however, in the interpretation of these images to bear in mind that there may be contributions to the image from (i) interposition of layers of differing refractive index, such as adsorbed protein layers, in the structure being examined and (ii) reflections from the other side of the cell if it lies close to the undersurface of the cell.

Much could be done to sort out these various effects if comparative studies were carried out using coverslips made of sapphire which, being of much higher refractive index than glass, gives completely different expectations of the sort of image to be formed. It is interesting to note that with a three or four component system and the values of refractive index that are normally to be expected there is a limited range of expected interferences: this in turn reflects that the finding of those degrees of phase retardation provides strong evidence that the gap refractive index is close to 1.34 and the plasmalemmal refractive index is close to 1.41. It should be borne in mind that the lateral resolution of IRM is relatively poor.

(ii) *Quantitative.* Quantitative interpretation of the IRM image requires optical modelling of the system and the fitting of the best model to the data. This is the main reason why very few quantitative studies of the gap dimensions to be revealed by IRM have been made. Furthermore, there have been difficulties in excluding errors arising from stray light. Fortunately the system carries an internal calibration in nearly every image. This is the reflection from a single interface, namely glass—medium. An early and simple quantitation was carried out by Curtis (11) using the relationship set out above. Much more recently Bailey and Gingell (18) have investigated the irradiances of close and of focal contacts using a mathematical system for modelling an object with up to five thin layers, that is, gap, adsorbed protein, plasmalemma, cortical cytoplasm and deep cytoplasm, and comparing calculated values with observed irradiances. It is curious hat no-one has yet taken advantage of varying the substrate refractive index as a method of testing the correctness or otherwise of the model being used to interpret the image.

7. DYES, FILTERS, SECTIONING AND PREPARATION

Several chapters in this volume give details of modern methodology for inducing contrast into biological material for microscopical examination by using dyes of various types, including those associated with antibodies, but a comprehensive inventory of the use of dyes and stains is outside the scope of this book. The use of absorbing chromogens for inducing contrast does of course rely on the uptake of these substances by differing physico-chemical entities in the specimen and the environment in which these entities exist can affect their properties profoundly.

7.1 **Dyes**

The technique of using dyes to render living material visible has been replaced by such optical techniques as phase-contrast and DIC but there are cases where the physico-chemical variability of the specimen can be exploited using fluorescent dyes. These are used in sufficiently low concentration not to harm the life processes yet they give out sufficient light to be visible against a black background. The older, non-fluorescent dyes achieved their colour or intensity effects by absorption of the light and had to be in considerable quantity to make any changes in the light visible to the eye. Such quantities were generally toxic.

With fluorescent dyes however the quantity of dye required is very low indeed and need not be at a toxic concentration. Ploem (19) mentions acridine orange at a dilution of 1:100 000 in phosphate-buffered saline (PBS) for macrophages and fibroblasts.

Until the recent regulations concerning animal experimentation came into being, it used to be possible to take a drop of blood from the tail of a living mouse carrying *Trypanosoma* and mix it with a drop of very dilute isotonic acridine orange. Illumination either by transmitted dark-field short wavelength blue light or by epi-illumination using the same wavelengths made the recognition of the swimming trypanosomes very spectacular. They would be bright apple-green and their nuclei could be seen as red. The erythrocytes were very dark and could be seen by using a very low illumination (in the epi-method) in a supplementary transmitted illumination at the same time. Too much UV (in the illumination) would of course destroy the parasites as would the more usual drying out problems.

Table 16. Use of filters for contrast.

1.	Examine, using bright-field microscopy, a specimen such as a well stained haematoxylin – eosin smear or a Masson-trichome-stained dermal section. Note the colour distribution using no filters.
2.	Repeat with a green filter in the light train. Note the reduction in intensity of all previously red features with an increase in contrast between them and any light green or blue areas.
3.	Repeat using other filters such as red or blue. Note for single-stained specimens the contrast will improve if a complementary coloured filter is used unless the specimen is very thick and over-stained in which case use a filter of the same colour.

A modern version of this very simple technique, avoiding the use of living animals, is the culture from frozen stock of *Trypanosoma brucei* in the multiple component medium M199 with Earle's salts (Flow Laboratories) supplemented by 10% fetal calf serum and glutamine, while protected by penicillin and streptomycin. Acridine orange at a concentration of 50 μg/ml with or without 5 μg/ml ethidium bromide will give, with quite low excitation requirements, good fluorescent, living, procyclic parasites (Hudson, personal communication).

Much more detail on the use of both absorptive and fluorescent dyes but with less emphasis on living tissues will be found in the chapters on histochemistry, immunocyto-chemistry and fluorescence later in this book (Chapters 4−6).

The quantitative assessment of staining within structures in cells can be done by, for example microdensitometry to prepare chromosome banding profiles as described in Chapter 9, Section 6.4.1. The measurement of fluorescence from the labelled antibodies featured in Chapter 4 is the subject of part of Chapter 6. The counting of coloured or tonally differentiated features within tissues is described in Chapter 7. These are all examples of the use of dyes for quantification in biological investigations.

7.2 **Use of filters**

A section of animal tissue such as, for example, the dermis stained by the trichrome method of Masson (ref. 1) will show a wide range of colour from blue to red. It is a most useful exercise (*Table 16*) to examine the same field of view using white then blue then green and finally red light. The portions of the tissue stained red, for example will show much darker in green light. Contrast is thus improved by the use of complementary coloured filters. Detail can sometimes be seen more easily in a very heavily stained specimen by the use of a filter of the same colour as the dye. The use of filters is well understood by photographers and in the taking of micrographs through the microscope but often seems to be ignored or forgotten in visual use of the microscope (see Chapters 3 and 8 for examples).

7.3 **Sectioning**

From a light microscopy point of view the purpose of sectioning is to provide a clear picture of as near a single plane as possible. That is, there should be minimum thickness in the z direction, as any detail in the image will be confused by out of focus information coming from both below and above the immediate plane of interest. There is in most working situations a compromise between the thickness with its problems and the contrast obtained in the x and y directions. Features must have enough thickness to enable adequate amounts of stain to be taken up to achieve sufficient absorption or even to achieve sufficient change of optical path in phase and DIC techniques. As a result very thin sections have to be stained to a greater extent than thicker ones to achieve the same degree of differentiation of tone or colour.

Knowledge of the plane of sectioning is of course very significant in the build up of the observer's understanding of the way the actual image observed gives information on the three dimensional structure of the specimen sectioned. More detail is given in the section on stereology in Chapter 7 (Section 3).

7.4 Preparation conditions

The microscopist must be continuously aware of the implications of the preparation of the specimen for microscopy in that, for example fixation may well provide light scattering boundaries by such processes as coagulation or precipitation. A well known domestic situation is that the white of an egg, white by virtue of its scattering of light, is transparent and invisible before it is coagulated by heat! Details on fixation are given in several of the histochemical procedures in Chapter 5 (see Section 2.2.2).

It can be a very salutary experience to compare the linear or area dimensions of a sample of living cells with those dimensions after the cells have been fixed and made as permanent microscope slide preparations. A sample of fresh cheek epithelial cells can be measured using a phase-contrast microscope and then another sample of the same type of cells, after fixing in mercuric chloride, staining with haematoxylin and eosin, dehydrating, clearing and mounting in Canada balsam, may be found to be 20% smaller. The statistical validation of this finding does require a large number of measurements from the two samples of cells before and after processing!

8. OTHER TECHNIQUES

To list all the techniques in use in both biological and materials science laboratories which might have potential for producing contrast in the images of living specimens is not possible. New developments are of course always occurring, and the very interesting one of scanning light microscopy is likely to bring major advances specifically to the study of thick biological materials. Brief mention is made below of two such techniques—firstly, dispersion staining, which has considerable value in identification of asbestos fibres and which could be applied to living cells. The second technique has already been discussed in Chapter 1, Section 2.6.1 in relation to depth of field.

8.1 Dispersion staining

The change of refractive index with wavelength and that degree of change as a property of the material through which the light is passing has already been discussed in Chapter 1. A technique known as dispersion staining has been developed, and that for identifying asbestos fibres by McCrone (20) is well known. A dispersion staining objective and a capacity for centring the aperture iris with respect to the condenser lens is required. High dispersion immersion oils are used with either ordinary white light or plane-polarized light to produce dispersion colours. An introductory kit containing test samples of asbestos is available from Triton Instruments.

The technique of exploiting dispersion by the glass lenses of the condenser has been used for examining such specimens as cheek epithelial cells (21). The cells are illuminated by an annulus below the ordinary uncorrected condenser. The condenser in which the dispersion is occurring is made to illuminate from positions well below its normal focusing position. Considerable skill is required to match the condenser to the situation.

8.2 Confocal, laser-scan and video systems

In so far as the removal of unwanted out-of-focus light improves the visibility of the required features, these methods clearly should be described in a chapter dealing with contrast techniques. The first two were mentioned in Chapter 1 under the section dealing with depth of field (Section 3) to which the reader should refer. The apparatus required is in its development stage but the technique looks very promising for a number of biological applications. It is particularly helpful, it seems, in removing flare light which often spoils the resolution and image recording in fluorescence microscopy. Thick transparent specimens are of course a common situation in biology and their thickness has been the reason why sectioning is so important. The preparation and processing required for sectioning often results in physical shrinkage as mentioned earlier which also greatly alters the chemistry of the tissues. The use of video microscopy for enhancement of contrast and visibility is dealt with in detail in Chapter 8.

9. CONTRAST TECHNIQUES FOR PROVIDING INFORMATION ABOUT A SINGLE SPECIMEN

A useful exercise would be to examine a single specimen using all the techniques available. The information contained in the images not only gives a picture of what the specimen 'looks like' but also provides information about its physico-chemical nature. Thus a technique may be the wrong one for a visual image but it will nevertheless give information for example about refractive index, absorptive power, lack of birefringence or other light−matter interaction.

The onion epidermis is a convenient specimen to use for this exercise because of its ease of preparation and because of its considerable inherent interest (*Table 17*). The section on polarized light microscopy (Section 5) explains why when the specimen is placed on the microscope stage it will have a specific orientation to both the observer and to the trains of light waves entering it. *Figure 6* gives a three-dimensional impression of what the individual cell will look like. Photomicrographs of the cells using four different contrast techniques are given in *Figure 7*.

Thus, in *Table 18* a variety of images of the same specimen are seen to be produced. These images are the results of various forms of light−matter interactions and can therefore be interpreted to give both qualitative and quantitative information about the physico-chemical properties of the specimen.

Data such as refractive index can be derived from measurements of o.p.d. seen, for example, in phase-contrast, and this in turn can give information on parameters such as concentration of proteins or mass of nuclei. The boundaries of change of refractive index are shown both by dark-field and phase-contrast by intensity changes while they are shown as apparent side lighted edges in DIC. The use of contrast produced by interferometric methods can give measurements of the optical path in the z direction

Table 17. Onion epidermis under a variety of contrast techniques.

1.	Prepare a piece of inner epidermis of the inner scale of a fresh onion and mount in water. Arrange to have the outer surface of the epidermis against the coverslip thereby keeping a check on its orientation to the light and to the eye. It may be useful to make two or three such slides as you may wish to use them in parallel.
2.	Prepare a chart as given in *Table 18* giving sufficient space for comments on the image obtained in each technique.
3.	Enter into the chart indications of brightness or darkness or relative intensity with surrounds or immediate neighbouring features. Remember to insert into the table equal brightness or darkness if there is nothing visible or there is no contrast.
4.	You may wish to amend some of the comments inserted into the *Table 18* but be prepared to think through why you disagree with what is entered.

and can show very fine differences in this direction. Again this technique shows gradients of optical path. Kulfinski and Pappelis (22) studying onion epidermal cells with interferometric techniques report that the nuclei have a mass per unit area of 1×10^{-14} $g/\mu m^2$ and a total nucleus weighs about 14×10^{-10} g.

Plane-polarized light will show areas of pleochroism while crossed polars will make contrast between isotropic and anisotropic regions. Further comparative accessories can make differences in the direction of birefringence distinct by colour. The highly orientated and crystalline nature of the cell walls of the onion skin specimen are clearly distinguished from the isotropic components of the cytoplasm and, by the use of the sensitive tint plate, the longitudinal walls are differently coloured from the transverse walls at any one orientation. The absorptive qualities of the nucleus and the vertical cell walls make the contrast in the bright-field image. There is no autofluorescence in the onion skin but, when stained with a fluorochrome, differences in the physico-chemistry of the specimen are made visible. The use of phase and particularly DIC make the cytoplasmic granules visible and they can be seen to be moving around the periphery of the cytoplasm. The dark-field image shows clearly the boundaries of refractive index and both the vertical walls and the nuclei can be seen bright against the dark background.

The horizontal walls of the cells are not made visible by any of the transmitted techniques, although perhaps some low birefringence is visible between cross polars. The horizontal walls are too transparent and transmitted light techniques are inappropriate. Incident bright-field is not a good way of showing them either because they do not show enough reflectivity. Dark-field incident light however, loses the confusing glare to some extent and they are then visible (*Figure 7f*) and seen to be sloping from a central ridge towards the long edges of the cells. It is possible that recent developments in confocal microscopy will further improve the visibility of such features.

The onion skin was chosen as an illustration here because it does show something different with a very wide range of techniques. Although the author has not yet tried the technique of IRM he does not expect to see a great deal more detail in the onion skin. The use of plastic replicas of the skin with subsequent metal coating or even of the vacuum sputter coating of the surface of the skin itself are older, alternative

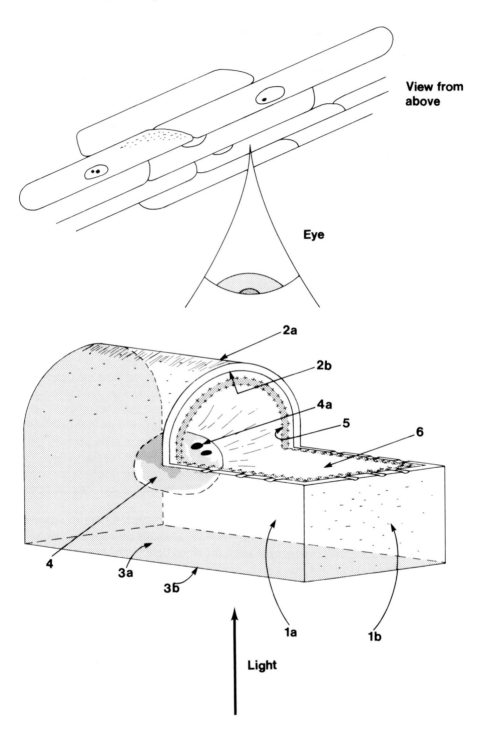

techniques and are used to increase the reflectivity and make for further provision of detail in the image, by incident light methods.

A word of caution is perhaps not inappropriate here. You may be very familiar indeed with the appearance of your specimen using one contrast technique and may be loath to change to another, but it is a valuable challenge to ask what does this specimen look like using another technique? Again you may be at first dismayed or at least very disappointed at its first appearance but if the principle which has been emphasized in this chapter is accepted, then the new image will be seen as giving some information on the physico-chemical nature of your specimen. Repeat the examination several times and if there is a consistent image produced then it is likely to be a useful one and worth learning to recognize. It may well then offer new opportunities for comparative studies of the specimen taken, for example from a range of experimentally tested animals or plants.

The use of two techniques together has been possible with the development of incident light fluorescence accessories attached to transmitted light microscopes. This can be very helpful in, for example locating fluorescent areas in respect of other, non-fluorescent cells or tissues. The techniques can be used simultaneously or consecutively, without disturbing the specimen, by merely moving a dark slide and turning up the voltage of the transmission source. See, for example Nielson *et al.* (23) using fluorescence and phase to show the association of calmodulin with lysosomes.

The specimen may still provide surprises even with the best contrast systems and the highest magnification in light microscopy. One phenomenon that must be a frequent occurrence, especially in specimens showing periodic structure such as diatom frustules, is the presence of Moiré patterns. These arise if there are two laminae of equal pattern superimposed one on top of the other. The features of their patterns can coincide to produce a super pattern. The comparison of electron microscope images with those taken with light microscopy have sometimes failed to demonstrate the presence of what was a well known pattern in the light image.

10. ACKNOWLEDGEMENTS

I would like to thank Professor Curtis for his contribution on IRM. Without the courses comprising the Summer Schools of the Royal Microscopical Society, my hands-on knowledge of microscopy would have been much less and without the patient and forbearing students on those courses over the years I would not have presumed to write the first two chapters. My grateful thanks to the publishers for their help with the art work and their encouragement throughout.

Figure 6. Diagrammatic three-dimensional appearance of onion epidermal cell in its relation to the light in a microscope and the observer's eye. (a) How the cells are seen from the top (see also *Figure 7*). **1**. Vertical walls in the (*z*) axial plane (a) short, (b) long. **2**. Horizontal wall in the plane of the stage of the microscope. Upper wall, (a) upper surface, (b) inner surface. **3**. Lower horizontal wall, (a) inside cell surface, (b) outer. **4**. Nucleus with (4a) nucleolus. **5**. Cytoplasm with granules. **6**. Vacuole.

Table 18. Onion bulb scale epidermis viewed under various light microscopy techniques.

Technique	Background	Onion skin components (see Figure 6)		Walls				Light–matter interaction
		Nucleus (4 and 4a)	Cytoplasm (5 and 6)[a]	Horizontal		Vertical		
				Top (2a)	Bottom (2b)	Short (3a)	Long (3b)	
Transmitted Bright-field	Bright	Only visible with aperture closed down				Only visible (dark) with closed down aperture	Only visible (dark) with closed down aperture	Map of absorption of light. Diffraction effect at edges with small apertures
Dark-field	Dark	Sky-grey				Bright at all orientations	Bright at all orientations	Scattering by refraction and reflection. Boundaries of refractive index
Phase (+ve) with green filter	Bright-green	Dark with halo round edge	Dark granules and edge with haloes			Bright (On either side of dark line)	Bright	Map of optical path in z directions = $t(n_o - n_m)$
DIC	Bright	Visible and 3D effect	Granules clearly visible. Movement			Bright (but see under polarized light)	Bright	Map of optical path in z directions = $t(n_o - n_m)$
Fluorescence see under incident fluorescence								
Polarized light plane	Bright	As bright-field				No change on rotation of specimen		No pleochroism

Technique					Interpretation
Polarized light (plane) crossed polars	Dark	Very slight white	Bright	Bright (at certain orientations only)	Map of birefringence and o.p.d. and orientation = $t(n_\parallel - n_\perp)$
Crossed polars + Sensitive Tint	Red 1		Blue (or yellow)	Yellow (or blue)	Indicates direction of birefringence $\pm (n_\parallel - n_\perp)$
Other[b]					
Incident Bright-field	Bright	Some brightness but excessive glare	Visible (?)	Visible (?)	Map of relectiveness
Dark-field	Dark	Clear contour of brightness	Grades into horizontal walls. Dark channels in wall clear.	Visible (?)	Map of surface light scattering features.
Fluorescence Primary	Dark				No primary fluorescence
Fluorescence Acridine orange	Dark		(If overstained yellow) (green)		Map of chemical affinity for acridine orange
Other[b]	Apple-green				

[a] Vacuole (6) within cytoplasm not rendered visible by any of these techniques. Implied by achieving plasmolysis of cytoplasm when mounting medium such as 2% sea salt in water has higher osmotic pressure than cell sap.

[b] Others might include transmitted AVEC–DIC (see Chapter 8, ref. 5) or incident light on sputter coated plastic surface replicas, which will give microtubules of cytoplasm and surface configuration of horizontal walls, respectively.

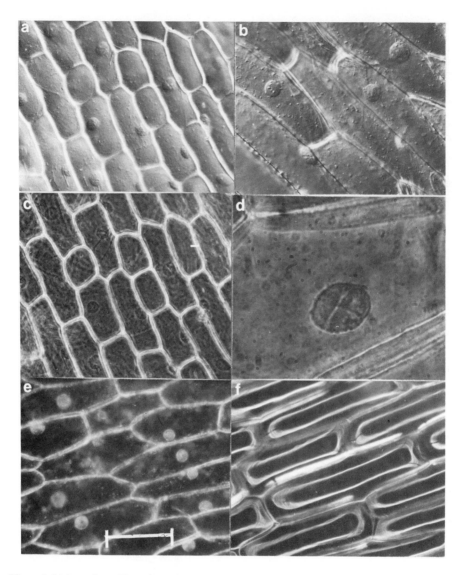

Figure 7. Living onion epidermal cells in different illumination systems. (**a**) DIC image shear in direction NE−SW. (**b**) DIC shear direction NW−SE parallel to long walls. (**c**) Phase-contrast of same field as (**a**). (**d**) Jamin-Lebedeff interference double focus Baker-Smith microscope with ×40 objective. (**e**) Dark-field transmitted light (scale bar = 100 μm). (**f**) Dark-field incident light.

11. REFERENCES

1. Humason,G.L. (1962) *Animal Tissue Techniques.* W.H.Freeman and Co., San Francisco.
2. Ross,K.F.A. (1967) *Phase Contrast and Interference Microscopy for Cell Biologists.* E.Arnold, London.
3. Barer,R. and Joseph,S. (1954) *Quart. J. Microsc. Sci.,* **95**, 399.
4. Barer,R. and Joseph,S. (1955) *Quart. J. Microsc. Sci.,* **96**, 1.
5. Barer,R. and Joseph,S. (1955) *Quart. J. Microsc. Sci.,* **96**, 423.
6. Frey-Wyssling,A. (1957) *Macromolecules in Cell Structure.* Harvard University Press, Cambridge, MA.
7. Ruthman,A. (1970) *Methods in Cell Research.* C.Bell and Sons, London.

8.   Schmidt,W.J. and Keil,A. (1958) *Die gesunden und die erkrankten Zahngewebe des Menschen und der Wirbeltiere im polarisation Mikroskop.* Carl Hanser Verlag, München.

9.   Roelofsen,P.A. (1959) *The Plant Cell Wall.* Gebruder Borntraeger, Berlin–Nikolessee.

10. Spencer,M. (1982) *Fundamentals of Light Microscopy.* Cambridge University Press, Cambridge, UK.

11. Curtis,A.S.G. (1964) *J. Cell Sci.,* **20**, 199.

12. Gingell,D. and Todd,I. (1970) *Biophsy. J.,* **26**, 507.

13. Gingell,D. and Todd,I. (1980) *J. Cell Sci.,* **41**, 139.

14. Gingell,D. (1981) *J. Cell Sci.,* **49**, 237.

15. Opas,M. (1988) *J. Cell Sci.,* **90**, 215.

16. Verschueren,H. (1985) *J. Cell Sci.,* **75**, 279.

17. Heath,J.P. and Dunn,G.A. (1978) *J. Cell Sci.,* **29**, 197.

18. Bailey, and Gingell,D. (1988) *J. Cell Sci.,* **90**, 215.

19. Ploem,J.S. (1975) In *Mononuclear Phagocytes in Immmunity, Infection and Pathology.* Furth,R.V. (ed.) Blackwells Scientific Publications, Oxford, UK.

20. McCrone,W.J. and Delly,J.G. (1973) *Particle Atlas.* Ann Arbor Scientific Publishers, Ann Arbor, MI, vol.2.

21. Taylor,R.B. (1980) *Proc. R. Microsc. Soc.,* **15**, 27.

22. Kulfinski,F.B. and Pappelis,A.J. (1971) *Phytopath.,* **61**, 724.

23. Nielson,T.B., Field,J.B. and Dedman,J.R. (1987) *J. Cell Sci.,* **87**, 327.

CHAPTER 3

Image recording

PETER J.EVENNETT

1. INTRODUCTION

No recorded microscopical image can compare in beauty and impact with the original, viewed directly at the instrument with constant minor adjustments to the fine focus. Yet recorded images are essential. Generally the need is for a record which is permanent, perhaps of an ephemeral specimen, a record which can be studied away from the laboratory, conveniently compared with other images, measured or analysed in various ways, presented in a lecture, or published in a report, paper or textbook.

This chapter will briefly consider recording microscopical images by drawing, but it consists principally of a consideration of techniques and equipment for photomicrography. Successful photomicrography depends on paying attention to a wide variety of choices concerning the microscope and its illumination, the film and its processing, and so on. The writer feels strongly that the correct approach to such a complex subject is through an understanding of these factors so that the reasons underlying the procedures can be appreciated, rather than the 'rule-of-thumb' approach. Our aim here has been to give just sufficient 'theoretical' information to enable the reader to understand the procedures and to make a serious attempt at 'troubleshooting'.

The reader impatient to take a photomicrograph may wish to begin with the practical schedules in Section 11, referring to the main text for clarification where necessary. Those who read the main text first will find that the practical Section 11 will summarize and help to keep the technical points in perspective.

2. DRAWING

Drawing is the earliest method of image recording, practised since the origins of microscopy. Drawing has its own peculiar advantages making it still relevant today, and it should not be seen simply as a cheap and non-technical substitute for more sophisticated methods; nor, with modern aids, does it require great artistic skill. A drawing can present an interpretation of a hypothetical ideal specimen, ignoring defects and combining features from several areas of examination, and can represent the synthesis of observations at both high- and low-powers and from several planes of focus. A specimen which would yield a less-than-perfect photomicrograph may thus be recorded in the form of a top quality drawing. In addition, a microscope used for drawing need not be equipped with expensive flat-field objectives.

An eyepiece with a graticule in its focal plane, carrying a grid of squares (such as in Figure 7A in Chapter 7) seen superimposed on the microscope image, provides the simplest assistance to freehand drawing. Used in conjunction with a grid of appropriate size in the form of an underlay to the drawing paper or lightly marked on the paper itself, this allows the outlines of the principal features to be drawn in correct proportion and relationship one to another. Finer details should be filled in later, freehand.

By the use of a special prism or simply a mirror placed at 45° just above the eyepiece, the microscope image may be projected onto a piece of paper where its outlines can be traced around, again in correct proportions, given that care is taken with the geometry to ensure that the image falls perpendicularly onto the paper. For this purpose the microscope should be fitted with a powerful light source and be used in a darkened room.

A camera lucida, or its modern successor, the drawing tube, provides the most comfortable approach to drawing. The camera lucida consists of a beam-splitting prism mounted above the eyepiece, with a mirror arranged so that the pencil-point at the drawing paper can be seen superimposed upon the microscope image. A drawing tube works similarly, but fits into the microscope usually beneath the binocular head enabling the pencil and paper to be seen through the eyepieces; it may contain adjustments for focus and for the size of the drawing. Using either of these methods, it is important to set the relative brightnesses of the microscope image and the illumination on the drawing paper, perhaps making adjustments as work progresses, so that both may be observed comfortably.

3. PHOTOMICROGRAPHY

Photomicrography is the recording of images from the microscope by means of photography. Note that (in English) photomicrography should be clearly distinguished from microphotography, which is the production of very small photographs.

Photography has been applied to the microscope since the earliest days of photography itself, and for the past 50 years has been the principal method of image recording. The advantages of photomicrography are that it is accurate, unbiased, repeatable, simple and produces high-quality results. But these advantages exist only provided good equipment is used carefully; photomicrography will also faithfully record deficiencies of specimen, equipment and technique, and may thus be even more demanding than visual microscopy. This chapter will cover principally the aspects of the use of the microscope and of photographic techniques which are involved in the production of good photomicrographs.

4. THE MICROSCOPE IMAGE FOR PHOTOGRAPHY

4.1 **Resolution**

A microscope is designed to be able to handle information about fine detail in the object (resolution) and present it in the form of an image, magnified sufficiently for the resolved detail to be detected by the eye; when necessary, the microscope may also increase contrast between features in the image. The objective lens forms a magnified primary image, which is usually further enlarged, first as it is transferred to the film, and subsequently by photographic enlargement, or projection of a transparency.

As shown in Chapter 1, the resolving power of a microscope is limited by fundamental laws such that the minimum resolved distance (d) is set by the wavelength (λ) of the

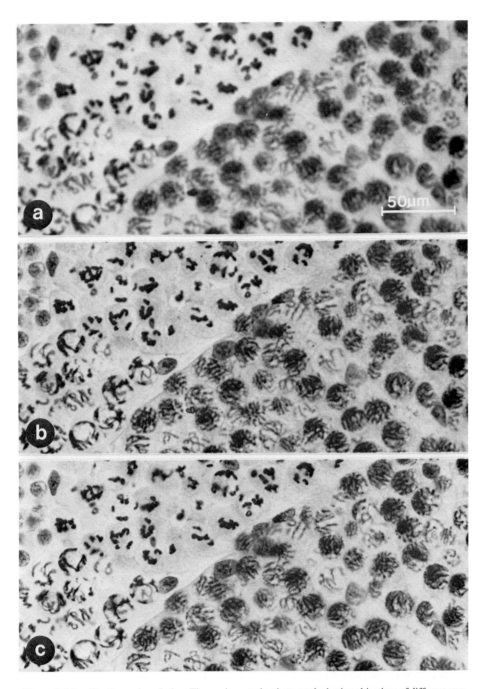

Figure 1. Magnification and resolution. Three micrographs photographed using objectives of different apertures and magnifications, all printed at a magnification of 400:1. (**a**) Objective 4/0.16. Final magnification 2500 × NA. The image is unsharp and obviously overmagnified. (**b**) Objective 10/0.25. Final magnification 1600 × NA. The image still lacks information in the fine details. (**c**) Objective 25/0.65. Final magnification 615 × NA. In this image the fine detail is sharply rendered. Scale bar = 50 μm.

radiation forming the image, and the numerical aperture (NA) of the objective lens thus

$$d = \frac{0.61 \, \lambda}{\text{NA}}$$

For practical purposes λ can be taken to be 0.5 μm (green light), and d may thus be simply calculated from the NA of the objective; the larger the NA the finer the detail resolved.

4.2 Resolution and the size of the print

The eye is capable of resolving about one-tenth of a millimetre (100 μm) at the normal nearest distance of distinct vision of 250 mm. If the final image from a microscope is presented to the eye as a micrograph, magnified so that the minimum resolved distance is 100 μm in the image, it will appear 'sharp'; an observer with good eyesight will be able to see its full content of detail. If it is of crucial importance to make the finest detail in the image clearly visible, then the final magnification should be a little more, bringing the resolved distance up to 200 or 300 μm, so that it can be seen comfortably by those with slightly imperfect sight or in poor viewing conditions. Any further enlargement constitutes 'empty magnification' and the resulting image is likely to be considered unsharp.

It can be shown using the formula in Section 4.1 that, for a micrograph intended to be viewed at normal reading distance, an appropriate total magnification (including photographic enlargement) lies between 500 and 1000 times the NA of the objective in use; this provides a soundly-based rule of thumb. This is illustrated in *Figure 1*. As an example, an objective of NA 1.0 has a minimum resolved distance of 0.3 μm using light of $\lambda = 0.5$ μm. A total magnification of 500:1 (500 $\times$ NA) will produce 150 μm detail in the image; magnification of 1000:1 will produced 300 μm. When a micrograph is in the form of a transparency for projection, considerations are complicated by the size of the image on the screen and by the viewing distance. For general use a magnification on the film of about 200 times the NA will be found to be satisfactory.

4.3 Producing a real image on film

Microscopes are designed principally for the eye. Rays leaving the eyepiece are parallel, so when they enter a normal relaxed eye they are converged to form an image on the retina. We 'see' a so-called 'virtual image' which appears to represent a magnified object situated at infinity; in these circumstances, the eyepiece operates in conjunction with the eye, to form a real image on the retina (a real image is one which can be received on a surface such as screen or film). When a microscope is to be used for photomicrography, the optical system must be modified in one of several ways so that it forms a real image on the film, at a finite distance from the eyepiece.

For visual use the primary image falls in the focal plane of the eyepiece thus producing parallel rays which enter the eye (see *Figure 2a*). If the primary image is lowered by refocusing the microscope, rays leaving the eyepiece will converge and form a real image in a plane suitable for accommodating the film (*Figure 2b*). This method of forming a real image requires no extra equipment, but it may produce an imperfect image since the conjugate distances between the object and the objective lens, and the

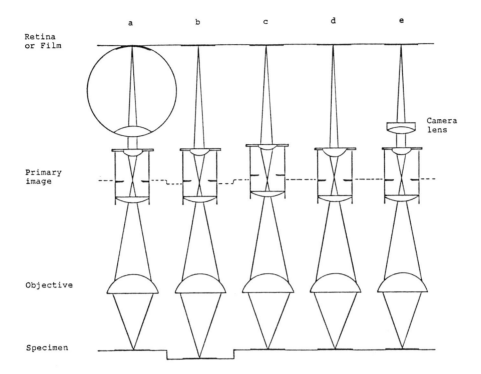

Figure 2. Methods of forming a real image. (**a**) Image formation with the eye. The microscope is adjusted so that the primary image falls in the first focal plane of the eyelens of the eyepiece; this means that parallel rays will enter the eye and be focused onto the retina by the eye. (**b**) The primary image is lowered by increasing the specimen – objective distance. This causes rays leaving the eyepiece to converge and form a real image in a plane suitable for accommodating the film, but performance of the objective lens will be impaired. (**c**) Specimen and primary image are in correct positions, but the eyepiece is raised to increase primary image – eyelens distance and form a real image on film. (**d**) Special projective lens or 'photographic eyepiece' designed to form a real image on film using normal settings of the microscope. (**e**) Normal eyepiece used in conjunction with a converging lens in a special photomicrographic camera attachment; optically similar to using the eye, (**a**) above.

objective lens and the primary image, are now longer and shorter, respectively, than those for which the spherical aberration correction was optimized in the design of the objective. This simple method may be adequate for recording images onto film of larger formats (where little refocusing of the microscope will be necessary because the distance from the eyepiece to the film is relatively long), or when using objectives of small NA (which are less sensitive to this effect), but it should be avoided for high-quality work. Where other methods are not available, an image may be formed on the film by raising the eyepiece in its tube, using a suitable ring or a piece of adhesive tape, while leaving focus of the microscope adjusted for visual use (*Figure 2c*).

Special lenses for photomicrography (strictly not 'eyepieces' since they cannot be used with the eye) are available from some manufacturers; these are designed to produce an image on film while maintaining optimum settings of the microscope (*Figure 2d*). The projective lenses of the large integrated photographic microscopes also fall into this category.

Finally, many photomicrographic systems mimic the action of the eye by including a converging lens above a normal eyepiece, arranged to form a real image (*Figure 2e*). While there may seem to be some logic in preferring to have a minimum of glassware between microscope and film (as in *Figure 2c* and *d*), in practice all commercial systems produce excellent results.

4.4 **Illumination**

Correct adjustment of the microscope for Köhler illumination as described in Chapter 1 is even more important for photomicrography than for visual observation, since photographic recording cannot 'make allowances' for defects as the eye does, and the results will be permanently recorded for all to see. This system of illumination was designed specifically for photomicrography, in order to provide a uniformly illuminated field of view when using a source of light of uneven intensity due to the structure of the filament of the lamp. When correctly set, the field will be uniformly illuminated, the illuminated field diaphragm will control the area of the field illuminated, and the illuminating aperture diaphragm will control the angle of the illuminating cone, matching it to the aperture of the objective in use. It is especially important for image recording that no surfaces which might collect dust should lie in planes which are conjugate with the final image; this will normally be true in a correctly designed and adjusted system. Maladjustment of the illuminating system will reduce resolving power, degrade the image by stray light, prevent the achievement of a uniformly illuminated field of view, and possibly allow the intrusion of the lamp filament and dust particles into the final image.

Some microscope illuminating systems include a fixed or removable diffuser in the form of finely ground or etched glass, designed to prevent the appearance of the filament in the final image. In addition, when a diffuser is present in the light path it will obviously be impossible to see the filament image, even in positions where it should be seen in correct Köhler illumination. Opinions differ on whether the use of such a diffuser detracts from the ultimate quality of the image; it can, however, be safely said that it is useful when using low-powered objectives. Because of their small NA these objectives have considerable depth of field, which may allow them to include an out-of-focus image of the filament in the final image even when the microscope is correctly adjusted. Depending on the position of the diffuser in the optical train, it is sometimes necessary to defocus the condenser very slightly, after setting it correctly, in order to avoid including the granular structure of the diffuser in the image.

5. THE MICROSCOPE FOR PHOTOMICROGRAPHY

In principle, any light microscope can be used for photomicrography; however, given some choice, certain features are desirable if good quality results are to be obtained consistently and conveniently.

5.1 **The microscope stand**

For preference, the microscope stand should be designed so that the viewing tube which carries the camera does not move with the focusing adjustments; fortunately most modern instruments are designed to support a photomicrographic camera. If it is necessary to use a stand where the focusing movements act on the limb or the body-tube, great care must be taken to see that the adjustments cannot drift downwards between setting focus and the end of the exposure, due to the weight of the camera. On most microscopes the stiffness of the coarse adjustment, the principal offender, can be adjusted by rotating the focusing knobs in opposite directions.

It is useful for the microscope to be equipped with a mechanical stage, enabling the slide to be moved delicately in order to frame the image. Where it is not possible to rotate the camera, or where this would involve the difficult task of rotating framing eyepieces accurately, in sympathy with the camera, it is helpful to have a rotating stage to enable the micrograph to fit best within the rectangular film frame.

The simpler camera adaptors can be mounted on the eyepiece tubes of most microscopes, which are of approximately standard diameter, but the use of inclined eyepiece tubes is best avoided. The ideal arrangement for a camera attachment is the trinocular head, which consists of a normal binocular viewing tube combined with a vertical camera tube to which the image may either be fully directed or shared with the viewing tube by means of a beam-splitting prism.

In recent years several large integrated photographic microscopes have been designed. These instruments, which combine the microscope and the photographic parts in one self-contained unit, are described further in Section 6.4.

5.2 The optical system

A photomicrograph cannot be of good quality if it is taken using a poor optical system. The quality of the objective lens is paramount, followed by the use of a suitable eyepiece or projective lens; the use of a well-corrected condenser can also have a detectable influence on the final image.

5.2.1 *Numerical aperture and magnification of objectives*

As described in Section 4.1 and in Chapter 1, objectives of larger NA are capable of resolving finer detail than those of smaller aperture. A glance at manufacturers' lists shows that the magnifications and apertures of objectives are not firmly linked. As an example, one manufacturer lists a 40/0.65, a 40/1.3, and also a 100/1.3. The first two will produce images of similar size, though that of the second will contain considerably more fine detail; the last two, oil-immersion objectives, should both resolve similarly well, but the 40 will include in its image a larger area of the object (at lower magnification) than the 100.

Objectives should be chosen for their aperture and hence their resolving power, rather than for magnification. Many microscopes are now fitted with some form of magnification-changer which enables the final magnification to be altered approximately two-fold and, where a print is to be made, extra photographic enlargement can be given in order to reach a desired final value. Of the three objectives quoted above, other characteristics being equal, the writer would recommend the 40/1.3 for high-quality photomicrography.

5.2.2 *Correction of aberrations*

Microscope objectives are available in several types according to their degree of correction for chromatic and spherical aberrations. These may be further subdivided according to their freedom from curvature of field, and are manufactured in a range of magnifications and NAs, in dry and immersion types.

Achromats are the simplest objectives, designed to bring red and blue rays to a common focus, which will be slightly different from that for green light. The image formed by these lenses may show slight colour fringes, in green or magenta depending on the setting of focus. Achromats are corrected for spherical aberration for green light only. They are relatively cheap, and suitable for visual microscopy; for photomicrography they should, if possible, be used with a monochromatic green filter such as the Wratten 58 (Kodak) or a green interference filter, with which they can perform well.

Fluorite objectives (so named because they formerly contained elements made from the mineral fluorite), or semi-apochromats, are more highly corrected for chromatic aberration than achromats. Because of this they can be made with larger apertures for a given magnification, and they produce a superior, more contrasty image. Their simple construction and high light-gathering power make fluorite objectives suitable for fluorescence microscopy, and they may be used for general photomicrography with success.

Apochromats are the most highly corrected lenses, approaching perfection in their correction for chromatic aberration, and with spherical aberration corrected for two colours rather than one. They are capable of producing high-quality images and are the most suitable objectives for photomicrography, particularly in colour. Because of the complexity of designing such a high-performance objective, many manufacturers deliberately arrange for lateral chromatic aberration to be corrected by a special 'compensating' eyepiece. For this reason, a fully corrected system consists of an objective together with its appropriate eyepiece. Objectives from different manufacturers, and sometimes even different types from the same manufacturer, are built to require eyepieces of different specification; some recent designs incorporate full correction in the objective itself. It is clear that the common practice of interchanging optical parts without due consideration will lead to poor results. When in doubt, the maker's recommendations should be sought.

5.2.3 *Dry objectives and coverglass thickness*

Objectives are designed to operate best when viewing the object through a specified thickness of material of a particular refractive index, and the larger the aperture of the objective the more important this is. If the objective is not used as recommended, its correction for spherical aberration will be inadequate and a poor image will result. Most dry objectives (i.e. those which operate with air between the front lens and the specimen) are designed for a 'No. 1½' coverglass of 0.17 mm thickness, and this figure is engraved on the lens mount. Some are made for uncovered specimens, marked Epi, 0, or simply ' − ', while others are made for use through the walls of culture vessels and may be adjusted for use through any thickness of wall up to about 2 mm.

Uncovered specimens are used in some kinds of work, for example in cytogenetics (see Chapter 9); in this field the Zeiss Epiplan 80/0.95 objective has been used with success, a lens more familiar to materials scientists than to most biologists.

Objectives with apertures of less than about 0.65 will generally be tolerant of minor errors in coverglass thickness, but dry lenses of large aperture (0.75−0.95) are not. These lenses are usually fitted with a 'correction collar' which adjusts the spacings of the components of the lens in order to maximize correction for spherical aberration. Even when the correct coverglass has been used on a preparation, an excess of mounting medium will increase the effective thickness, and make adjustment necessary.

For work of the highest quality, ensure that coverglasses are of the correct thickness for the objectives in use, normally 0.17 mm, and use the minimum of mounting medium; when sections are being used, they should if possible be mounted directly on the underside of the coverglass.

To adjust the correction collar:

(i)   Focus the image of the specimen.
(ii) Move the collar a little in one direction and refocus.
(iii) Consider whether the contrast of the image is improved.
(iv) If it is, move the collar further in steps in the same direction, refocusing each time.
(v)   When the image begins to deteriorate, move the collar back again until the best setting is found.

Figure 3 demonstrates the importance of correct setting of the correction collar. If a specimen is suspected of having too thick a coverglass or too much mountant and a

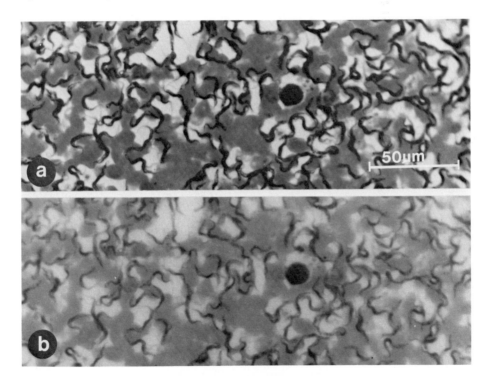

Figure 3. Correct setting of the correction collar for adjusting correction for spherical aberration of a dry 40/0.95 objective. Printing conditions for both micrographs are identical. (a) Correct adjustment, collar set for coverglass thickness 0.17 mm. (b) Incorrect adjustment, collar set for 0.23 mm, an error of about 0.05 mm (50 μm) in coverglass or mountant thickness. Scale bar = 50 μm.

lens with correction collar is not available, use an oil-immersion objective — see Section 5.2.4.

5.2.4 *Immersion objectives*

It is necessary to use an immersion objective if a NA of 1.0 or greater is required. Most immersion objectives operate using a specially prepared oil, though water-immersion and glycerol-immersion objectives are available, as are also those which are adjustable for use with any of these immersion media.

While one may wish to avoid where possible the inconvenience of using immersion objectives for routine visual microscopy, they provide considerable benefits for photomicrography. Apart from their greater aperture and thus superior image quality, the use of immersion objectives, particularly oil-immersion, almost completely avoids the difficulties due to inaccurate coverglass thickness described above. Since the optical properties of mounting medium, coverglass and immersion oil are all closely similar, the use of more of one and less of another will have no detectable effect. There is, then, a strong case for using a 40/1.0 oil-immersion objective for photomicrography, rather than a dry 40/0.95.

Immersion oils are not all identical, so it is advisable to follow the lens manufacturer's recommendation. Equally important is to ensure that oils are not mixed; if a lens carries traces of oil from previous use, and is used with another kind of oil, considerable loss of clarity may result. See Section 5.2.8 on cleaning lenses.

5.2.5 *Depth of field*

Depth of field is the distance in the object, measured along the axis of the microscope, within which object features appear acceptably sharp in the image. Depth of field decreases with increasing NA; objectives of large aperture have a depth of field of less than 1 μm, hence only a very shallow slice of the specimen can be in focus at any one time. This slice may be considerably thinner than the section or other object, and a number of different views or 'optical sections' can be produced from any one area. Because of this, an experienced microscopist will constantly adjust the fine focus control while viewing.

For photomicrography one particular plane of focus must be selected for each exposure, and there may be advantage in taking a 'through-focus series' of an important and difficult specimen, adjusting the fine focus in small increments between exposures, from one side of focus to the other.

It is of course possible to use an objective of smaller aperture in order to increase depth of field, but the resolution will then suffer, and the useful magnification will be lower. It is an inescapable conflict in conventional light microscopy that the large apertures which provide the finest resolution are the most restrictive of depth of field. Nevertheless, when photographing specimens requiring increased depth of field, resolution may well already be severely limited by other factors such as the thickness of the object, and large aperture objectives will thus be unnecessary.

5.2.6 *Curvature of field*

Many objectives produce an image in which the central parts and the periphery cannot be simultaneously in focus. For visual microscopy this is not normally a handicap, but such images are unsuitable for photomicrography. Using relatively thick specimens the problem may not appear too serious, since some feature is likely to coincide with the curved field of the lens and thus appear sharp, but, particularly with the thin sections which are used more and more nowadays, the images may be unacceptable.

In order to avoid this problem, manufacturers supply objectives specially designed to minimize curvature of field, usually designated by the syllable 'Plan' in their name, such as Planachromat and Planapochromat. Plan objectives are specially recommended for photomicrography, always used, of course, with an appropriate eyepiece as recommended by the manufacturer.

5.2.7 *Types of condenser*

Condensers which are more highly corrected for spherical and chromatic aberrations are especially recommended for photomicrography; their superior quality permits greater precision of adjustment, higher contrast due to less stray light, and a more uniformly illuminated field. Their benefits may be particularly noticeable in critical colour photomicrography.

Condensers range from the simple two-lens Abbe type, uncorrected for chromatic and spherical aberrations, through aplanatic types corrected for spherical aberration, to the achromatic-aplanatic design which gives the highest performance. Because of their aberrations, the simpler condensers form an unsharp and colour-fringed image of the illuminated field diaphragm, making accurate setting for Köhler illumination impossible. Moreover, because of their restricted aperture, they are incapable of fully illuminating large aperture objectives and thus realizing their full potential resolving power.

The more highly corrected condensers are generally designed to be optically united with the slide by a drop of immersion oil when apertures of greater than 1.0 are in use. Immersion of the condenser, though messy, is recommended for top-quality results in photomicrography when large-aperture objectives are used.

When using objectives of lower magnification (generally < 10), it may be difficult to illuminate the whole of their relatively large field using a standard condenser. This problem often tempts microscope users into one of the cardinal sins of microscopy, that of lowering the condenser from its correct position, with deleterious effects on image quality. For use with low-powered objectives, some condensers have a swing-out top lens and some microscopes are provided with auxilliary lenses to be inserted beneath the condenser. Others include an extra lens system in the microscope base, between the lamp collector and the condenser. All of these systems enable the condenser to throw a larger image of the illuminated field diaphragm onto the specimen. As an example, the condenser on a microscope used by the writer provides an illuminated field at the object of about 1.5 mm diameter with its top lens in position, and approaching 5 mm with the top lens swung out. In cases where the condenser does not have a swing-out top lens, it may be possible to unscrew the top component and achieve a similar effect. If low-magnification photomicrography forms an important part of a laboratory's work, a condenser of long focal length, specially designed for this kind of work, will be found invaluable.

5.2.8 *Cleaning lenses*

The best way to have clean lenses is to avoid dirtying them. Some optical glasses are considerably softer than window glass, and their antireflection coatings are similarly vulnerable to abrasion; cleaning an optical surface inevitably carries some risk of damage to the fine finish. Some useful points of advice are:

(i)   All optical parts should be covered when not in use; even a simple polythene bag placed over a microscope will significantly reduce the need for cleaning.

(ii)   Take great care not to contaminate dry objectives with immersion oil when removing the slide.

(iii) Avoid contamination with Canada balsam or other mounting medium from a freshly prepared specimen, since this can be particularly difficult to remove, especially if it is allowed to harden.

(iv)   Never stand an immersion lens 'upside-down' on the bench without first removing any oil which might run inside the mount.

(v)   Never be tempted to dismantle an objective for cleaning; unless you have the skills of the manufacturer's optical department it will be impossible to reassemble it correctly.

Lenses are, however, easily contaminated despite all precautions, and the consequent effects on the image are so serious that cleaning must occasionally be undertaken. It is difficult to offer universal advice on cleaning lenses, since manufacturers and other authorities differ in their opinions on the subject, particularly with regard to the choice of solvents.

Before the glass makes contact with any cleaning material, dust should be removed; optical glass may be relatively soft, and easily damaged by abrasive particles from the atmosphere. If a can of compressed air is available, use this; failing this use a blower brush, first blowing and then lightly brushing the surface. *Never* blow with the mouth, since a drop of dried saliva inside an objective lens will be extremely difficult to remove.

Next, wipe the lens lightly with lens tissue, each time using a fresh part of the surface of the tissue, and one which has not been contaminated with grease by contact with the fingers. Breathe lightly on the lens, in effect depositing a thin film of distilled water, and wipe again carefully. Keep the lens tissue packet closed when not in use, carefully avoiding contamination. Use *only* lens tissue for cleaning microscope lenses; domestic paper tissues do not have the same fine texture and may scratch the lens.

If contamination still remains, and the lens appears to require more drastic treatment, the use of a solvent must be considered. Xylene, ethanol, ethanol/ether mixture, proprietory 'microscope lens-cleaning fluid', light petroleum distillate and other fluids have all been recommended. Equally, most of these have been rejected by other authorities, because of possible damage, particularly to the cement which may secure the lens in its mount, and also to the paint and even to the plastics of which some lens mounts are now made. If you have the recommendation of the lens manufacturer, follow this; otherwise select a solvent from the list above. The best advice is to use any solvent extremely sparingly; several applications with a dampened tissue are more likely to be effective than a heavy dousing which carries the risk of contaminated solvent entering the inside of the lens. Apply the solvent with the tip of a roll of lens tissue, changing the surface frequently. Finish by cleaning with dry tissue, perhaps after breathing on the surface.

6. THE PHOTOMICROGRAPHIC CAMERA

Camera equipment for photomicrography ranges from the single-lens reflex (SLR) camera as used in normal photography, together with a simple adaptor, through the purpose-built photomicrographic camera which may be attached to any microscope, to fully integrated systems in which the camera is built into or onto the microscope itself. Most microscope cameras are built to use standard 35 mm film, though adaptors are usually available for film of larger formats, including the Polaroid range of 'instant-print' materials. Each system has its own advantages, and all are capable of producing top-quality results within their own limitations.

6.1 Choice of size of film

Thirty-five mm film is used for a large proportion of general photography, and because of this is readily available in a wide range of types, with good processing facilities easy to find. For these and several other good reasons, 35 mm film is the obvious choice when the final result is to be a transparency for projection, and it is probably the most popular format for all forms of photomicrography.

The optical system of the microscope sets a severe limitation on the information content of the image to be recorded, and thus on the degree of enlargement of the final image from the film (see Section 4.2). Because of this even 35 mm film, formerly considered in general photography to be 'miniature', is fully capable of recording the detail in the average photomicrograph.

A large-format camera must necessarily be used for Polaroid 'instant-print' films, since the prints from these materials are the same size as the image which falls on the film. These films are available for prints of 3¼ × 4¼ inches or 5 × 4 inches, and are considered further in Section 8.4.

A microscope fitted with a large-format camera may be arranged to use conventional roll-film for images in a range of sizes up to 6 × 9 cm, or cut-film up to 5 × 4 inches, films in both of these types being available in black-and-white and colour. Despite the reasons given above for the adequacy of 35 mm film, micrographs from these larger formats may show a slight superiority because they require a smaller degree of enlargement for a print of a given size. Furthermore, any blemishes on the larger film will be less obtrusive than similar blemishes on smaller negatives. Film in the larger formats is used when work of the highest quality is required. Cut film, used in single sheets, also enables film from individual exposures to be processed separately, so that the results can be assessed immediately, and exposure or development conditions modified as necessary.

Film in the larger formats is naturally more expensive for each exposure than 35 mm, though this may be partially compensated by exerting greater care in its use. A significant disadvantage of recording on large-format film is that longer exposures will be required than for 35 mm film of the same speed and in the same circumstances. This is because the light from the eyepiece must be spread over a considerably larger area of film, leading to difficulties when recording images of low brightness in dark-ground or fluorescence microscopy, for example.

6.2 **The 35 mm single-lens reflex camera**

The SLR camera using 35 mm film offers the possibility of taking photomicrographs at minimal cost; a camera may already be used in the laboratory for other tasks, but even if it has to be purchased along with a microscope adaptor and some accessories, it is unlikely to cost more than about £500. Suitable adaptors are supplied for use with SLR cameras from Leitz, Minolta, Nikon, Olympus, Pentax, Zeiss, and other major manufacturers. Most modern cameras incorporate a system for determining exposure by measuring the light which will fall onto the film—known as 'through-the-lens metering'. They also have the advantage that their handling is familiar to many camera users, the complete system is relatively cheap, simple and easily understood, and uses a minimum of optical and mechanical parts; moreover, the camera can easily be transferred from one microscope to another.

The 35 mm SLR is unfortunately less convenient in use for photomicrography than more advanced equipment in several ways; viewing the image and focusing are more difficult, the controls are situated on the camera itself and are thus less accessible, and the range of shutter-speeds may be more restricted. The SLR also lacks the many refinements found on more elaborate photomicrographic equipment. The most important limitation of the SLR is that the mirror and shutter mechanisms may cause vibration of the complete camera—microscope system, leading to unsharp results. Paradoxically, better results will often be gained by using longer exposure times (1 sec or even more) so that any mechanical vibrations, which are generally of short duration, will take place for only a small proportion of the exposure. Nevertheless, with sufficient attention to technique, a SLR camera can produce micrographs indistinguishable from those taken with advanced equipment. Using a remote shutter-release together with a motor-driven winder to advance the film, thus reducing the need to reach up to the camera for each exposure, the writer has used such a system for more than half of the micrographs he has taken in recent years.

6.3 **The purpose-built photomicrographic camera**

This is the category of equipment most likely to be used for routine photomicrography. Examples are the Leitz MPS 12 and Orthomat, Nikon Microflex UFX, Olympus PM 10 and Zeiss MC 63 (*Figure 4*), which cost in the range from £1000 to £7000. Supplied by the major microscope firms, some photomicrographic cameras can be used only with microscopes from the same manufacturer, while others may be used on microscopes from other manufacturers also. This adaptability is principally dependent on whether the camera is designed to attach to an eyepiece tube of standard diameter, or has a larger fitting peculiar to a particular range of microscopes. The special fitting restricts interchangability, but allows some special facilities to be fitted, and also provides greater mechanical stability and freedom from vibration.

Compared with the SLR camera, purpose-built photomicrographic cameras offer greater convenience and a wide range of facilities. Viewing of the image is more

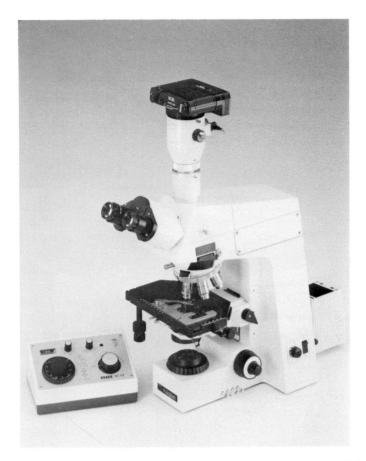

Figure 4. A purpose-built photomicrographic camera, the Zeiss MC63, attached to a Zeiss Axioplan microscope. The camera control box, on the left of the microscope, enables film-speed, exposure-time and reciprocity corrections to be made. The shutter-release button is at the top right. Photograph courtesy of Carl Zeiss (Oberkochen) Ltd.

comfortable, and focus easier to establish precisely. An exposure metering system is built in; some models offer the choice of spot or integrated readings, and exposure control is by means of a shutter designed to minimize transmission of mechanical vibrations to the rest of the system. Film advance is usually by means of an electric motor, and the controls are conveniently presented on the panel of a unit designed to stand on the bench. This control panel may include means of setting automatic correction for reciprocity failure, and other facilities.

6.4 The integrated photographic microscope

These instruments offer maximum convenience in photomicrography, allowing photomicrographs to be made without interrupting routine work. They cost from around £12 000, depending on the optical equipment included, examples including the Nikon

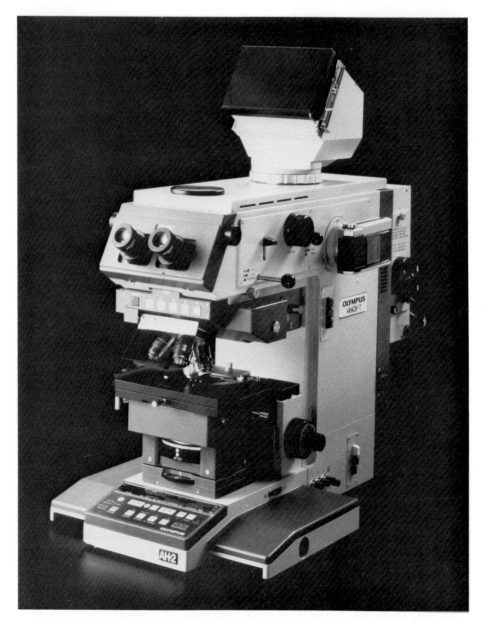

Figure 5. An integrated photographic microscope, the Olympus Vanox-T. This instrument is fitted with two 35 mm camera bodies, one on each side, and a large-format camera, here shown with a Polaroid back, on the top. A television camera may be fitted to the opening shown just above the eyepieces. Photograph courtesy of Olympus Optical Co. (UK) Ltd.

Microphot, Olympus Vanox, Reichert Polyvar and Zeiss Axiophot.

These instruments incorporate one or more interchangeable camera bodies or film magazines with automatic wind-on of 35 mm film, or adaptors for larger format film, and the complete unit is built into a sturdy metal box protecting its components from dust and vibration. Focusing is done through the normal eyepieces and exposure is controlled by a sophisticated automatic electronic system offering facilities similar to those described in Section 6.3. In addition, a wide range of optical equipment for the various contrast techniques is available for these highly versatile instruments. An example of such an instrument—the Olympus Vanox—is shown in *Figure 5*.

6.5 The bellows camera

This type of camera offers great versatility at the expense of convenience and speed of use. It is designed to use large-format film and is especially valuable for work of the highest quality, and for photomacrography (see Section 9), rather than for routine photomicrography. It consists of an extremely heavy and rigid stand carrying a shutter and a focusing screen connected together by an extensible light-tight bellows, both the shutter panel and the screen being movable up or down the stand as required. The focusing screen can be removed and replaced by standard holders for roll-film, cut-film or Polaroid instant materials. A reflex housing may be fitted between the bellows and the film holder, enabling the image to be composed and focused on a viewing screen in a more convenient position.

The microscope stands on the base of the apparatus, and the shutter panel is lowered close to the eyepiece, a light-trap being fitted between the shutter and the microscope tube. Since the image is projected over a relatively long distance from the eyepiece to the film, a standard eyepiece normally provides good image quality, and it should not be necessary to use a special photographic eyepiece or other device. The size of the final image increases with projection distance, hence adjustment of the bellows allows magnification to be varied continuously. This feature can be useful when a final result of a particular exact size is required.

7. FOCUSING AND EXPOSURE DETERMINATION

Whatever kind of photomicrographic equipment may be used, it is of the greatest importance that the image should be precisely in focus on the film, and that the correct exposure is given. The problems of focusing and exposure metering are covered in the next sections.

7.1 Focusing

There are three methods of viewing the image to achieve correct focus; choice will depend on the apparatus in use.

(i) The simplest and most direct method of focusing is found in the bellows camera, designed to use film of the larger formats. Here, the image is focused onto a ground-glass screen positioned in the film plane, the screen being subsequently removed and exchanged for the film holder. The screen generally has a clear central spot within which a relatively dim image may more easily be seen and critically focused using a magnifier. If the screen has no clear area, one can be provided simply by making a pencil cross on the ground surface of the glass, and cementing a suitable microscopical coverglass over the area using Canada balsam or similar medium.

(ii)   Using a SLR camera mounted on a simple adaptor, focusing can be carried out
        on the camera's normal viewing screen in a similar way. While this will suffice
        for elementary work, the image may be very dim, and no advantage can be
        taken of any of the screen's focusing aids such as microprisms or split-image
        'rangefinder'. A great improvement can be made, if the camera has inter-
        changeable screens, by changing to a clear screen with central cross-lines, and
        viewing the much brighter aerial image. In this case, since there is no ground
        glass to define the plane in which the image should fall (by diffusing it to fuzzi-
        ness when it falls in any other plane), it is essential to use an adjustable accessory
        magnifier to fix the plane of focus of the eyes onto the cross-lines on the screen.
        This is often incorporated into an accessory right-angle finder.

(iii)   In the third method, the image is focused outside the camera, in a plane arranged
        to be conjugate, that is in focus simultaneously, with the final image. Some
        purpose-built photomicrographic cameras have a beam-splitting prism to direct
        the image into a side tube containing a screen and a focusing telescope. The area
        which will be included in the photograph is framed by a graticule on the screen,
        and the image focused on the central cross-lines which are arranged to be par-
        focal with the final image on the film. Especially where a photomicrographic
        camera system is designed for use with a specific model of microscope, the focus-
        ing system may operate through the normal binocular head using a framing
        eyepiece containing a graticule and cross-lines. Such a system relies on correct
        adjustment so that the camera and visual images are in focus simultaneously;
        this is normally preset in manufacture but on some models it is a user adjustment.

When focusing is to be done on a clear screen with cross-lines or through a framing
eyepiece, first defocus the microscope image so that the eye can be focused critically
on the cross-lines using the magnifier or eyepiece adjustment, without disturbance from
the image. Then bring the image into focus, moving the head very slightly from side
to side so that defocus can be more readily recognized due to parallax; at correct focus
there will be no relative movement between the image and the cross-lines. Finding focus
is relatively easy using high-magnification, large-aperture objectives; with low-power
lenses great care must be taken if a good micrograph is to result, since accommodation
of the eye can have a serious effect on the plane of focus chosen. It can help considerably
if a small telescope or pair of binoculars focused on infinity is used between the eye
and the eyepiece when setting final focus, since this procedure restricts the depth of
field of the system. Older photomicrographers whose eyes have lost the ability to
accommodate will find less difficulty here.

At this stage the microscope image should have been optimized, and considerations
of the requirements for photography follow. Nevertheless, since focusing is of such
critical importance to good photomicrography, and is so easily upset due to minor
movements of the microscope and specimen, it should be reset as the very last step
before releasing the shutter.

7.2 Exposure determination

The manufacturer of film quotes a nominal 'speed' for the material, which must be set on the exposure meter of the equipment. This is an expression of the quantity of light, known as the 'exposure', which must fall on the film for best results; different kinds of film require different exposures. If too little light is received, negatives will be too pale and the final prints or transparencies too dark, and vice versa. Exposure is the product of the intensity of the light and the time for which it falls on the film (the exposure time). For a given degree of exposure there is thus a reciprocal relationship between intensity and time, that is if the light is twice as bright it should fall on the film for half the time, and so on. When exposure times are abnormally long or short, this relationship breaks down ('reciprocity failure', see also Section 8.3.5), and a greater-than-expected exposure must be given; many photomicrographic cameras are designed to take account of this effect.

Exposure is determined by light-sensitive devices which produce an electric current or alter their electrical resistance in response to the light falling on them. Most modern SLR cameras have an exposure metering system which measures light 'through-the-lens', that is they measure the intensity of the light which falls onto the film; purpose-built photomicrographic equipment cameras use an essentially similar system. The result is indicated in terms of the recommended shutter-speed, by a meter or an electronic display system, and use of this speed should result in a correctly exposed micrograph. The shutter speed may be set manually, or the system may control it automatically.

Exposure metering systems are designed to give correct exposure with 'average' subjects. Where the distribution of light and dark areas within the image is not 'average', incorrect exposure can be expected. A well-stained biological section which fills the field of view may be correctly exposed if the meter reading is followed without modification. A specimen which does not fill the field and is surrounded by a bright background is likely to be under-exposed, while images in which only parts of the field are bright, against a dark background, as with fluorescence or dark-ground microscopy, are likely to be over-exposed if a simple metering system which measures the integrated brightness over the whole field is used without correction. Some SLR and photomicrographic cameras are designed with 'spot-metering', which enables the meter reading to be taken from a suitably representative small portion of the image.

If the camera does not incorporate a light-metering system, a separate exposure meter may be used, such as the Leitz Microsix or Zeiss Ikophot M. Attachments are also available to adapt some normal photographic exposure meters for microscope use. With such meters, readings may be taken from above or in place of an eyepiece, from the camera viewing tube, or from the film plane of a large-format camera.

Exposure meters must be set according to the nominal 'speed' of the film (see Section 8.1.1). Most systems now carry a scale calibrated in ISO (International Standards Organisation) or ASA (American Standards Organisation) speed ratings; others may be marked in arbitary units and must be calibrated by experiment. Even with the former, setting the nominal speed of the film as quoted by the manufacturer will not necessarily result in correctly exposed micrographs; the system should be calibrated to determine the most suitable setting for the equipment, type of specimen, contrast mode, film−developer combination and colour filter in use (schedules for initial calibration are given in Sections 11.2 and 11.3).

7.3 **Exposure control**

Exposure time is controlled by a shutter built into the body of the camera. An SLR camera generally has a 'focal-plane shutter'—a blind with an adjustable slit which moves across the frame just in front of the film. Combined with the movement of the mirror in such a camera, the shutter is likely to cause some vibration of the microscope with consequent blurring of the image. A purpose-built photomicrographic camera will have a shutter which is designed to avoid transmitting vibration to the system, and for this reason alone is much to be preferred. Shutter speeds may be controlled manually, semi-automatically with a knob which sets the light-meter reading against a datum or an indicator light, or fully automatically where the shutter remains open until sufficient light has reached the film.

A camera shutter is adjustable only over a certain range of speeds. With some risk of vibration and specimen movement, and extended exposure due to reciprocity failure, long exposure times caused by dim images may be tolerated; excessively bright images which call for shutter-speeds shorter than the agility of the system will allow will have to be reduced in intensity. Unlike a camera lens, a microscope has no diaphragm intended for altering the brightness of the image, even though this may occur as a side effect of adjusting a diaphragm. All of the microscope's optical adjustments are concerned with the *quality* of the image; its *intensity* can be altered only by adjusting the voltage on the lamp (but for the effect of this on colour film see Section 8.3.3), or reduced using colour filters (Section 8.2.1) or neutral-density filters (Section 8.3.3) in the light path.

8. PHOTOGRAPHIC CONSIDERATIONS

Most of the needs of the photomicrographer can be met by a relatively small selection from the wide variety of photographic films and developers available on the professional and amateur markets. The aim of the following paragraphs is to assist with the rational choice of materials so that once a satisfactory system is established, attention can be concentrated on the microscopy, where most of the difficulties lie.

8.1 **The photographic process**

All photographic processes, whether for negatives of prints, black-and-white or colour, are based on the special property of certain salts of silver, whereby they are light-sensitive. These salts are the silver halides, principally the chloride and bromide, and photographic materials consist of an emulsion of these salts in a layer of gelatin, coated on a base of transparent plastic, paper or, now rarely, glass. When light falls on a silver halide crystal, the energy carried by about 10 photons converts some of the salt to a tiny speck of metallic silver; crystals thus affected constitute the *latent image*. When treated with a mild reducing agent (known as a *developer*) the whole of each crystal carrying a latent image is preferentially reduced to a grain of metallic silver in a relatively short time; prolonged treatment with developer or exposure to light will result in the ultimate conversion of all the halide to silver. After a predetermined time, the developer is removed and the unchanged halide is dissolved with sodium or ammonium thiosulphate, known as the *fixer*. Those areas on which light has fallen will be black due to the silver grains, against a clear background; the image will thus be in the form of a negative. A positive print is made from the negative by a further similar process; transparencies are positives resulting directly from their special reversal processing.

Colour film responds to the action of light in the same way as black-and-white, but contains substances which are converted to coloured dyes during processing (see Section 8.3).

8.1.1 *Film speed*

Since about 10 photons must fall on one silver halide crystal to render it developable, it follows that large crystals will collect their requirement of light in a shorter time than smaller crystals. A film composed of large crystals will thus require less exposure for a satisfactory image; or to put it in more familiar photographic terms, films which are coarse-grained will be 'faster'. This situation is complicated by the incorporation into the emulsion of sensitizing dyes, which absorb energy from light of longer wavelengths and transfer it to the silver halide, thus widening the colour sensitivity of the film from blue only, through blue and red (orthochromatic) to the full visible spectrum (panchromatic).

Film speeds have been expressed using many systems; the one standardized by the ISO (similar to that of the ASA), will be used here. This is an arithmetic scale on which a film rated at ISO 100 has twice the speed (i.e. requires half the light) of a film rated at ISO 50. Films regarded as slow at the time of writing have ISO ratings up to about 50, medium speed films from 50 to about 200, and fast films up to about 1600. It should be understood that the speed ratings quoted by film manufacturers are only recommendations to provide useful guidance in normal circumstances. When establishing a new routine in photomicrography it is strongly recommended to use a test film with the exposure metering system set at a range of ISO values on both sides of the nominal, combined with the keeping of careful records of all important factors.

8.1.2 *Grain*

After development the image is in the form of grains of silver about 0.5 μm across. These are obviously invisible except with a microscope, but because they are randomly arranged rather than uniformly distributed, clumps of grains exist, which give rise to the 'graininess' commonly seen in the faster films. However, in photomicrography, since the resolving power of the microscope itself imposes a limit on the total final magnification, such that the film will not normally be enlarged more than 5-, or at the most, 10-fold, the graininess of the film is seldom a serious limitation.

8.1.3 *Contrast*

Control of contrast is particularly important in photomicrography, since the contrast-range in images from the microscope is generally less than that in 'domestic' photography. Apart from being an inherent property of a particular film, slower films being generally more contrasty, contrast also depends on the type of developer used, its temperature, and the development time.

8.2 **Black-and-white photomicrography**

All the major manufacturers produce films suitable for photomicrography, and for general photomicrography the slower, more contrasty films such as Ilford Pan F and Kodak Panatomic X will be found satisfactory. Their manufacturers offer recommendations on development, and it will normally be found preferable to follow a procedure leading to a contrast-index (similar to the older concept of gamma) of about 1.0. Two particularly useful developers for increasing contrast are Kodak's HC-110 and D19.

Kodak Technical Pan Film 2415 has proved to be very suitable for photomicrography, since its contrast can be altered over a wide range by choice of development conditions. HC-110 will give a contrast index from about 1.0–2.7, and D19 up to 3.6. This adaptable film is available in 35 mm and 120 sizes, and finds other uses in the laboratory for document copying and making lecture slides, for example; with its special developer, Technidol, it can produce relatively low contrast (0.5–0.75), and is thus also suitable for high-quality pictorial work.

As an example, and to provide a starting-point for black-and-white photomicrography, try Kodak Technical Pan 2415 film rated at ISO (ASA) 50. Develop in Kodak HC-110, dilution D (see the instructions provided with the developer) for 6 min at 20°C.

For lower contrast, rate Technical Pan at ISO 25 and develop in Kodak Technidol Liquid Developer for 9 min at 20°C, and for high contrast, rate Technical Pan at ISO 125 and develop in D19 for 4 min at 20°C.

The differences between the negatives and the prints resulting from these three procedures will be obvious (see *Figure 6*), and is should not be difficult to select the one which suits a particular combination of specimen, microscope and contrast technique. Intermediate values of contrast can be achieved by adjusting development time, bearing in mind that longer or more energetic development will increase both contrast and effective film speed. Similar procedures can be established using other fine-grained films.

8.2.1 *Using colour filters in black-and-white photomicrography*

Colour filters can significantly improve black-and-white photomicrographs in several ways. A microscope will resolve finer detail when light of a shorter wavelength is used. Thus if light of longer wavelengths, especially red, is removed from the light with which an image is formed, very slightly higher resolving power will result; this can be done using a blue-green (or cyan) filter, sometimes known for obvious reasons as 'minus red'. It might be thought that further advantage would accrue from using light of only the shortest visible wavelength, blue, and this is indeed true in theory but not in practice. Apart from the extremely dim appearance of blue light, which would make focusing difficult, most microscope lenses are designed to operate best in the centre of the visible spectrum.

Lenses are never quite perfect; they suffer from a number of aberrations, many of them wavelength dependent. The lens designer arranges for spherical aberration to be best corrected for green light, the colour to which our eyes are most sensitive; as the wavelength of the imaging light departs from about 550 nm, the image quality may deteriorate.

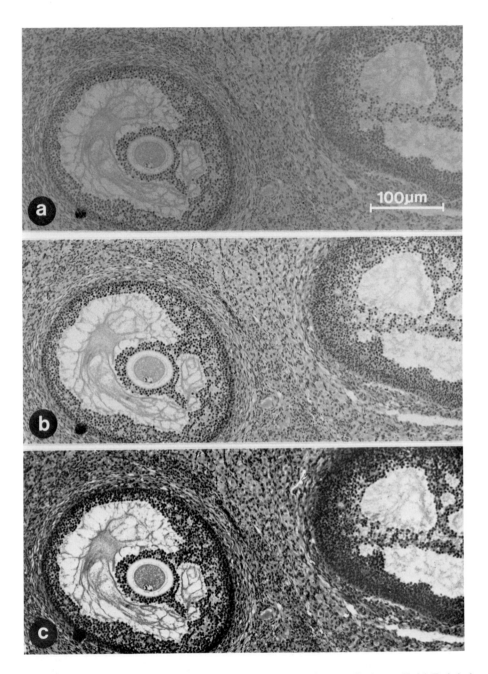

Figure 6. The effect of film developer on contrast. These micrographs were all taken on Kodak Technical Pan film, and developed as shown below. Printing conditions for all micrographs are identical. (**a**) Low contrast; developed in Technidol, 9 min at 20°C. (**b**) Medium contrast; developed in HC-110, Dilution D, 6 min at 20°C. (**c**) High contrast; developed in D19, 4 min at 20°C. Scale bar = 100 μm.

Achromatic objectives are not corrected for axial and lateral chromatic aberrations for all wavelengths, and they thus form an image with slight colour fringes visible on critical examination. If these lenses are provided with almost monochromatic light by the use of a suitable colour filter, their inadequate correction becomes irrelevant. Semi-apochromats (sometimes known as fluorite lenses) and apochromats, both of which are well corrected for chromatic aberrations, may also sometimes benefit slightly from the use of light of a narrow wavelength band.

Perhaps the principal reason for the use of colour filters is for the control of contrast in images of coloured specimens. Filters make it possible to alter the way in which colours in the image are rendered as shades of grey in the photographic record. Take for example, in visual microscopy, a red-stained specimen against a clear background. When a red filter is inserted there will be less contrast between the object and the background; both will appear red, and will be recorded in similar tones of grey in a photomicrograph. Conversely if a blue-green filter (cyan or minus red, see above) is inserted instead of the red filter, the red object will appear dark and in high contrast against a cyan background, since it does not transmit blue or green light (that is why it normally appears red!). Thus, filters of a colour similar to the object will lighten the tone of that object in a black-and-white photograph, and filters of the complementary colour will darken the object (see *Figure 7*). Colours and their complementaries are shown in *Table 1*.

Colour filters are available from most of the major microscope manufacturers, and are built into some more advanced microscopes. They are available from Kodak (Wratten filters), from other photographic firms such as Ilford, and from microscopic manufacturers.

Care should be taken about where filters are inserted into the optical system. Obviously they should be placed in the illuminating rather than the imaging parts of the system wherever possible, unless they are of high quality and designed for this application. Some microscopes are fitted with a filter tray situated just beneath the condenser, an ideal position for a filter since it is unlikely that fingermarks, dust and other blemishes on its surface will be seen together with the image. The design of many microscopes invites the user to place the filter into the opening in the base, beneath the condenser, through which the light emerges. At this level the filter is often close to the illuminated field diaphragm of the microscope, which is conjugate with the specimen and thus the final image, and blemishes may then intrude on the image. Care should be taken to avoid problems due to this.

Table 1. Colours and their complementaries.

Colour	Complementary colour
Red	Cyan
Green	Magenta
Blue	Yellow
Cyan	Red
Magenta	Green
Yellow	Blue

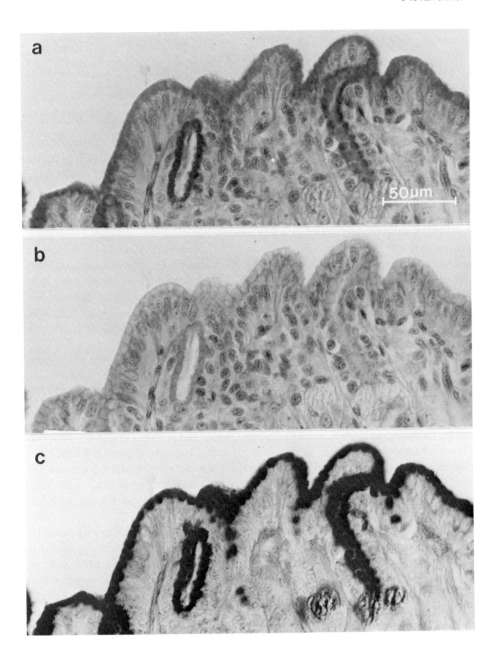

Figure 7. The effect of colour filters on contrast. The specimen, a section of frog stomach, is stained with periodic acid−Schiff and Light Green, photographed on Kodak Technical Pan film, and developed for medium contrast. The appearance of the micrographs is closely similar to that seen directly with the eye. (**a**) No filter. The tips of the cells, densely stained magenta, appear slightly darker than the rest of the specimen, which is in shades of green. (**b**) Red filter. The magenta areas now appear colourless and the green-stained nuclei are dark. (**c**) Green filter. The green-stained nuclei cannot now be seen, while the magenta-stained regions are shown in stronng contrast. Scale bar = 50 μm.

In summary, the following advice is given.

(i)   A narrow-band green filter such as the Wratten 58 or a green interference filter should be used to improve the optical performance of the microscope, except when it will excessively affect contrast.

(ii)   Green stains should be avoided, since they will be hardly detectable in the preferred green illumination. For preference use magenta or red stains, which give good contrast in green light.

(iii) Strongly coloured filters may require the film-speed setting on the exposure metering system to be altered, since the spectral sensitivity of the light-measuring device of the camera is unlikely to be exactly the same as that of the film. This should first be investigated with a trial film (see Section 11.2).

8.3 Colour photomicrography

The recording of colour is based on the fact that the sensation of any colour can be produced by an appropriate mixture of the three primary colours, red, green and blue. Conventional modern colour films consist of three layers of emulsion, each sensitive to one of these colours. In most films these layers also contain the precursor molecules for dyes of appropriate colours, dyes being formed when these molecules react with the oxidation product of a special developer when it reacts with the silver halide; the silver is later dissolved away.

8.3.1 *Colour prints or transparencies?*

Colour films can be designed to produce either negatives which can be printed onto paper (negative films), or transparencies for projection (reversal films). It is obviously convenient to produce micrographs in the form of prints since these can be viewed easily, presented as posters, used to illustrate theses and reports, and so on. Reversal films have important applications in lecturing, but this accounts only partly for the great popularity of this kind of film in photomicrography. Several excellent processes are available for producing prints from transparencies, making reversal film, in the view of many experienced photomicrographers, the more useful type for use with the microscope.

Many users of the microscope wish to produce prints and are familiar with the rapid and cheap processing services now available. Since disappointment with the results is common, a brief explanation of the problems involved is necessary here. It is an unfortunate fact that it is more difficult to achieve top quality results on negative film than on reversal film, when using commercially available routine processing in each case. Reversal film is processed by a standardized and closely controlled system, and the final transparency is the very piece of film which was exposed in the camera; given competent processing, the final result should be a faithful record of the image as it fell onto the film. Negative film undergoes an additional process—printing onto paper— and it is here that variations from the intended colours may occur.

Most routine processors serving the mass market use machines which automatically compensate for what appear to be colour casts by inserting filters into the light path. For most domestic photography this automatic process works well, since in an 'average' photograph there is an approximately equal distribution of all three primary colours. This assumption, on which the operation of the machines is based, is generally not true for photomicrography; features may be stained in only one or two colours, and are often seen against a clear white background. As an example of the problems this causes, if a blue-stained object is photographed against a white background, the automatic printing machine will add yellow in order to counteract the excess of blue which it detects, resulting in a dull-grey object against a yellow background. The inspector at the end of the production line, being unfamiliar with scientific images, allows it to pass, to the disappointment of the discriminating microscopist.

It must be said, however, that excellent photomicrographic prints can be produced using negative film, provided the prints are made individually by an operator who understands the result required; they will then unfortunately be considerably more expensive than routine prints or transparencies. If prints are being made 'by hand' rather than automatically, it will be helpful to provide the printer with a transparency or a print showing the correct colour rendering, or at least with the information that the background should be rendered, for example, as white.

In addition to the reasons given above, most transparency films are more contrasty, and capable of recording slightly finer detail than negative films of similar speed, and are still preferred to prints by many publishers. Furthermore, the transparencies themselves are valuable since they can be viewed by projection, while negatives are useful only as a stage in the production of prints.

8.3.2 *Prints from transparencies*

Several systems are available for making prints from transparencies, from Agfa, Ilford, Kodak and other manufacturers. These are offered by commercial processors, but are also simple enough for use in the normal laboratory darkroom without the necessity for a large amount of specialized equipment. Ilford's Cibachrome has proved to be well suited to printing photomicrographs, partly because it tends to be rather more contrasty than other processes. It is very tolerant of minor errors in exposure and colour balance, and once a procedure has been established for printing from correctly exposed transparencies taken on a particular kind of film, excellent prints can be made with very little effort. Printing in the laboratory has the further advantage that the colour balance can be varied in difficult cases, under the microscopist's supervision, until an acceptable result is achieved. Processing requires only three solutions, and a finished print can be produced in only 12 min after exposure. Kodak's Ektachrome 22 paper, processed in their R3000 solutions is also recommended for laboratory use.

It is true that prints made from transparencies by any of these processes will be considerably more expensive than machine-made prints from negative film. However, experience shows that prints will normally be required from only a selection of the transparencies, rather than from the whole film, thereby reducing the potential cost.

8.3.3 *Colour temperature*

White light, so called, consists of a distribution of energy across the spectrum of wavelengths of visible light, from violet through to red. But lights which we generally perceive as 'white' are not all alike. For example, the light of the midday sun contains a greater proportion of its energy in the form of the shorter wavelengths (blue), than the light from the tungsten filament of a conventional electric-light bulb, which contains

Table 2. Colour temperature of some light sources.

Light source	Colour temperature (K)
Tungsten-filament lamp run at rated voltage	2900
Tungsten – halogen lamp run at rated voltage	3200
Photoflood lamp	3400
Daylight (average)	5500
Electronic flash	6000

Table 3. Wratten colour-correction filters.

Colour temperature of light source	Colour temperature balance of film		
	3200	3400	5500
2900	82B	82C + 82	80A + 82B
3200	None	82A	80A
3400	81A	None	80B
5500	85B	85	None
6000	85B + 81A	85B	81A

more in the longer wavelengths (red). This property of light is expressed in terms of colour temperature since it depends on the temperature, expressed in Kelvin, of the body emitting the light (see *Table 2*).

Our eye – brain system is adept at compensating for differences in colour of ambient light; the paper on which this book is printed is perceived as 'white', whether it is examined by daylight, electric light, or even by candle light. Colour film cannot compensate similarly, and records these differences faithfully, giving results which are more blue when exposed by daylight, and more red when exposed by the light of tungsten-filament lamps. Because of this, colour films are manufactured with their colour rendering balanced for a particular colour temperature of light. Daylight film is balanced for a colour temperature of 5500 K; artificial light (or tungsten) film is balanced for 3100, 3200 or 3400 K. Since many light sources do not conform exactly to these values, or we may wish to use a more readily available daylight film with a microscope fitted with a tungsten-halogen lamp, the use of colour correction filters is frequently necessary to ensure correct rendering of colour. *Table 3* indicates the use of filters from the Kodak Wratten range, probably the most comprehensive and best known colour correction filters.

Because colour temperature depends on the supply voltage at the lamp, care should be taken to adjust this to the specified value, and to ensure that this setting is reproduced for every micrograph. It will be found useful to measure the supply voltage at the lampholder itself, rather than at the power supply, since the resistance of the inter-connecting cable can result in a considerable voltage drop, leading to a reddish colour cast.

At its full rated voltage the lamp may be too bright for comfortable viewing and for the shortest exposure time available. In these circumstances, neutral-density filters should be used to absorb all colours approximately equally, reducing the intensity of the light without altering its colour. Not all 'grey' filters are entirely neutral in their effect: for preference use the filters with a mirror-like, partially-transmitting metal coating, supplied for this purpose by the microscope manufacturers.

Several modern photographic microscopes incorporate systems for ensuring consistent colour temperature. One approach is to stabilize the voltage applied to the lamp at a value considered to be correct, at all times or only when set for photomicrography. In this case, changes of brightness (and hence exposure time) are made by means of neutral-density filters built into the illuminating light path of the microscope. Another approach is to use a colour temperature meter, available as a separate instrument or as an integral part of some photomicrographic cameras, altering the voltage at the lamp and/or adding correction filters until the meter indicates that the light is of the correct colour temperature.

8.3.4 *Correcting colour casts*

Even when every care has been taken to ensure that the colour temperature of the light source corresponds to the requirement of the film, the resulting transparencies may still show a colour cast. This can be due to slight colorations in the glasses used in the illuminating and imaging systems, and also to more obvious causes such as yellowing of the mounting medium. Such minor colour casts should be corrected by the use of pale Colour Compensating (CC) filters, again available from Kodak and other manufacturers, in the colours red, green, blue, yellow, magenta and cyan. The transparencies should preferably be viewed on a colour-corrected light box, and CC filters inserted alone or in combination, until the colour rendering appears correct. Using these filters in the light path of the microscope should neutralize the colour cast, though it is not certain that the film will respond to the change in colour of the illumination in exactly the same way as the eye.

Colour films record UV light as if it were blue. Some microscope lamps produce a significant proportion of UV radiation, and this can cause a bluish haze in the micrograph. Wratten filters 2A, 2B and 2E all absorb UV, and use of one of these extremely pale yellow filters will avoid difficulty due to this cause.

Some manufacturers recommend the use of a filter made of didymium glass, especially when photographing specimens stained in eosin. This filter has the property of emphasizing pink colours, and may in certain circumstances considerably improve colour rendering.

8.3.5 *Reciprocity failure and colour balance*

Colour film shows reciprocity failure when exposed for abnormally short or long times, requiring adjustments to its exposure as does black-and-white film (see Section 7.2). But colour film suffers an additional problem: the three layers of emulsion, sensitive to the three primary colours, lose speed at different rates, leading to incorrect colour rendering. In general, artificial light films, which are made to give correct colour balances when exposed for longer times than daylight films, are recommended for photomicrography when light levels are low.

Correction of colour balance can be made when printing, but for transparencies correction filters should be used at the time of exposure. Each kind of film suffers reciprocity failure in its own way, so the film manufacturer's technical information should be consulted for advice on filtration. Since it may be impossible to correct completely for colour imbalance due to reciprocity failure, exposure times of longer than about 5 sec are best avoided by the use of a faster film or a more intense light source.

After exposing a few test films, keeping careful records of all important operating conditions, it should be possible to arrive at a standard setting of the lamp brightness and a combination of filters to compensate for the peculiarities of a particular combination of microscope, film and technique, so that micrographs with correct colour rendering can be produced on every occasion (see Section 11.4).

8.3.6 *Choice of film speed*

Colour films are available with speed ratings ranging from ISO 25 to 1600 or 3200. Highest image quality and contrast are provided by the slower films (up to ISO 100), and these are recommended for general photomicrography where ample light is usually available. Where extra speed is required, as for dark-ground, phase-contrast or fluorescence microscopy, films of ISO 200 and even 400 may be used without serious loss of quality. In these cases, consideration should also be given to the technique often known as 'push-processing' in which the effective speed of a transparency film is increased by modified processing (extending the time in the first developer). Film speeds may be doubled or quadrupled by this method, particularly useful for ISO 200 and 400 films, and which is offered without extra charge by some processors. Apart from increasing the effective film speed, push-processing has the advantage for some forms of photomicrography of increasing contrast; its disadvantages of increasing graininess, and decreasing maximum density leading to slight greying of the background in dark-ground or fluorescence microscopy, will not normally be found unacceptable.

8.3.7 *Fluorescence photomicrography*

Recording images from the fluorescence microscope presents special difficulties since the fluorescing features frequently occupy only a small proportion of the photographic frame, are rather dim, and may fade during prolonged exposure to the exciting illumination. These factors result in inaccurate estimation of exposure, and long exposure times further prolonged by reciprocity failure. Using the best microscopical techniques, however, good results can normally be expected.

Except for extremely low-magnification work, an epi-fluorescence system should be used, in which the specimen is illuminated by light emerging from the objective. Epi-fluorescence has the advantages that unabsorbed illumination is directed away from the observer, the illumination falls directly onto the surface being observed, and alignment is simplified since the objective acts as its own condenser. The quantity of light transmitted by a lens is proportional to the square of its NA; in epi-illumination, since the objective lens acts also as the condenser, image brightness is related to the fourth power of its NA. Image brightness is reduced, however, according to the square of magnification. Manufacturers thus offer objectives designed especially for fluorescence microscopy, with large NA (frequently achieved by oil-, glycerine- or water-immersion) combined with low magnification, and the use of these is particularly advantageous for photomicrography.

Where a spot-metering system is available, this should be used so that the exposure is determined principally from the fluorescing areas of the specimen. If spot-metering cannot be used, the exposure time should be reduced from that indicated by the meter, according to the proportion of the field which is not fluorescing. This will inevitably lead to some incorrectly exposed results, and in all cases it is advisable to take several frames of film, using exposure times covering quite a wide range.

Since the level of light resulting from fluorescence is likely to be extremely low, fast films of ISO 200 or 400, or possibly even higher, should be used. Further speed can usefully be obtained by 'push-processing' (see Section 8.3.6). Many microscope camera systems contain a beam-splitting prism which divides the light leaving the objective, so that it enters the eyepieces and the camera simultaneously. In fluorescence microscopy it is advantageous to set the prism, where possible, so that all the available light is directed to the eyepieces for observation and all to the camera for photomicrography. This can double the quantity of light reaching the film, yet may reduce the exposure time required by considerably more than a factor of two when the consequences of fading and reciprocity failure are taken into account. With certain photomicrographic cameras, the long exposure times required in fluorescence photomicrography may permit a significant quantity of light to enter the camera through the eyepieces or camera viewing tube, fogging the film, perhaps superimposing an image of the window or room lights, or causing incorrect exposure determination. Some microscopes are fitted with a shutter in the viewing tube to prevent this; if no such shutter is fixed, this possibility should be considered in the event of problems.

Colour film balanced for use in daylight is most frequently used for fluorescence photomicrography and should be tried first, but because of the variety of fluorescing substances which may be studied, some users may find their specimens better rendered on tungsten-balanced film. When black-and-white results are required, try Kodak Technical Pan film for brightly fluorescing specimens, or an ISO 400 film if this proves too slow.

8.4 Polaroid films

The Polaroid Corporation (USA) produces a wide range of films designed to reduce the delay, due to processing, between making the exposure and seeing the result. These can be extremely valuable in photomicrography, particularly for specimens which are not permanent, since they permit immediate assessment of the result, with the possibility of a repeat exposure after appropriately altering the conditions. Polaroid films make photomicrography a practical possibility for many laboratories which lack facilities for developing and printing conventional materials, and their greater cost can be offset against considerable savings in staff time.

Most of the Polaroid films are supplied in the larger formats, designed to produce prints, and one type produces a negative as well as a print; 35 mm films are also available, producing black-and-white or colour transparencies.

8.4.1 *Polaroid 'instant-print' films*

Polaroid 'instant-print' films are sensitive to light due to the familiar properties of silver halides. Each individual film incorporates a 'pod' of developer, which becomes split and spread evenly across the film as it is withdrawn from the camera between a pair of rollers, and by a special process a positive print on paper is produced simultaneously. After the development time quoted in the instructions (usually <1 min), the packet is torn open to expose the print. The print from some types of film must be coated with the protective lacquer provided with the film.

These films produce prints which are the same size as the camera image, and are available in the larger formats of 8.3 × 10.8 cm (3¼ × 4¼ inches) and 13 × 10.5 cm (5 × 4 inches). The smaller format is supplied in an eight-exposure film pack, and the larger in individual sheets. In order to use either kind, a Polaroid adaptor is required, supplied by the manufacturers of most photomicrographic cameras. If the camera is already fitted with a 5 × 4 inch 'international back', a standard adaptor available from Polaroid can be used.

Polaroid produce a wide range of black-and-white and colour films, with speeds ranging from ISO 50 to ISO 3000. The writer has used Type 665 black-and-white and type 668 daylight colour, both ISO 80, for photomicrography; the former produces excellent negatives as well as prints.

Since the instant prints resulting from Polaroid materials do not allow correction for errors in exposure or colour balance to be made at the printing stage, it is important to get these factors correct at the time of exposure. The colour materials are particularly sensitive, and care should be taken to follow the recommendations given in the manufacturer's instructions. In order to establish the correct exposure most economically when using a non-automatic system, a stepped exposure can be made on one sheet of film, in the traditional way for large-format photography. The film-pack adaptor has a light-tight slide to prevent light entering when the pack is removed from the camera. This can be used to cover portions of film successively, after increments of exposure have been given. The packet in which each 5 × 4 inch film is fitted can be used similarly.

To demonstrate the principle, suppose that the camera shutter is marked in the usual series of steps: 1/15, 1/30, 1/60, 1/125, 1/250 sec.

(i) Withdraw the slide fully and expose for 1/250 sec.

(ii) Insert the slide about one-quarter of its travel and expose again at 1/250 sec. At this stage, step one has received 1/250 sec and the rest of the film 2 × 1/250 = 1/125 sec.

(iii) Insert the slide until it is about half-way home and expose for 1/125 sec. The second step will now have received 1/125 sec and the rest 2 × 1/125 = 1/60 sec.

(iv) Insert the slide until only one-quarter of the film remains to be exposed, and expose this for 1/60 sec. This last step will have received 2 × 1/60 sec = 1/30 sec.

(v) Process the film and evaluate the results.

The print should be in the form of four stripes of graded density. If they are all too dark, repeat the series using increased exposure times; if they are too light, repeat with shorter times.

8.4.2 *The Polaroid SX70 camera*

This is a SLR camera which produces 3¼ × 4 inch colour prints using its own special film, the processed prints being ejected from the camera immediately after exposure by a motorized system. A microscope adaptor is available for this camera, incorporating an automatic exposure control device. The SX70 provides a simple and relatively inexpensive approach to photomicrography for laboratories which are not equipped for conventional processes, and where an instant result is required.

8.4.3 *Polaroid transparency films*

Polaroid manufacture two 35 mm films of interest in photomicrography: Polapan black-and-white ISO 125, and Polachrome colour ISO 40. These are packed in standard cassettes of 12 or 36 exposures, each supplied with a processing pack for use in the special Polaroid processor; after removal from the camera the complete film is processed, a procedure which takes less than 2 min.

Polachrome is unusual in that it uses the additive colour process, in which the image is built up from separate red, green and blue components in the form of fine lines, rather similar to a colour television picture. This process inevitably results in a considerable loss of light when viewed, and because of this the transparencies should not be judged 'in the hand'; when projected they are very good. A high-contrast version of Polachrome has just been introduced, which promises to be particularly suitable for rapid photomicrography.

In addition to their rapid processing films, Polaroid have recently introduced a colour transparency film for conventional processing, known as Professional Chrome. This is supplied as individual sheets, each in its own light-tight exposure packet, for use in the standard Polaroid 5 × 4 inch (13 × 10.5 cm) camera back with which many laboratories are already equipped. Since the films are already packaged for use, there is no need to load individual dark slides as with conventional cut-film. Moreover, since this film is completely interchangeable with standard 5 × 4 Polaroid film, test exposures may be made using the instant process without any alteration of the equipment. Transparencies in the 5 × 4 format are large enough to be appreciated without a magnifier when displayed on a light-box, and one can imagine that this film will find applications for photomicrographs used in teaching. Large-format transparencies such as these are also excellent as a stage in the production of high-quality colour prints.

8.5 **Storing unexposed film**

Film is damaged by dampness, warmth and light, and should be kept in its original container, unopened, as long as possible. Where film is to be kept for some months, it is a great advantage to store it in its original packing in a refrigerator or freezer, at 13°C or lower. After removal from the refrigerator, it should be allowed about 1 h to warm up to room temperature before the packet is opened, if damage due to condensation on the film is to be avoided.

Storage in cool conditions is advantageous for all films, but it is essential for those designated 'professional' if good colour balance is to be maintained; colour balance alters as film ages, and these films are sold at the point when optimum balance has been reached.

9. PHOTOMACROGRAPHY

The term photomacrography is generally used to refer to photography in which the image on the film ranges from life size (1:1) up to the lower regions of the conventional microscope's magnification ($\sim$25:1). The procedure is often considered to be difficult, and many laboratories are poorly equipped for this work; only a limited range of equipment is manufactured, and much of it lacks the sophisticated automation of modern high-power microscopes.

9.1 Imaging systems for photomacrography

Apart from the use of a conventional camera with 'close-up' facilities, which will not be considered here, there are essentially three methods of producing low-magnification images: the stereomicroscope, the purpose-built 'macroscope', and specially designed photomacrographic objectives used on a long-bellows camera. The first of these is the most readily available, the second the most expensive, and the third combines the highest image quality with least convenience in use. For occasional use in transmitted light, the photographic enlarger may also be found suitable.

9.1.1 Stereomicroscopes for photomacrography

While a stereomicroscope is in fact a compound microscope, its photographic results being thus more properly called photo*micro*graphs, it will be considered here since it operates within the 'macro' range of magnifications. The stereomicroscope has an advantage over the other systems since it may be the very instrument with which the specimen is being observed at the scientist's bench, and photographs can thus be taken conveniently at intervals during the study. In the past, microscopists have been discouraged from recording images from stereomicroscopes since the image quality, while acceptable for visual use, was inadequate for photography. Instruments of modern design are much improved, producing excellent images, and most manufacturers provide adaptors enabling conventional photomicrographic equipment, 35 mm SLR or larger cameras to be used without difficulty.

A stereomicroscope conventionally consists of two optical systems, mounted with a small angle between them, and directed at a common point on the object. Because of this angle and hence the different viewpoints, slightly different images are seen by the two eyes, leading to the perception of depth by the brain. Some stereomicroscope camera adaptors are reversible so that the left- and right-eye views may be recorded in sequence; twin adaptors may also be available permitting simultaneous recording of a stereo-pair. Examined in a suitable viewer, these pairs of images demonstrate the three-dimensional nature of the object most vividly; this is a technique which deserves greater popularity. As a consequence of the angled view through each of the optical pathways of a stereomicroscope, the images will show very slight distortion, and the edges may be slightly out of focus. In a system where both optical trains share a common objective lens, each will make use of only one side of the aperture of the lens, possibly of lesser optical quality than the centre. Some specialized designs include a third objective lens optimized for photography, or they enable part of the system to slide sideways so that the centre of the objective is used for photography; some camera attachments incorporate an iris diaphragm in each optical path for increasing depth of field. It is worth considering all these factors when purchasing a stereomicroscope which will be used for photography.

9.1.2 *Macroscopes*

These instruments generally form part of a system of stands, illuminating equipment and cameras specially designed for convenient photomacrography. They incorporate a zoom magnification-changer and a diaphragm for control of depth of field. The Wild Makroskop looks superficially like a stereomicroscope, with its binocular viewing system, but has a single imaging system beneath, as does the Zeiss Tessovar.

Macroscopes have a long working distance, facilitating illumination and manipulation of the specimen, but as a partial consequence of this their NA and hence their resolving power are limited. This is not a serious handicap for most applications, where their great convenience is overriding, but it may become evident in the very highest quality work, particularly in transmitted light.

9.1.3 *Photomacrographic objective lenses*

These lenses are specially computed for use without an eyepiece, to project an enlarged image onto film, with a relatively long camera extension. Examples are the Leitz Photar, Nikon Macro Nikkor, Olympus Macro Zuiko and Zeiss Luminar lenses, all of which will provide macro images of the highest quality. Some of these lenses are offered as part of their manufacturer's 35 mm camera system and thus fitted with a camera lens mount; others have a standard microscope objective lens thread, but are quite unsuited for use as conventional objectives.

When a 35 mm negative or transparency is required, it is most convenient to use the macro lens on the manufacturer's bellows or extension tubes; stands and illuminators are usually also available. For use with larger format film or Polaroid materials, the lenses may be fitted to the lens panel of a bellows camera such as the Leitz Aristophot or Nikon Multiphot, as described in Section 6.5.

Most photomacrographic objectives are fitted with an iris diaphragm, often calibrated in arbitrary units each representing a halving of the quantity of light passing through. As with camera and microscope lenses, smaller apertures provide increased depth of field but also poorer resolving power. The aperture should not normally be reduced more than one or two divisions from maximum; if smaller apertures are used, the principal advantage of these lenses, their fine resolving power, will be lost.

9.1.4 *Photographic enlargers and photomacrography*

A photographic enlarger may in some cases be used to produce macrographs of specimens in transmitted light, within certain constraints on the size of object and magnifications obtainable. The specimen should be fitted into the negative carrier and, if possible, masked around the area of interest to reduce stray light in the darkroom. The image can then be projected onto a piece of cut film on the baseboard, as if a print were being made, or into a camera from which the lens has been removed; in the latter case the camera's exposure-measuring system and shutter may possibly be used. Either negatives or transparencies can be made in this way. If the laboratory is equipped for printing from colour transparencies, for example by the Cibachrome process, first class results can be obtained by printing a transmission specimen such as a biological section directly onto paper with a photographic enlarger, as if the specimen were a transparency.

9.2 Illumination for photomacrography

Illumination of the object presents the greatest challenges in photomacrography, and only brief advice can be offered here. For solid specimens to be photographed by reflection, light may be provided by spotlights, by illuminators in which light is directed from an intense source via fibre-optic light-guides, or by flash. The effects of the lights on the positions of the shadows and modelling of the object should be carefully considered; it is generally desirable to use more than one source, one acting as a main light and the other, less intense, for filling-in the shadows. Many objects are best photographed with the almost shadow-free illumination provided by a light-tent. A simple tent can be improvized by making a collar of thin white paper or other suitable material, several times the diameter of the object, and encircling the object with it. Even a single light source, directed through the paper, will be diffused and reflected to provide shadow-free lighting. The light-tent is particularly useful for objects with shiny surfaces, and also for work with flash where it is not easy to assess the extent of the shadows. Useful light-tents for small objects may also be made from halved eggshells (white) or ping-pong balls, paper cups, and so on.

Achieving uniform illumination in photomacrography by transmitted light presents particular difficulties, and ideally requires special equipment such as the complex diascopic illuminators provided by Leitz and Nikon for their bellows camera equipment. These provide a uniform field of illumination similar to the Köhler method for microscopy, but require careful readjustment for every change in magnification. The transmitted-light illuminating bases supplied with stereomicroscopes and macroscopes provide convenient illumination over a more limited range of magnifications, and will be found generally useful. In an emergency, acceptable results can be obtained by supporting the specimen on clean glass, some distance from a piece of white card uniformly illuminated by means of tungsten lamps or flash.

9.3 Exposure determination in photomacrography

Determination of exposure often causes difficulty in photomacrography because of the wide variety of specimens and lighting and imaging systems encountered. A camera with through-the-lens metering is especially useful for 35 mm work. When using large-format and Polaroid films, exposure should be determined by means of the stepped-exposure (see Section 8.4.1), or by using a meter which measures from the film plane. Long exposure times are likely when using large image sizes, particularly at higher magnifications, and the effects of reciprocity failure and the possibility of disturbing vibrations should be borne in mind.

If flash is to be used for 35 mm photography, a camera with 'off-the-film' metering of flash, such as the Olympus OM2, OM4 and several others, is strongly recommended. In this system the flash is extinguished as soon as sufficient light has been detected falling on the film, yielding correctly exposed results even in the difficult circumstances of photomacrography.

10. THE FINISHED MICROGRAPH

10.1 **Keeping records**

Record keeping should begin with the preparation of a sheet on which details of all exposures on a film are recorded. Ideally such a sheet should provide each film with a unique serial number, perhaps incorporating the date or year, give the name of the operator, describe the type of film and development procedure, and carry at least the following information about each exposure, where appropriate: exposure number, specimen, objective used, eyepiece, camera factor or magnification-changer value, contrast method, details of any filters used, lamp voltage setting, film speed setting and any exposure adjustment, integrated or spot exposure measurement, and any special notes which may be of value later.

10.2 **Storing negatives**

Negatives should be kept free from damp, heat, dust and risks of abrasion and finger-prints in suitable envelopes; colour negatives should be kept in the dark to avoid risk of fading. Probably the most convenient storage system is the loose-leaf file, one example of which takes pages of approximately 26×31 cm, each accommodating seven six-exposure strips of 35 mm film. The pages are available in translucent or completely transparent material, the latter being preferred since it permits critical examination of the image without removing the film. The record sheets made at the time of exposure may be punched and filed along with the negatives, and together with a contact-print sheet. Also available are similar loose-leaf files for roll-film negatives, and negative bags for individual six-exposure strips of 35 mm film and for cut-film negatives in various sizes.

10.3 **Storing transparencies**

Transparencies must be stored in the dark, in similar conditions to negatives (Section 10.2). They may have already been mounted by the processor; unmounted strips should be stored in the same way as negatives until mounted for viewing. Mounted transparencies, usually in 50×50 mm mounts, can be stored in the boxes in which they may have been returned from processing, in purpose-built slotted boxes, in special translucent hangers in a filing cabinet, or in magazines suitable for direct insertion into a projector. Alternatively, the 'Slidex' system offers semi-stiff translucent file pages which can be used directly on a special projector, and provides a useful storage and projection system where funds permit.

No ideal system of organizing transparencies is known to the writer, and the system adopted will depend on how the slides are to be used. Transparencies may be filed chronologically, all those from each film being kept together, numbered and indexed so that they can be referred to the appropriate record sheet. Alternatively they may be filed according to subject, or assembled into groups required for a particular lecture.

Problems arise when a slide is required for use in more than one lecture, and in this case it will be helpful to have duplicates, either copied from the original or by multiple exposure at the microscope. All slides should carry a unique number in addition to other information, and they should always be returned to the files after use. This is a counsel of perfection and difficult to maintain, but its adoption will avoid much frustration and wasted time searching for a missing slide.

10.4 Mounting transparencies

Transparencies from 35 mm film are normally mounted in 50 × 50 mm mounts for projection or viewing. Many types of mount are now available, with or without protective coverglasses. Mounts without glass are cheap, quick to use, and avoid interference with the image from the four potentially dirty surfaces of a pair of coverglasses; they offer no protection from contamination and abrasion, but are adequate particularly for slides which are stored in projector magazines and handled infrequently. Apart from protection, coverglasses serve to hold the film flat in the projector's focal plane. Because of the difficulty of achieving focus across the whole of the image on a curved piece of film, Leitz supply a special curved-field (CF) lens for their projectors especially for projecting unglazed slides.

Mounts fitted with coverglasses are not without their own problems, however. Experience shows that the glasses are frequently dirty when supplied, requiring careful cleaning, and they develop a film of contamination on the inside after some years. Moreover, the film may stick to the glass, usually on the film-base side, particularly after use in an undercooled projector. The rainbow-like Newton's rings, interference colours due to close contact between the film base and coverglass, may also be troublesome; special glasses with a slightly unflat surface help to avoid this problem.

After mounting transparencies it is essential to label them, and mark them to ensure that they are correctly oriented when projected; there are seven incorrect ways, and only one correct way, of orienting a slide for projection! The conventional way of marking slides is to apply a coloured spot, either self-adhesive or with ink, facing the observer, in the lower left-hand corner of the mount, when the slide is correctly oriented for viewing when held in the hand. When the slide is inserted into the projector, the spot will be at the top right corner, under the thumb of the right hand.

10.5 Storing prints

All prints should be marked with the serial number of the film and the exposure number, as well as a brief description. This information should be recorded on the back of the print or its mount, using a soft pencil or ballpoint pen. Avoid using fibre- or felt-tip markers with water-soluble inks, since these are slow to dry on many kinds of modern photographic paper, and may become transferred to the face of another print beneath. To avoid any risk of this, and when packing prints to send by post, arrange them face-to-face and back-to-back. Each print should be marked with its magnification (see Section 10.6).

'Working prints' will frequently be kept unmounted, in envelopes or files, but those which are to be displayed on noticeboards, used in discussions or teaching, or submitted for publication, should be mounted on card, after careful trimming. Mounting protects prints against damage and improves their appearance, and may be done using conventional dry-mounting tissue in a heated press, with a spray adhesive, or with a pressure-sensitive adhesive sheet.

10.6 Determination and expression of magnification

Most observers do not find it helpful to be told the magnification of a micrograph in the form of a number. If this is all that is quoted, in order to gain any meaningful information about the dimensions of the objects depicted it is necessary to measure them on the micrograph and then divide this dimension by the magnification factor, taking care to avoid errors of several orders of magnitude in the process. In the case of a projected transparency, the frequently heard statement that 'this is a magnification of 100' is clearly nonsense, since the projector has further enlarged the image perhaps 50-fold.

A scale bar drawn on a printed micrograph is very much to be preferred, since it immediately demonstrates what a dimension of, say, 100 μm in the specimen looks like. Furthermore, a bar drawn on a micrograph will automatically alter in length if the image is enlarged or reduced in preparation for publication. Similarly, for a transparency it is helpful to quote the width of the field of view.

The information for drawing the scale bar or for determining the field width is easily obtained from one extra exposure on the film—a photograph of a stage micrometer, taken at the same magnification as the micrograph, and subsequently printed at the same degree of enlargement as the micrograph. The print of the micrometer scale can be used as a 'ruler' for measuring features in the image, and as the basis for drawing an accurate scale bar.

Where it is necessary to quote the magnification of a micrograph, this is nowadays more frequently being expressed as a ratio, such as '100:1', similar to the scale of a map, a form which is recommended for all enlarged or reduced real images. The magnifying power of an objective is thus expressed in this way since the lens produces a real primary image. This style is recommended rather than '×100', which should be used to refer to the angular magnification of a virtual image; eyepieces and magnifiers are thus marked, for example, ×10.

11. PRACTICAL SCHEDULES

11.1 Taking a micrograph using a 35 mm camera system

(i)   Select a film of suitable speed: black-and-white, daylight or tungsten-balanced colour transparency, or colour negative.

(ii)   Check the camera for dust, hairs, fragments of film etc., and load the film.

(iii)   Advance the film to frame number one, checking where possible that the camera rewind knob is rotating, indicating that the film is being transported.

(iv)   Set the film speed (previously verified by experiment) on the control of the exposure measuring device.

(v)   Prepare a suitable record sheet for notes about each exposure.

(vi)   Clean the slide thoroughly.

(vii)   Switch the microscope on, select the objective to be used, move the specimen to fit the frame and set up correctly for Köhler illumination.

(viii)   Where necessary, ensure that the beam-splitting prism is set so that light may reach the camera, and that any blanking slide is removed from the system.

(ix)   Defocus the image of the specimen and adjust the eyepieces or camera focusing system until the cross-lines (or other markings) are seen comfortably in sharp focus. Refocus the image.

(x)   If using colour film, adjust the lamp voltage to the predetermined best setting and insert any necessary colour-correction filters (see Sections 8.3.3 and 8.3.4).

(xi)   Consider the distribution of bright and dark areas within the photographic field. If not 'average', use spot-metering or apply a correction so that more exposure than indicated is given when there is a clear bright background, and less for a dark background or when using fluorescence microscopy.

(xii)   Read the exposure meter and check that the indicated speed falls within the range of the equipment (see camera manual).

(xiii) If indicated exposure time is too short, insert neutral-density filters until correct.

(xiv)   Focus the image precisely. Check adjustments of condenser focus, illuminated field diaphragm and illuminating aperture diaphragm.

(xv)   Move the specimen slightly using the mechanical stage, while viewing through the focusing tube or eyepieces. Look for any feature which does not move with the specimen and remove it by appropriate cleaning; where this proves impossible, try to put it out of focus by very slight defocusing of the condenser.

(xvi)   Refocus precisely once again; check that the stage and focusing movements are not drifting out of adjustment.

(xvii) Press the shutter-release, avoiding touching the microscope or bench during the exposure.

(xviii) Wind on the film (where this is not automatic).

(xix)   Note optical and specimen details on record sheet.

(xx)   For particularly important micrographs repeat after refocusing and after adjusting for slightly more and slightly less exposure.

(xxi)   At end of film, rewind into cassette.

(xxii) Process film as soon as possible.

11.2 Initial calibration of exposure metering system

In this section, Kodak films and developers are named in order to give specific examples; similar products from other manufacturers will be equally suitable.

(i)   Load the camera with Kodak Technical Pan Film 2415.

(ii)   Choose a well-stained section; set up the microscope optimally for bright-field illumination, using a 10, 25 or 40 objective, and select an area which fills the field of view.

(iii)   Remove any colour filters or neutral filters from the light path.

(iv)   Turn the lamp brightness to a level suitable for comfortable viewing, and note the reading of the meter (voltage or current) on the power supply. Keep brightness at this level throughout the calibration sequence, if possible.

(v)   Set the film-speed adjustment of the equipment to ISO 12 or, if it is not marked in ISO speed ratings, to a value at the slow-speed end of the range.

(vi)   If the equipment indicates exposure time, check that this is not too short or too long for the capability of the equipment. If necessary, readjust lamp brightness appropriately.

(vii)   Make a sequence of exposures, at ISO 12, 25, 50, 100, 200 and 400, carefully noting all relevant microscopical and photographic data on the record sheet. If the equipment is not marked in ISO ratings, use successive major scale divisions such that the full range is covered in about six exposures.

(viii) Expose one blank frame (e.g. without the specimen).

(ix)   Repeat steps (v) to (viii) for each colour filter which may be used for contrast enhancement.

(x)   Remove the film and develop in HC-110 developer, Dilution D, for 6 min at 20°C. See instructions packed with the developer for an explanation of the dilution.

(xi)   When dry, examine negatives and rank them for printability; if in doubt, make prints (or have them made) and assess these. Choose the optimum setting for use with and without the various filters.

(xii)   Transfer frame numbers on margin of film to record sheet (your first exposure might not be on frame number one) and write comments on exposures on record sheet.

The above procedure may then be repeated using Technical Pan developed for lower contrast results using Technidol Liquid Developer for 9 min at 20°C, and for higher contrast using D19 for 4 min at 20°C. Where necessary it should also be repeated with other imaging modes and kinds of specimen likely to be encountered in the laboratory's work.

11.3 Initial calibration of a system without exposure measurement

Follow steps (i) to (iv) in Section 11.1, making particularly careful notes on all settings of the equipment, so that the conditions can be exactly reproduced on the next occasion.

(v)   Set the camera shutter to 1 sec and make an exposure.

(vi)   Repeat for successive exposures, halving in duration down to one-thousandth of a second (or the shortest available), keeping careful notes on the record sheet.

(vii)   Expose one blank frame (e.g. without the specimen).

(viii) Repeat steps (v) and (vii) for each objective and colour filter likely to be used.

(ix)   Follow Section 11.2 from step (x) to the end.

Photomicrography without means of measuring exposure is tedious and uncertain; it is likely to be satisfactory only when the requirements are for repeated micrographs in closely standardized conditions.

11.4 Achieving correct colour rendering in transparencies

(i)   Select a film and load the camera. As a starting point choose one balanced for tungsten light, and of slow to medium speed, such as Ektachrome 50 Professional (Tungsten) (ISO 50) or Ektachrome 160 (ISO 160), both balanced for 3200 K, or Fujichrome 64 T (ISO 64), which is balanced for 3100 K.

(ii)   Set up the microscope for correct Köhler illumination using a typical well-stained specimen with mounting medium which has not yellowed with age.

(iii)   Set the lamp rheostat carefully so that the lamp is running exactly at its rated voltage.

(iv)   If the microscope has a tungsten-halogen lamp, make a series of exposures as in Section 11.2(vii) with no colour-conversion filters in the light path. If a Wratten 2A, 2B or 2E UV-absorbing filter is available, make another series using this. For colour work, where processing is standardized, five different exposures should be adequate: one at the manufacturer's ISO film speed rating, and one each at twice, four-times, one-half and one-quarter this rating; intermediate values may be used for greater precision if desired. If a conventional tungsten-filament bulb is used, see (vii) below.

(v)   Results can be evaluated at this stage, (viii) below, or a more detailed check on colour rendering carried out (vi).

(vi)   Repeat (iv) using a gradation of bluish (82, 82A, 82B) and yellowish (81, 81A, 81B) filters. If a satisfactory film-speed setting has already been determined, one exposure on each side of this value will be sufficient.

(vii)   When using a conventional tungsten bulb, a Wratten 82B filter should be used in (iv) above, together with one of the UV-absorbing filters, and in (vi) the range from no filter, 82, 82A, 82C, 82C + 82 and 82C + 82A should be tried.

(viii)   After processing evaluate the results, preferably viewing them on a specially colour-corrected light-box when they can be seen simultaneously and compared, or by projection.

(ix)   Choose a suitable ISO rating for future use from the results from (iv); an intermediate value may be chosen where two adjacent exposures are almost correct. If results using the UV-absorbing filter are noticeably less blue than those without, indicating that significant UV may reach the film, this filter should be used routinely.

(x)   The results from (vi) or (vii) should permit a suitable light-balancing filter to be chosen.

(xi)   If even the best result still shows a slight colour cast, examine the transparencies through a selection of colour-compensating filters until the colour appears correct. Make another series of trial exposures, adding this CC filter to the filter pack, and evaluate again.

12. FURTHER READING

Kodak Technical Publications (1977) *Closeup Photography and Photomacrography*. Eastman Kodak Co., Rochester, New York.

Kodak Technical Publications (1988) *Photography Through the Microscope*. Eastman Kodak Co., Rochester, New York.

Loveland,R.P. (1970) *Photomicrography: A Comprehensive Treatise*. Wiley, New York, 2 Volumes.

Morton,R.A. (1984) *Photography for the Scientist*. Academic Press Inc., London, 2nd edition.

Thomson,D.J. and Bradbury,S. (1987) *An Introduction to Photomicrography*. Royal Microscopical Society Microscopy Handbook No. 13, Oxford University Press, Oxford.

White,W. (1987) *Photomacrography: An Introduction*. Focal Press, Boston.

CHAPTER 4

Immunohistochemistry

MICHAEL G.ORMEROD and SUSANNE F.IMRIE

1. INTRODUCTION

Immunohistochemistry utilizes antibodies to localize specific products in tissue sections. Briefly, a tissue section is incubated with a labelled antibody, the section is then washed and the site of reaction of the antibody is identified by visualizing the label.

A variety of histochemical techniques have been used for many years to identify certain constituents of cells. These methods lack the specificity obtained from an antibody and it is this property which enables a worker to map precisely the distribution of a particular product in a tissue. Aside from its power as a tool in research, immunohistochemistry has found an increasing role in diagnostic histopathology with particular application to the diagnosis of tumours. The range of applications of the method has been advanced immeasurably by the development of technology for the production of monoclonal antibodies. This has produced a wide range of antibodies, many of which have specificities which would have been difficult, if not impossible, to obtain from conventional antisera.

The initial sections (2−5) of the chapter give the background to the techniques. The later sections describe the procedures in detail. Section 2 briefly defines an antibody and its structure, gives a method for its purification and describes the difference between polyclonal sera and monoclonal antibodies. The target for an antibody is called an antigen; the possible effects of tissue processing on the antigen are considered in Section 3. A wide variety of different labels may be employed; these are described and compared in Section 4 which also includes methods for conjugating enzymes to antibodies. Section 5 describes and compares in general terms the different methods used for applying antibodies to tissue sections.

Section 6 contains the details of the experimental methods used with enzyme-labelled antibodies. The use of controls and problem solving are discussed in Section 7 and ways of visualizing two antigens simultaneously in Section 8.

Although the chapter concentrates on the use of antibodies, a similar technique can be applied using any reagent which shows a similar specificity. For example, lectins can be used to identify certain groups of carbohydrates. A recent exciting development is in the use of DNA probes for specific sequences in genes or in mRNA. This is described in Section 9.

Most of the chapter describes techniques for staining sections cut from tissue. The same methods can also be applied to cytological preparations and a short section (Section 10) on staining smears is included.

Section 11 contains a brief discussion on quantitation and Section 12 describes a staining tray—the only special piece of equipment needed.

For further reading, several books on immunohistochemistry have been published. Most of these give details of the application of the technique to particular problems (1–5).

2. ANTIBODIES

Antibodies, collectively called immunoglobulins (Igs), comprise approximately 20% of the proteins in human plasma. They are produced by plasma cells and can exist in millions of different forms. During an immune response, the foreign body (antigen) stimulates division of those plasma cells responsible for producing an Ig reactive with that particular antigen. This enables the animal to produce large numbers of specific antibodies. Any animal in a normal environment continually undergoes immune responses and its plasma will contain antibodies directed against thousands of different antigens.

2.1 Immunoglobulin structure

The basic structure of an Ig is a dimer, each half containing two polypeptide chains, one called heavy (containing ~440 amino acids), the other light (containing ~220 amino acids). The chains are linked by disulphide bonds (*Figure 1*). The heavy chain has an invariant (constant) region, which determines the class of the Ig, and a variable region. The latter, in conjunction with the variable region on the light chain, forms the site which binds the antigen. Light chains can be of two classes, kappa or lambda.

The major classes of Igs are listed in *Table 1*. The IgG class is further subdivided (IgG1, IgG2, etc.). During an immune response, the bulk of the Igs in the plasma are IgGs. Consequently an antiserum will contain predominantly Igs of this class.

Immunoglobulins are sometimes digested enzymatically to produce fragments which are still reactive with the antigen. Pepsin digests the constant part of the heavy chain below the 'hinge' region leaving a dimeric fragment called F(ab')2 (see *Figure 1*). Papain creates two monomeric fragments, Fab, plus the constant region of the heavy chains, Fc. Fab fragments are sometimes used in place of the whole Ig.

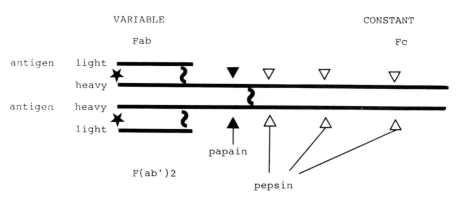

Figure 1. A simplified representation of an Ig molecule showing the antigen binding site and the sites of cleavage of papain and pepsin. ~ represents a disulphide bond.

Table 1. The major classes of immunoglobulins.

Name	% Total Ig in blood	Comment
IgG	80	Major antibody in immune sera
IgA	15	Found in sero-mucous secretions
IgM	5	Produced early in immune response, pentameric
IgD	<1	Present on lymphocyte surfaces
IgE	<1	Responsible for allergic reactions

2.2 Polyclonal antisera

To raise an antiserum, a group of animals are injected with the purified antigen together with a non-specific stimulant of the immune response (an adjuvant). A protocol can be found in ref. 6. If the immunization is successful, the antigen will have stimulated a variety of lymphocytes, each of which will have undergone several divisions to produce a clone of antibody producing plasma cells. For this reason, the term 'polyclonal' antiserum is sometimes used.

The site on an antigen with which an Ig reacts is called an epitope. An antigen may contain several epitopes and an antiserum will contain Igs directed against each of them. Furthermore different Igs directed against the same epitope may have different binding affinities and also may see a slightly different part of the epitope. An antiserum will therefore contain a variety of Igs reactive with the antigen but with different specificities and affinities. Consequently, two antisera are never identical.

Apart from a high concentration of antibodies directed against the appropriate antigen, an antiserum will also contain other Igs which were present before immunization and may contain antibodies reactive with impurities in the original preparation of antigen. These Igs may give undesired reactions requiring their removal.

2.3 Monoclonal antibodies

Many of these problems are avoided by the use of monoclonal antibodies. These are made by a cloned line of cells which produce a single Ig. The cells are derived by fusing spleen cells from an immunized animal with a line of drug-sensitive myeloma cells. After the fusion, the cells are incubated in the presence of the drug. The only cells to grow will be hybrids resulting from a fusion of a normal spleen cell (which is drug-resistant) and a myeloma cell (which brings immortality to the hybrid). The hybrids are cloned and clones producing specific antibodies are selected and re-cloned. The resulting cell lines (called hybridomas) each produce indefinitely a single Ig. A full description of these methods can be found in ref. 7 or in another volume in this series, *Antibodies: A Practical Approach, Volume I* (D.Catty, ed.).

2.4 Purification of antibodies

Sometimes it is necessary to purify an antibody before attaching a label to it. This is comparatively easy if a monoclonal antibody contained in the supernatant from a hybridoma culture is used. An ammonium sulphate precipitation followed by exclusion chromatography would suffice. If a monoclonal antibody from an ascitic fluid or a polyclonal antiserum is used, it is desirable to separate the specific antibody from the

Table 2. Affinity purification of anti-rabbit immunoglobulin.

1.	Dissolve 2.5 g of ammonium sulphate in 10 ml of rabbit serum. Stand at 4°C overnight.
2.	Centrifuge and discard the supernatant.
3.	Dissolve the precipitate in water and dialyse against 0.02 M phosphate buffer, pH 7.0.
4.	Place the solution on a column of DEAE−Sephadex A50 (Anion-exchanger, Pharmacia) equilibrated in 0.02 M phosphate buffer, pH 7.0. Elute IgG in the same buffer.
5.	Couple rabbit IgG to cyanogen bromide-activated Sepharose (Pharmacia) according to the manufacturer's instructions. Resuspend the slurry of Sepharose in PBS.
6.	Pour IgG−Sepharose into a small (7 mm diam × 150 mm) column. Pass anti-rabbit IgG serum through the column.
7.	Wash the column with PBS until there is no detectable absorption at 280 nm.
8.	Elute the specific anti-rabbit Ig from the column with 4 M potassium thiocyanate. Collect 1 ml fractions and monitor optical density at 280 nm.
9.	Bulk the fractions containing protein and dialyse against PBS. (Do not leave the sample in potassium thiocyanate as this can denature the Ig.)

other Igs. The best method is to immobilize the antigen on a solid support and to use affinity chromatography. A suitable procedure is given in *Table 2*.

2.5 Specificity of antibody reactions

An important property of an Ig is the affinity with which it binds to its antigen. The concentration needed to produce a desired end point clearly depends on this affinity. The 'strength' of a polyclonal antiserum is directly related to the concentration of antibody multiplied by its affinity constant; the higher the affinity, the greater the working dilution. For the reasons given below, antibodies of high affinity generally show greater specificity.

Antibodies are used as reagents because of their high specificity. However it is important to realize that they can also give rise to non-specific reactions. These can have three causes.

(i)   From impurity antibodies. This only arises with a polyclonal antiserum and should be detected by the use of appropriate controls (see Section 7). It is less likely to be a problem if the antiserum can be used at a high dilution.

(ii) From cross reactions. These arise if two molecules have similar, but different, structure. Some antibodies may recognize both molecules but often the undesired reaction will be of lower affinity and will only be a problem if the antibody is used at too high a concentration. This emphasizes the desirability of using antibodies of high affinity.

(iii) From two molecules sharing the same epitope. This possibility will not be revealed by the usual controls and would normally only be shown by a detailed immunochemical study.

2.6 Storage of antibodies

For antibodies and antisera obtained commercially, the supplier's instructions should be followed. Other reagents should be stored without first diluting them. It is advisable to aliquot them into suitable quantities and to store the aliquots frozen. Once an aliquot is thawed it should not be re-frozen as thawing and freezing will denature Igs.

If an antibody is used rarely, it may be stored in amounts sufficient for one experimental run and an aliquot thawed when necessary. If this involves microlitre quantities of reagent, the aliquot should be covered with a small quantity of glycerol before freezing to prevent it freeze-drying.

If a reagent is in regular use, it may stored in larger aliquots. We have found that most antisera have a shelf life at 4°C of at least a month, some considerably longer. For such reagents we regularly make them 0.01% in sodium azide to prevent bacterial contamination. If reagents are ever contaminated, they should be immediately discarded.

3. EFFECT OF TISSUE PROCESSING ON ANTIGENS

Before applying an antibody, a section of tissue must be prepared. Before deciding how to handle a tissue the effect of any processing on the antigen of interest must be considered.

Conventionally, when sections are to be stained with haematoxylin and eosin, tissue is fixed, often in formalin, and embedded in paraffin wax. After sections have been cut, the wax must be removed by immersion in xylene, or a similar solvent, and brought to water through ethanol. This treatment yields sections of high quality. It also alters the proteins in a tissue so that many antigens no longer react with the appropriate antibody. In this case, alternative methods of fixation and processing may be tried but frequently sections must be cut from frozen tissue.

3.1 Choosing conditions for processing

When using a new antibody, it is important that the optimum conditions for fixation and processing are determined. While these can only be established empirically, it is possible to lay down some guidelines.

A set of fixatives is chosen. In this laboratory we would select formalin, methacarn, 90% ethanol, Bouin's, chloroform—acetone and formol—calcium followed by chloroform—acetone. These fixatives are then tested on a set of frozen sections. Several frozen sections are cut from the tissue which contains the antigen of interest and one of the sections immersed in each of the selected fixatives for 5 min. The sections are stained with the antibody using the selected method and the result read. If all the fixatives chosen appear to destroy the antigen, the experiment is repeated using more gentle fixatives (e.g. paraformaldehyde—lysine—periodate). Occasionally, the staining must be carried out without prior fixation; this will give sections with poor morphology.

Recipes for various fixatives, including those mentioned above, with some indications of their use are given in *Table 3*.

If the antigen survives one or more of the fixatives, tissue is fixed in the optimum fixative and embedded in paraffin wax. Sections are cut and stained. If the processing has destroyed the antigen, this is repeated using a paraffin wax with a low melting point (45° as opposed to 58°C). At this stage it should be possible to select the optimum conditions for a particular antibody.

Table 3. Recipes for some common fixatives.

A. *Fixatives generally used on tissue subsequently processed into blocks of paraffin wax.*

1.	10% formol−saline.	100 ml of 40% formaldehyde, 9 g of NaCl in 900 ml water.
2.	Methacarn.	60% methanol, 30% chloroform, 10% glacial acetic acid. Tissues are usually fixed at room temperature overnight and then transferred to 70% alcohol.
3.	Modified methacarn.	Use inhibisol in place of chloroform.
4.	Carnoy's fluid.	60 ml of ethanol, 30 ml of chloroform, 10 ml of acetic acid.
5.	Bouin's fluid.	75 ml of saturated picric acid, 25 ml of 40% formaldehyde, 5 ml of acetic acid.
6.	B5.	60 g of mercuric chloride, 21 g of sodium acetate trihydrate in 900 ml of water. Add 100 ml of 40% formaldehyde before use.

When using this fixative, after sections have been cut, mercury must be removed before performing an immunohistochemical stain.

(i)   Take sections to water.
(ii) Immerse for 5 min in Lugol's iodine (2 g of KI, 1 g of iodine in 100 ml of water).
(iii) Wash in water.
(iv) Immerse for a few seconds in 5% (w/v) sodium thiosulphate.
(v)   Wash in water.

B. *The following are usually used on frozen sections.*

7.	Formol−calcium.	100 ml of 40% formaldehyde, 100 ml of 1 M $CaCl_2$, 800 ml of water plus a few chips of marble (or $CaCO_3$). Store at 4°C. Fixation is followed by a further fixation in chloroform−acetone.
8.	Chloroform−acetone.	50:50 v/v. 5 min fixation at 4°C. Often used to fix frozen sections prior to using monoclonal antibodies to distinguish lymphocyte phenotypes.
9.	Ethanol−acetic acid.	95% ethanol, 5% glacial acetic acid. 1 min at 4°C for frozen sections.
10.	Ethanol.	Various percentages of ethanol at 4°C or room temperature for a predetermined time.
11.	Periodate−lysine−paraformaldehyde.	(i) Prepare a 3.6% w/v solution of paraformaldehyde by dissolving 2 g in 0.14 M sodium dihydrogen phosphate, 0.11 M NaOH at 70°C.

(ii)   Filter, cool and add 2.5 ml of 1 M HCl.
(iii) Prepare a solution of lysine by adjusting the pH of a 0.2 M solution of lysine−HCl to 7.4 by addition of 0.1 M dibasic sodium phosphate.
(iv)   Dilute to 0.1 M lysine by addition of 0.1 M phosphate buffer, pH 7.4.
(v)   Just before use mix 1 part of paraformaldehyde solution to 3 parts lysine solution and add solid sodium periodate to a concentration of 10 mM.

This is a gentle fixative which is suitable for labile antigens such as the H2 antigen in mouse tissues.

The antigens found on the surfaces of lymphoid cells, such as the histocompatibility antigens and the subset specific markers, are often unstable. Some care must be taken to preserve their integrity. To demonstrate the necessary steps, the procedure we use

Table 4. Preparation of sections for staining for lymphocytic markers.

A. *Freezing the tissue*

1. Cover small pieces of tissue (maximum size 3 × 3 × 15 mm) with OCT embedding compound (obtainable from BDH) on a slice of cork.
2. Freeze in isopentane pre-cooled in liquid nitrogen.
3. Store the tissue and cork in plastic ampoules in liquid nitrogen.

B. *Cutting sections*

1. Cut sections, 8 μm thick, on a microtome mounted in a cryostat and mount as usual on glass slides.
2. Dry sections at 37°C for 1 h. They may now be used immediately or stored.
3. To store sections, wrap them in plastic film (the type sold for 'cling' wrapping food) and store at −20°C. Before use, bring to room temperature and remove film.

C. *Fixation*

1. 5 min in formol−calcium (see *Table 3*).
2. Dip in cold acetone.
3. 5 min in chloroform−acetone (50:50 v/v) at −20°C.
4. Dip in cold acetone.
5. Wash twice in PBS.

---

to prepare sections for staining labile markers on the surface of lymphocytes in human tissue is given in *Table 4*.

It should be noted that the various epitopes on an antigen might be affected differently by a fixative. If several monoclonal antibodies to the same antigen are available, each should be optimized separately.

Sections of calcified tissue, such as a tumour metastasized to bone, are normally decalcified with 5% formic acid or 0.5 M EDTA before histological staining. If the antigen has survived fixation in formol−saline, it will usually also survive the process of decalcification.

3.2 Revealing hidden antigens

Sometimes treatment of a section of fixed tissue with a proteolytic enzyme will 'reveal' an apparently destroyed antigen. The mechanism by which partial digestion of protein on a section permits a previously inhibited reaction of one of the proteins with an antibody is not properly understood. Possibly, after fixation, the epitope in question is unaffected but surrounded by a matrix of crosslinked proteins whose removal allows the antibody access.

The use of a proteolytic enzyme adds to the procedure an extra variable which is difficult to control precisely. As a general practice we would not recommend it. However, in diagnostic pathology, sometimes the only available tissue has already been fixed in formalin and embedded in paraffin wax and these may not be the optimal conditions for a particular antigen. In these circumstances there is little choice but to try this approach. A suitable recipe is given in *Table 5*.

The effect is demonstrated in *Figure 2* which shows a stain for glial fibrillary acidic protein on a section of formalin-fixed human cerebellum embedded in paraffin wax using an indirect method. In (b) the section was pre-treated with pronase according to the protocol in *Table 5*; (a) was untreated. The primary antibody was a rabbit polyclonal antiserum raised by Mr N.Bradley (Institute of Cancer Research) and used at a dilution of 1/100. The secondary antibody was a peroxidase conjugated swine anti-rabbit (Dakopatts) used at a dilution of 1/100. Colour was developed using diaminobenzidine (DAB), the section counterstained with Mayer's haemalum (see Section 6.9.2) and the coverslip mounted in DPX (see Section 6.9.3). The antibody picks out the astrocytes, the fine fibrils are the cellular processes delineated by the stain. The stronger staining on the section pre-treated with pronase can be seen.

4. CHOICE OF LABEL

The labels used to visualize an antibody fall into three main classes: fluorescent, enzymatic and gold with silver enhancement. It is also possible to use a radioactive label followed by autoradiography. This is not a normal immunohistochemical procedure and the technique will not be discussed here.

Table 5. A method for treatment of sections with pronase.

This is used on sections from fixed tissue prior to applying the first antibody.
1. De-wax the section and take it through alcohol to water.
2. Incubate the section in PBS at 37°C for 5 min.
3. Incubate the section in PBS, 50 μg/ml pronase at 37°C for 20 min.
4. Wash in running tap water for 5 min.
5. Wash twice in PBS.
6. Apply antisera according to the desired protocol.

After this treatment, the sections are very fragile and must be handled with care.

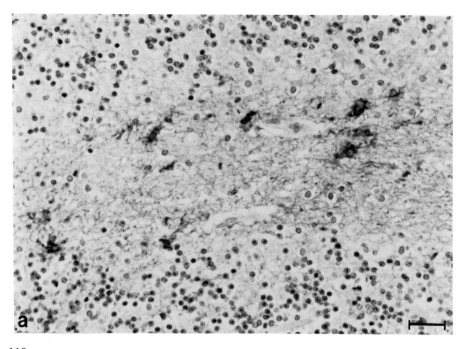

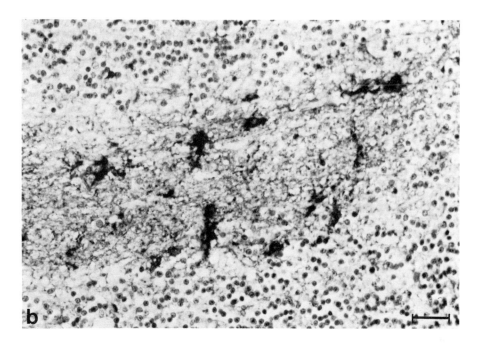

Figure 2. Human cerebellum stained for glial fibrillary acidic protein showing the effect of pre-treatment with a protease. The bar represents 30 μm.

4.1 **Fluorescent labels**

An ideal fluorescent label has a high quantum yield, good separation between the wavelengths of excitation and emission, a wavelength of maximal absorption close to a strong line from a mercury arc lamp (used for fluorescence microscopy) and an emission wavelength suitable for photographic film and the human eye. In practice, the two substances which are in common use are fluorescein and rhodamine which can both be attached to protein by reaction of lysine residues with the isothiocyanate (see *Table 6*). All manufacturers of fluorescent microscopes sell suitable filters for these compounds. Recently, phycoerythrin, a fluorescent protein found in red algae, has been used. At present, few standard laboratory microscopes are equipped with filters optimized for this reagent. (See Chapter 6 for further details on filters.)

The advantage of using a fluorescent label is its speed. Once the slide has been incubated with labelled antibody, the coverslip can be mounted and the result read. The disadvantages are the need for specialist equipment and that one has literally to work in the dark. The architecture of the tissue and the cellular morphology are not revealed. Although the earliest immunohistochemistry employed antibodies labelled with fluorescein, this method is being superseded by methods employing enzymes except when two different antigens are to be revealed on the same section (see Section 8).

4.2 **Enzymatic labels**

When an enzyme is used as a label, it is visualized by means of a reaction which gives an insoluble coloured product. An ideal enzyme has a low molecular weight (for ease

Table 6. To label an immunoglobulin with fluorescein.

1.	Dissolve 10 mg of IgG in 2 ml of 0.5 M carbonate−bicarbonate buffer, pH 9.5.
2.	Add 10 mg of fluorescein isothiocyanate adsorbed to Celite (Sigma) and stir gently at room temperature for 4 h.
3.	Remove Celite by centrifugation.
4.	Pass the supernatant over a small (7 mm diam × 150 mm) column of Sephadex G25 (Pharmacia) equilibrated in 0.05 M phosphate buffer, pH 7.8. Collect fractions in the void volume.
5.	Dialyse against three changes of the phosphate buffer (2 litres) over a period of at least 48 h.
6.	Read optical density at 280 nm and 495 nm.
7.	Calculate the ratio of fluorescein to protein.
8.	Store at 4°C.

of attachment to Ig) and a high turnover number (to give a high yield of product), is absent from normal tissue and can be used to give a product which is insoluble in water, ethanol and xylene (so that the coverslip can be mounted conventionally).

4.2.1 *Horseradish peroxidase*

Horseradish peroxidase (HRP) was the first such enzyme to be used and remains a popular choice. It fulfills the above criteria except that it is found in normal tissue, particularly granulocytes, erythrocytes and cells of the myeloid series. Usually it is necessary to block the activity of enzyme endogenous to the tissue. The substrate is hydrogen peroxide and the product oxidizes a chromogen. It is commonly used with DAB which gives a brown precipitate at the site of reaction. An advantage of this stain is that, for all practical purposes, it is permanent. Care must be taken not to confuse the reaction product with endogenous brown pigment.

4.2.2 *Alkaline phosphatase*

The substrate for this enzyme is usually a naphthol phosphate with a diazonium salt. The phosphatase releases the naphthol which couples with the diazonium salt to form a precipitate. Used in conjunction with Fast Red it gives a red precipitate which dissolves in ethanol and xylene so that the coverslip must be mounted in an aqueous medium. Alkaline phosphatase is found in several tissues including bone marrow, breast, endothelium, kidney, placenta and intestine. That found in the intestine is a different isoenzyme and is the more robust, surviving many procedures for processing which destroy the enzyme at other sites. The red colour catches the eye and we have found this label particularly useful when trying to identify rare cells (e.g. micrometastases in bone marrow smears) (8). It has the disadvantage that the reaction product fades over a period of months.

4.2.3 *Glucose oxidase*

Glucose oxidase is found in bacteria and is absent from mammalian tissue. The substrate is oxidized in the presence of a tetrazolium salt and the hydrogen acceptor, phenazine methosulphate. Upon reduction, the tetrazolium forms a coloured precipitate which, if the nitroblue derivative is used, is blue. If used when labelling two antigens on the same slide, it makes a pleasing contrast with the product from alkaline phosphatase. It is necessary to use methyl green if a counterstain of the nuclei is required.

Table 7. To conjugate alkaline phosphatase to immunoglobulin by the glutaraldehyde method.

1.	Dissolve 5 mg of alkaline phosphatase in 1 ml of 0.1 M phosphate buffer, pH 6.8.
2.	Dialyse against two changes of 2 litre, 0.1 M phosphate buffer, pH 6.8.
3.	Add 5 mg of Ig in 1 ml of PBS, pH 6.8.
4.	Add 0.15 ml of 0.1% glutaraldehyde.
5.	Stand at room temperature for 3 h.
6.	Dialyse against two changes of 5 litre PBS.
7.	Store at 4°C.

4.2.4 *Galactosidase*

The enzyme β-galactosidase is extracted from *Escherichia coli* and is readily conjugated to other proteins. The optimal pH for the bacterial enzyme (7.0−7.5) differs from that of human β-galactosidase (5.0−5.6) so that, if the correct buffer is used, there is no need to block the endogenous enzyme. Furthermore, the latter will be inactivated by heating above 55°C so that tissue embedded in paraffin wax will contain no active enzyme.

4.2.5 *Conjugating enzymes to antibodies*

Several manufacturers now produce a range of conjugated second antibodies of high quality. Producing conjugates in the laboratory is not recommended if commercial reagents are available.

Before conjugating an enzyme to an antibody, it is preferable if the antibody is first purified (see Section 2.4). This avoids the presence of extraneous labelled proteins which might result in non-specific staining.

There are three methods of conjugation in general use employing either glutaraldehyde, periodate or a bifunctional reagent. In the first, a mixture of Ig and enzyme can be lightly crosslinked using glutaraldehyde. We have used this method satisfactorily with alkaline phosphatase. A suitable protocol is given in *Table 7*.

Horseradish peroxidase has a high content of carbohydrate; aldehyde groups can be formed on sugars with vicinal hydroxyl groups by treatment with periodate. These will react with lysine residues on the Ig to give stable bonds after reduction. Some years ago, we used this method regularly but later found that an alternative method was more reliable and gave more stable conjugates. This uses a bifunctional disulphide *N*-succinimidyl 3-(2-pyridyldithio) propionate (SDPD), to link proteins together with disulphide bonds via their lysine residues (protocol in *Table 8*).

4.3 **Colloidal gold**

This technique employs colloidal gold particles onto which an antibody has been absorbed. It was originally developed for electron microscopy where it has the advantage that gold is electron dense and that different antibodies may be labelled with gold particles of different sizes, enabling two or three antigens to be localized simultaneously. Used in light microscopy, the gold particles have to be visualized by a silver precipitation. The method has proved to be more sensitive than those employing enzymatic labels.

Since immunogold is usually used in an indirect method and several manufacturers produce suitable reagents, these are usually best purchased. Streptavidin absorbed onto

Table 8. To conjugate peroxidase and immunoglobulin by the SDPD method.

1.	Dissolve 10 mg of HRP in 1 ml of 0.1 M phosphate buffer, 0.1 M NaCl, pH 7.5.
2.	Pass it through a small column (15 mm diam × 150 mm) of Sephadex G25 (Pharmacia) equilibrated in the same buffer. Collect fractions in the void volume.
3.	Dissolve 1.2 mg of SDPD in 1 ml of ethanol.
4.	While stirring gently, add 250 μl of SDPD solution to the solution of peroxidase. Stand at room temperature for 40 min.
5.	Pass it through a column of Sephadex G25 as in step 2.
6.	Pass 5 mg of affinity purified antibody through a column of Sephadex G25 as in step 2.
7.	While stirring the solution of antibody, add 10 μl of SDPD solution. Stand at room temperature for 1 h.
8.	Pass through a column of Sephadex G25 equilibrated in 0.1 M acetate buffer, 0.1 M NaCl, 25 mM dithiothreitol (DTT), pH 4.5. Stand at room temperature for 1 h.
9.	Pass through a column of Sephadex G25 as in step 2 (0.1 M phosphate, 0.1 M NaCl, pH 7.5).
10.	Mix the solutions of antibody and peroxidase. Stand at room temperature overnight.
11.	Store at 4°C.

colloidal gold is also obtainable and can be used in conjunction with biotinylated antibodies.

The amount of protein absorbed onto the surface of colloidal gold particles is affected by the pH, particle size, ionic concentration and the concentration of protein. The first of these, pH, is critically important and needs to be close to the isoelectric point of the protein being absorbed. Methods for absorbing proteins onto gold are described in the article by J.Roth in ref. 2, volume 2, and the manufacturer, Janssen Pharmaceutica, has produced an excellent booklet giving detailed recipes for different types of protein.

4.4 Selecting a label

There is no right or wrong choice of label. Selection should be guided by the tissue to be studied. Alkaline phosphatase would probably be a poor choice if a study is to be made of the gut since there is a high concentration of the enzyme in this tissue. Peroxidase is usually best avoided if the tissue contains large amounts of endogenous peroxidase or brown pigment (which can be confused by the colour produced by a commonly used substrate), particularly if the antigen is likely to be affected by the procedures necessary to eliminate these. Another important factor is the availability of labelled reagent, since for most applications it is usually easier and cheaper to buy rather than make labelled antibody. The ultimate choice may come down to the personal preference of the individual worker.

5. METHODS OF APPLICATION

The simplest method of detecting an antigen in a tissue section is to apply a labelled antibody (the so-called direct method). More commonly the primary antibody is left unlabelled and the label is attached to a different reagent which is then used to detect the primary antibody. For example, if the primary antibody is a mouse IgG, it may be detected by a labelled goat anti-mouse IgG antibody (the indirect method).

Of the methods described below, the direct method gives the fastest result. The indirect method combines ease of use with acceptable sensitivity while the enzyme—anti-enzyme

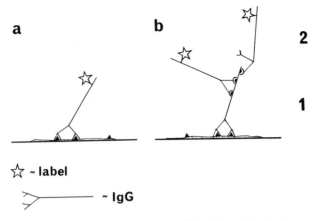

Figure 3. The (**a**) direct and (**b**) indirect methods of visualizing the reaction site of an antibody on a tissue section. 1, primary antibody; 2, labelled second antibody.

and immunogold methods are the most sensitive. As in many things, there is no one 'correct' method and the choice is often governed by the personal preferences of the investigator.

5.1 The direct method

The label is attached directly to the antibody (*Figure 3a*). To do this, the Ig must first be purified. The advantage of the method is its speed—it has only one step. The disadvantages are that it is less sensitive than other methods and that each antibody has to be labelled separately involving a considerable amount of work and often wastage of a valuable reagent.

5.2 The indirect method

A second antibody is raised to the Igs of the species from which the antibody of interest was obtained. Sections are incubated with the first antibody, washed and then incubated with the labelled second antibody (*Figure 3b*).

The major advantage is that, for a series of first antibodies, only one preparation of labelled second antibody is needed. This creates less work and also is more economical with the primary antibody since some reagent is always lost during a chemical procedure. The indirect is also more sensitive than the direct method because more than one second antibody molecule can react with each first antibody.

In a particularly sensitive variant of the indirect method, the second antibody is absorbed onto particles of colloidal gold (see Section 4.3).

5.3 Enzyme–anti-enzyme methods

This is a variation of the indirect method which is used to increase sensitivity. The label used in immunohistochemistry is frequently an enzyme. For ease of description, it is assumed that the first antibody has been raised in a rabbit. An antibody to the enzyme

115

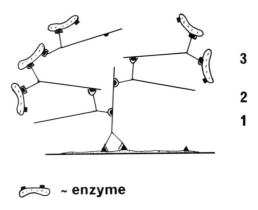

~ **enzyme**

Figure 4. The enzyme–anti-enzyme method. 1, primary antibody; 2, second (linking) antibody; 3, enzyme–anti-enzyme complex. Note that the primary and the anti-enzyme antibodies must be raised in the same species.

is raised also in a rabbit. A further requirement is an unlabelled antibody (e.g. raised in a goat) to rabbit IgG.

As is outlined in *Figure 4*, the section is incubated in the first (rabbit) antibody, washed and then incubated in an excess of goat anti-rabbit IgG serum and washed again. A solution of complexes of enzyme–anti-enzyme is prepared by adding rabbit anti-enzyme serum to a solution of enzyme. This is added to the slide and, after incubation and washing, a colour reaction for the enzyme performed.

Because Igs are dimers, they can react with two separate molecules of antigen. If an enzyme carries more than one epitope and the enzyme and antibody are mixed in the correct concentrations, a crosslinked network of enzyme and antibody will be formed. The purpose of the goat anti-rabbit IgG serum is to link the enzyme–anti-enzyme complexes to the first antibody.

The advantage of this method is its greater sensitivity since even more molecules of label can be added to each molecule of first antibody. It also avoids having to link covalently an enzyme to the second antibody. The disadvantages are the need for an additional reagent (the anti-enzyme serum) and the additional time required because of the extra step in the sequence of reactions.

If the enzyme used is HRP the method is called PAP (peroxidase–anti-peroxidase); alkaline phosphatase–anti-alkaline phosphatase (APAAP) is also frequently used.

Figure 5 shows a human lymph node stained for the leukocyte common antigen using (a) the indirect method with alkaline phosphatase and (b) the APAAP method. Tissue was fixed in modified methacarn and embedded in paraffin wax. The primary antibody was a murine monoclonal at a dilution of 1/200. The secondary antibody in (a) was a rabbit anti-mouse alkaline phosphatase used at a dilution of 1/200. In (b) rabbit anti-mouse Ig was applied at 1/20 followed by mouse APAAP at 1/100. The reagents were obtained from Dakopatts. The colour was developed using Fast Red TR; the counterstain was Mayer's haemalum (Section 6.9.2) and the coverslip mounted in glycerin jelly (Section 6.9.3). The stain demonstrates that the antibody reacted with the surface of the lymphocytes and a comparison of (a) and (b) demonstrates the greater sensitivity of the APAAP as compared to the simple indirect method.

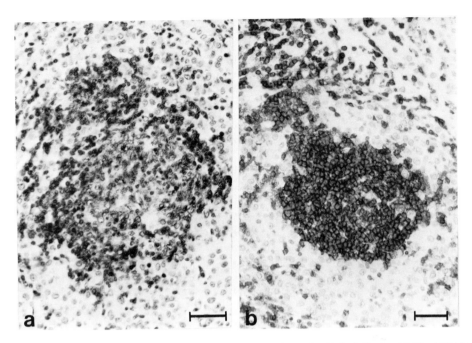

Figure 5. Human lymph node stained for leukocyte common antigen using (**a**) the indirect and (**b**) the APAAP method. The bar represents 200 μm.

5.4 Systems using biotin−avidin

Avidin, a protein extracted from egg white, has four binding sites of high affinity for biotin which is found in liver. Biotin can be covalently bound to either the first or second antibody which is then visualized using labelled avidin (*Figure 6*). Avidin has an isoelectric point close to 10 and is positively charged in neutral buffers. It is therefore likely to bind negatively charged molecules in a tissue. Streptavidin, isolated from *Streptomyces avidinii*, also has four binding sites for biotin but has an isoelectric point close to 7.

For enzyme labelling, rather than attach an enzyme directly to avidin or streptavidin, the enzyme may be biotinylated and unlabelled avidin used as a bridge. For increased sensitivity, this system may be used in a manner analogous to the enzyme−anti-enzyme technique. Complexes of biotinylated enzyme−avidin are preformed applied in place of the labelled avidin—the so-called ABC method (avidin−biotinylated peroxidase complex). It is claimed that larger complexes can be created than in the enzyme−anti-enzyme method thereby giving greater sensitivity.

Because of the presence of biotin in liver, particular care should be exercised if avidin or streptavidin are used on sections of this tissue.

A method for biotinylating proteins is given in *Table 9*. The molar ratio between biotin *N*-hydroxysuccinimide and protein needed to give optimal labelling will depend on the protein used. The details given here were used by Dr J.H.Westwood (Institute of Cancer Research) to label a mouse monoclonal antibody and gave an average of six biotin molecules per molecule of Ig.

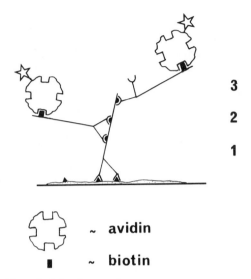

Figure 6. The biotin−avidin method. 1, primary antibody; 2, biotinylated second antibody; 3, labelled avidin.

5.5 **Other methods**

The avidin−biotin system can be mimicked by attaching a small molecule (hapten) to the first antibody and using a labelled antibody to the hapten. Common haptens are dinitrophenol (DNP) and arsinilate. This method can be useful when labelling a section with two antibodies from the same species (see Section 8).

The methods above may be used in a variety of combinations. In attempts to increase sensitivity, antibodies must be piled on antibodies. Except in special circumstances, this is generally neither necessary nor desirable. If an antigen has apparently been destroyed in fixed tissue, its presence may be revealed by using a highly sensitive method. However, the increased sensitivity will also show weak cross-reactions and amplify any slight non-specific staining. Whenever possible, it is preferable to select the correct conditions for processing a tissue.

6. EXPERIMENTAL METHODS

Work sheets for the different methods are presented below. They may vary in detail from those of other authors. There is no one correct method and variation can be introduced as long as certain guidelines are followed.

In the following methods, all incubations and treatments are at room temperature unless otherwise stated. In many laboratories, this can vary from 18 to 24°C and this will lead to considerable variation in the rate of reaction. If standardization is desired, and particularly if quantitative measurements are to be made, the temperature should be controlled. If a cold room is used, the times of incubation should be lengthened.

Recently, it has been reported that times of incubation with antibody solutions can be reduced from 1 h to 1 min by use of a microwave oven (9). This relies on the speed of the antibody−antigen reactions at the high temperatures induced by localized heating by the microwaves.

Table 9. Biotinylation of an immunoglobulin.

1.	Dissolve 150 g of biotin *N*-hydroxysuccinimide in 15 litres of *N*-dimethyl formamide (DMF).
2.	Dissolve 1 mg of Ig in 1 ml of PBS.
3.	Mix the two solutions and stir them gently for 4 h at room temperature.
4.	Dialyse at 4°C against two changes of 1 litre PBS over at least 20 h.
5.	Store at 4°C.

Table 10. The indirect method.

1.	Bring the section to water or buffer.
2.	When using an enzyme conjugate, if necessary block endogenous enzyme.
3.	Rinse in PBS and wipe any excess from the slide with a paper tissue. This ensures that the antiserum is not diluted on the slide.
4.	Place 100 μl of the first antibody, diluted appropriately[a], over the section. Incubate for 1 h at room temperature in a moist chamber.
5.	Wash the section with PBS, 0.5% bovine serum albumin (BSA).
6.	Wash several times in PBS, 0.01% detergent (BRIJ or Tween 80).
7.	Wash with PBS and wipe excess from the slide.
8.	Place 100 μl of the second (conjugated) antibody, diluted appropriately, over the section. Incubate for 1 h in a moist chamber at room temperature.
9.	Repeat steps 5 and 6.
10.	If using an enzymatic label, wash the slide in an appropriate buffer and develop the colour. Mount the coverslip.

[a]Dilute antibodies in either PBS, 0.5% BSA or, preferably, in PBS, 5% serum; the serum being obtained from the same species as the second antibody.

6.1 A general method

A generalized method for applying antibodies to a section is given in *Table 10*. This is for an indirect method but can be adapted for all the other methods above. It is to be used together with the detailed protocols for each type of label in the methods listed in the rest of this section.

The purpose of most of the additives to the buffers is to prevent Igs sticking non-specifically to the section and to remove the last traces of unreacted Ig after incubation. Protein is added to prevent the antibody absorbing onto the surface of the section and the addition of a small amount of detergent is intended to reduce hydrophobic interactions between Igs and proteins in the section.

In our laboratory, if large numbers of sections are being stained, for washing, they are placed in racks and gently agitated in a staining trough. If there are only a few sections, solutions for washing are kept in wash bottles. When a section is washed, the slide is held at a slant and a stream of solution directed just above the section. The sections will be washed several times and care must be taken not to wash the section off the slide. This also emphasizes the need to prepare sections of high quality.

6.2 Choosing the correct dilution of antibody

When using a new antibody, a number of sections should be cut from a tissue known to contain the antigen. They should be stained using a set of dilutions which should be in geometric not arithmetic progression (e.g. neat, 1:2, 1:4. 1:8, etc.). It is best to start with large steps (e.g. in fives) to find the approximate range and then to repeat the experiment using doubling dilutions.

The optimal dilution is that which just stains a section close to the maximum strength. With some antisera, at high concentrations, non-specific staining of the section may be observed. In this case, a dilution has to be sought at which the non-specific staining has been diluted out while the specific stain is still acceptably strong. This should not be a problem when using a monoclonal antibody.

6.3 Fluorescent labels

Some fixatives (e.g. glutaraldehyde) make a tissue autofluorescent. This should be checked before starting. After following the procedure in *Table 10*, coverslips should be mounted in an aqueous, non-fluorescent mountant such as Hydromount (from National Diagnostics) or 9:1 (v/v) glycerol:40% formaldehyde. Another mountant can be made up as follows.

(i)   To 3 g of analytical grade glycerol add 1.2 g of polyvinyl alcohol (PVA) (Goshenol from Polaron) and stir.

(ii)   When the PVA and glycerol have mixed completely, add 3 ml of water, stir and leave for 4 h at room temperature.

(iii)   Add 6 ml of 0.1 M Tris−HCl buffer, pH 8.5, and keep at 50°C with occasional agitation until the PVA has dissolved.

6.4 Peroxidase

The chromogens used with peroxidase often produce a brown colour which might be confused with brown pigment in the section, for example, that found in red blood cells in tissue fixed in formol−saline. It is therefore often necessary to bleach the section before starting any other procedure. This is done by immersing the slide in H_2O_2 (we use 7.5%) for 5 min and then washing well in water. A check should be made to see if this has a deleterious effect on the antigen being studied.

6.4.1 *Blocking endogenous enzyme*

Before placing antibody on the section, incubate the slides in 2.3% periodic acid for 5 min, wash in water, rinse in 0.03% freshly prepared potassium borohydride and wash in water. Another method often used is to incubate the slides for 30 min in 0.3% H_2O_2 in methanol.

If these treatments destroy the antigen, an alternative is to incubate the slides in PBS, 0.1% phenylhydrazine for 5 min.

6.4.2 *Chromogens for peroxidase*

From the large number of possible chromogens for use with peroxidase only about five have found general use in immunohistochemistry. Their recipes are given below.

Diaminobenzidine with or without enhancement. This is the most commonly used chromogen. It gives a brown precipitate which is insoluble in water, ethanol and xylene.

Table 11. Protocol for silver enhancement of a peroxidase/diaminobenzidine stain.

The section is first stained using a suitable method with a peroxidase labelled antibody and the colour developed with DAB. The colour may then be enhanced using the method below.

1. Wash the sections in distilled water.
2. Immerse for 5 min in 2.5 mM gold chloride, pH 2.3. Wash in distilled water.
3. Immerse for 5 min in 0.1 M sodium sulphide, pH 7.0. Wash in distilled water.
4. Immerse for 2−6 min in the silver reagent (see below). Wash thoroughly in distilled water leaving the slide in water for 10 min between washes.
5. Immerse in 1% v/v acetic acid for 15 min changing the acid once during this time. Wash in water.
6. Counterstain in Mayer's haemalum and mount.

Silver reagent

Make up the following solutions by dissolving each given quantity in 100 ml twice distilled water.

Solution A	sodium carbonate	5.08 g
B1	ammonium nitrate	0.83 g
B2	silver nitrate	0.82 g
B3	dodeca-tungstosililic acid	3.97 g

Add 1 ml of solutions B1, B2 and B3 to 1 ml of water. Add 5 μl of 40% (v/v) formaldehyde solution. Add this solution to 4 ml of solution A with vigorous mixing.

DAB has been suspected as a carcinogen and should be handled with care.

(i)   Just before use dissolve 100 mg of DAB in 100 ml of 0.1 M Tris−HCl buffer, pH 7.2.
(ii) Add 100 ml of water containing 70 μl of 30% H_2O_2.
(iii) Immerse sections for 5 min. Wash thoroughly in water.
(iv) Counterstain in Mayer's haemalum and mount coverslips in DPX (see Section 6.9).

It is possible to intensify the stain using either imidazole or silver. For the former, make the DAB solution above 0.01 M in imidazole before use.

For silver enhancement, the method which is used in this laboratory is given in *Table 11*. The result of a silver enhancement is shown in *Figure 7*. Sections are of human colon fixed in formalin and embedded in paraffin wax. They were stained for carcino-embryonic antigen (CEA) by the indirect method. The primary antibody was rabbit anti-CEA raised in this laboratory and used at a dilution of 1/4000. Secondary antibody and colour development as *Figure 2*. In (a) there was no further treatment; in (b) the colour was enhanced with silver. CEA is located on and in the epithelial cells. The staining is barely visible at this dilution of primary antibody without enhancement but is clearly demonstrated after the silver reaction.

Hanker−Yates Reagent. This consists of 1 part *p*-phenylenediamine−HCl to 1 part (w/w) pyrocatechol. It gives a blackish-brown precipitate insoluble in water, ethanol and xylene.

(i)   Just before use dissolve 150 mg of Hanker−Yates Reagent in 100 ml of 0.1 M Tris−HCl buffer, pH 7.6, and add 120 μl of 30% H_2O_2.
(ii) Incubate sections for 15 min. Wash in water.
(iii) Counterstain in Mayer's haemalum and mount in DPX.

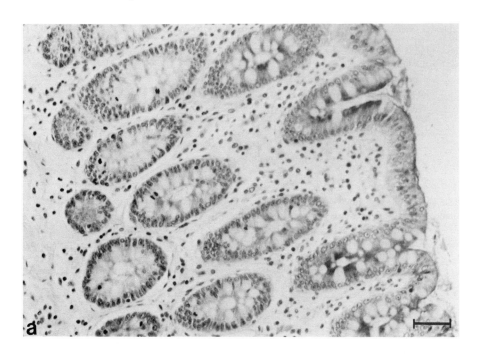

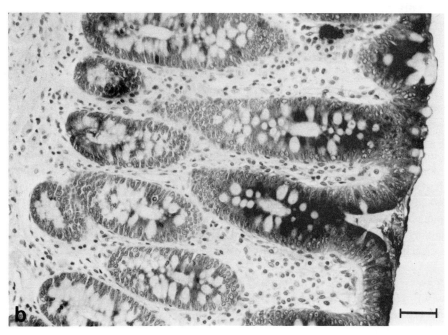

Figure 7. Human colon stained for CEA showing the effect of silver enhancement on the product of the peroxidase reaction. The bar represents 50 μm.

3-Amino-9-ethyl carbazole. This gives a red precipitate soluble in ethanol.

(i)   Dissolve 2 mg of 3-amino-9-ethyl carbazole in 0.5 ml of DMF in a glass tube.
(ii)   Add 9.5 ml of 0.2 M acetate buffer, pH 5.
(iii)   Just before use add 5 μl of 30% H_2O_2. Incubate sections for 15 min.
(iv)   Counterstain in Mayer's haemalum and mount in glycerine jelly.

4-Chloro-1-naphthol. This can be used if a blue colour is required. The precipitate is soluble in xylene. We have not found a suitable nuclear counterstain since methyl green is soluble in water.

(i)   Dissolve 20 mg of 4-chloro-1-naphthol in 40 ml of 20% methanol in Tris−saline, pH 7.6, by heating to 50°C.
(ii)   Before use add 15 μl of 30% H_2O_2. Incubate sections for 10 min.
(iii)   Mount in glycerine jelly.

Tetramethyl benzidine. This is an alternative blue stain, insoluble in ethanol and xylene.

(i)   Dissolve 5 mg of tetramethyl benzidine in 2 ml of dimethyl sulphoxide.
(ii)   Add to 50 ml of 0.02 M acetate buffer, pH 3.3, containing 20 μl of 30% H_2O_2 immediately before use. Incubate sections for 15 min.
(iii)   Counterstain in methyl green and mount in DPX.

6.5 Alkaline phosphatase

6.5.1 *Blocking endogenous enzyme*

Before applying the first antibody, immerse the slides in 20% acetic acid for 5 min. Wash well in water. This will destroy all alkaline phosphatase activity including that in the intestine. If it also destroys the antigen, alkaline phosphatases other than intestinal can be inhibited by making the substrate solution 1 mM in levamisole (increased to 2 mM for frozen sections of tissues rich in alkaline phosphatase such as placenta and kidney). This takes advantage of the fact that the alkaline phosphatase used for preparing conjugated antibodies is extracted from calf intestine.

6.5.2 *A cautionary note*

Phosphate buffer may inhibit alkaline phosphatase. If the last antibody has been washed off in phosphate buffer (as recommended in the protocol in *Table 10*), the sections should be washed thoroughly in distilled water before developing the colour.

6.5.3 *Chromogens for alkaline phosphatase*

Fast Red. This is the usual chromogen for use with alkaline phosphatase. It gives a red precipitate soluble in ethanol.

(i)   Dissolve 5 mg of sodium naphthol AS BI phosphate in a few drops of DMF in a glass tube.
(ii)   Add to 5 mg of Fast Red TR salt in 10 ml of veronal acetate buffer, pH 9.2.
(iii)   Add levamisole if required. Filter.
(iv)   Incubate slides for 1 h and wash in water.

(v)   Counterstain with Mayer's haemalum and mount in glycerine jelly.
(vi)   The intensity of the reaction may be increased if the substrate is renewed after 30 min.

Fast Blue salt. This gives a blue product soluble in ethanol.

(i)   Dissolve 5 mg of Fast Blue BB salt in 10 ml of 0.1 M Tris buffer, pH 9.0, and add 5 mg of sodium naphthol AS BI phosphate in DMF as above.
(ii)   Add levamisole if required. Filter.
(iii) Incubate for 15 min replacing the substrate after each 5 min. Wash in water.

There are problems finding a counterstain for this dye since haemalum stains the nuclei blue and methyl green is soluble in water while the blue precipitate is soluble in xylene.

New Fuchsin. This produces a red precipitate insoluble in ethanol and xylene.

(i)   Mix 250 μl of 4% New Fuchsin in 2 M HCl with 250 μl of 4% $NaNO_2$ and leave to stand in the cold for 5 min.
(ii)   Add to 40 ml of 0.2 M Tris$-$HCl buffer, pH 9.0, and add 10 mg of sodium naphthol AS TR phosphate dissolved in 0.2 ml of DMF.
(iii) Add levamisole if required. Filter. Incubate sections for 10 min.
(iv)   Counterstain with Mayer's haemalum and mount in DPX.

6.6 Glucose oxidase

The chromogen usually used with glucose oxidase is nitroblue tetrazolium which gives a dark blue precipitate insoluble in ethanol and xylene. It should not be necessary to block endogenous enzyme.

(i)   Dissolve 335 mg of β-D-glucose and 33.5 mg of nitroblue tetrazolium in 50 ml of 0.05 M Tris buffer, pH 8.3.
(ii)   Heat at 37°C for 1 h in the dark and add 8.3 mg of phenazine methosulphate (this compound may be carcinogenic and should be handled with care).
(iii) Incubate sections for 1 h at 37°C in the dark. Wash in water.
(iv)   Counterstain in 0.1% methyl green and mount in DPX.

6.7 Galactosidase

If the reagents below are used, the product is blue, stable in ethanol and xylene. It should not be necessary to block the endogenous enzyme.

(i)   To 7 ml of PBS, 1 mM $MgCl_2$, pH 7.0, add 0.5 ml of 50 mM potassium ferricyanide and 0.5 ml of 50 mM potassium ferrocyanide.
(ii)   Add 10 mg of 5-bromo-4-chloro-3-indolyl-β-D-galactosidase previously dissolved in a drop of DMF (this solution may be stored frozen for 2 months).
(iii) Incubate sections in the above for 1 h at 37°C. Wash in water.
(iv)   Counterstain in 0.1% methyl green and mount in DPX.

6.8 Immunogold

The gold is visualized by silver precipitation by a chemical process similar to that used

Table 12. Silver enhancement of colloidal gold.

Before adding the first antibody.

1. Wash the section in water.
2. Immerse the section for 5 min in Lugol's iodine (see step 5, *Table 3*). Wash in water.
3. Rinse in 2.5% (w/v) sodium thiosulphate in water.
4. Wash in TBS, 1% Triton X-100.

Apply antibodies as described in *Table 10* substituting Tris-buffered saline (TBS) for PBS and using antibody absorbed on immunogold in the final step. The immunogold method is very sensitive and can bring up any slight background staining. Before applying an antibody, it may be necessary to incubate the section for 10 min in a normal serum (from the same species as the second antibody).

To develop the stain.

1. Wash in water.
2. Incubate in the silver enhancement solution in subdued light (e.g. a dark room safe light) for 40−60 min.
3. Wash in water.
4. Counterstain. Take through ethanol to xylene (or histoclear) and mount.

Silver enhancement solution. 20 ml of 1 M citrate buffer (1 M citric acid, 0.5 M trisodium citrate), pH 3.5; 33 ml of 30% gum acacia, 15 ml of silver lactate (0.11 g in 15 ml); 15 ml of hydroquinone (0.85 g in 15 ml); 17 ml of water.

to develop photographic film. If the tissue has been fixed in formalin and embedded in paraffin wax, it is necessary to pre-treat the sections with Lugol's iodine or some other oxidizing agent (10). The reason for this is not clear since it is not necessary when other conditions of fixation and processing are used. A suitable protocol is given in *Table 12*.

6.9 Some general procedures

6.9.1 *Coating the slides*

During an immunohistochemical stain the slides are washed frequently. It is important that the section adheres well to the glass slide. Several different procedures are used to coat slides to improve adhesion. Four of those commonly used in this laboratory (gelatine−formaldehyde, gelatine, albumin and poly-L-lysine) are given in *Table 13* which also indicates the circumstances under which they are used.

6.9.2 *Counterstaining*

It is general practice to counterstain the nuclei in a section in order to reveal the architecture of the tissue. If the immunohistochemical stain is red, black or brown, either haematoxylin or Mayer's haemalum is generally used. We use the latter. If the immunostain is blue, then methyl green is preferred.

Mayer's haemalum.

(i) Dissolve 1 g of haematoxylin in 1 litre of distilled water using gentle heat.
(ii) Add 50 g of aluminium potassium sulphate, heat if necessary.
(iii) Add 0.2 g of sodium iodate, mix well and leave overnight.
(iv) Add 1 g of citric acid. Mix well.

Table 13. Solutions used for coating slides.

A. *Gelatine–formaldehyde.* Often used for frozen sections.
1.   Mix 100 ml of 1% gelatine (warm gently to dissolve) and 100 ml of 2% formaldehyde.
2.   Immerse slides for 3 sec and dry at room temperature.
3.   Store at room temperature and use as required.

B. *Another gelatine based adhesive.* Sometimes used for frozen and soft wax sections.
1.   Melt 15 g of gelatine in 500 ml of warm water.
2.   Dissolve 1 g of chrome alum in 220 ml of distilled water.
3.   Mix the two and add 70 ml of glacial acetic acid and 300 ml of 95% ethanol.
4.   Store at room temperature.
5.   Coat slides as in step 1 above.

C. *Albumin.* Routinely used for ordinary paraffin wax sections.
1.   Dissolve 2.5 g of egg albumin, 0.25 g of NaCl in 50 ml of distilled water (warm to 37°C).
2.   Add 50 ml of glycerin, 0.05 g of thymol.
3.   Coat slides just before use.

D. *Poly-L-lysine.*
1.   Immerse slides in 100 μg/ml of poly-L-lysine (high molecular weight, Sigma) in water.
2.   Dry and store.
3.   For slides to be stained by *in situ* hybridization of DNA, use poly-L-lysine at 1 mg/ml (see *Table 14*).

(v)   Add 50 g of chloral hydrate.

(vi)   Immerse the slide in the haemalum for 5–20 min (depending on the desired strength of nuclear stain) and wash thoroughly in running tap water.

(vii)   Dip in a saturated solution of lithium carbonate which renders the stain blue and wash in tap water.

Methyl green. Use a 0.1% solution of methyl green in distilled water. The stain will dissolve out in water and the coverslip should be mounted in a non-aqueous mountant.

6.9.3 *Mounting the coverslip*

If the reaction product is insoluble in ethanol and xylene, the slides are brought from water through ethanol to xylene. The coverslips are mounted in a natural or synthetic resin such as DPX. If not, they are left in water and the coverslips mounted in glycerine jelly or a similar water-based mountant.

DPX consists of 10 g of distrene 80, 5 ml of dibutylphthalate and 35 ml of xylol. It is usually purchased ready made up.

To make glycerine jelly:
(i)   dissolve 10 g of gelatine in 60 ml distilled water using gentle heat;
(ii)   add 70 ml of glycerine and 0.25 g of phenol and mix well;
(iii)   aliquot into 10 ml batches and store in the cold;
(iv)   before use melt in a water bath; avoid shaking as this creates air bubbles.

7. CONTROLS AND PROBLEM SOLVING

There are two types of problem encountered: the unwanted presence of stain and the unexpected absence of stain. That there is a problem will be revealed by the appropriate controls.

7.1 Controls

Two types of control are needed—positive and negative. For each antibody, a block of tissue known to contain the antigen should be selected and a large number of sections cut. One of these should be included in every run in order to monitor the strength of the staining reaction.

Each run should also include an experimental section on which the first antibody has been omitted. This checks for non-specific staining by the reagents used to detect the primary antibody. If the direct method is being used, this control would be omitted. If an enzyme stain is being used, a section which is developed for colour only is included to monitor endogenous enzyme activity.

These controls, although necessary, do not check for non-specific staining by the primary antibody. When using a polyclonal antiserum, an aliquot of serum can be absorbed with the original antigen. This should reduce the specific and reveal non-specific stain. It will not demonstrate the presence of cross-reacting antigens. It is unnecessary to do this control with every run but it should, if the antigen is available, be performed when first using a new antiserum and on one or two key sections. This type of control is meaningless with a monoclonal antibody.

The test section itself may act as a control. The distribution of an antigen is probably known; for example, an antibody against T lymphocytes should not stain epithelial cells. The correct structures should be well stained and other structures quite clean. If the 'wrong' cells are stained then non-specific staining can be suspected. In particular, stromal cells and muscle cells tend to give a 'dirty' background stain if the washing procedures are inadequate.

7.2 Problem solving

Problem solving is a matter of applying simple logic. The controls discussed above should pinpoint which part of the procedure is in error. Each step should be carefully considered in turn. Frequently the source of the problem is something quite trivial. For example, it is advisable to use slides with ends of frosted glass and to mark the slide clearly in pencil. Without this, it is sometimes difficult to recognize one side of the slide from the other and hence stain the wrong side of the slide.

There are three types of incorrect staining—under-, over- and non-specific staining. No staining at all on the positive control usually suggests that a reagent has been inadvertently omitted or the wrong reagent used. If a large number of sections have been stained with different antibodies, perhaps the incorrect second antibody has been used on that section (anti-rabbit on a section stained with a mouse antibody).

Weak staining suggests that one of the more labile reagents has deteriorated. For example, if a peroxidase stain is in use, it is important that the solution of H_2O_2 is fresh. Sodium azide is frequently added to buffers to prevent bacterial growth. This compound inhibits many enzymes and could cause difficulties if a buffer containing azide has been used during the development of the chromogen.

Over-staining is often caused by accidentally diluting one of the reagents incorrectly.

Temperature can occasionally cause a problem. Most routine work is carried out at room temperature which can vary as much as 10°C in a laboratory without air-conditioning. Enzyme reactions proceed much faster at higher temperatures and a procedure worked out during a chilly day might over-produce chromogen during a heat wave.

If the volume of antibody solution is insufficient to cover the section properly, any evaporation during incubation will concentrate antibody at the edge of the section. This will cause overstaining. In the extreme, if solution actually dries onto part of the section, a heavy background stain will result.

Non-specific staining will increase if the slide is over-stained. Apart from this, if this is a general problem, attention should be paid to the procedures used for washing the slides and, in particular, to the protein in the buffer. If a problem arises with a particular primary antibody, it may help if the section is pre-incubated for 15 min with PBS containing 5% serum from another species (see note at the foot of *Table 10*). If a particular secondary reagent causes a problem, it should be discarded and another purchased from a different source.

8. DETECTING TWO ANTIGENS ON THE SAME SECTION

The application of two antibodies to the same section requires some care. If both antibodies have been raised in different species, an indirect method may be used with two differently labelled second antibodies selected to ensure that there is no cross-reaction between them. If the two antibodies are from the same species, as will often be the case when working with murine monoclonal antibodies, it may be possible to carry out a complete immunohistochemical stain for the first antigen followed by a stain for the second using a different enzyme. (The formation of the coloured precipitate will often prevent the first antibody reacting with the second set of reagents.) If this does not work, then either a direct method should be used with the antibodies each conjugated to a different label or a distinguishing compound must be attached to each antibody. For example, one antibody could be biotinylated and then detected using labelled avidin and the other reacted with DNP and detected with a labelled anti-DNP.

The correct order of application of the reagents must be established empirically. If antibodies from two different species are used, good results are obtained usually by applying both first antibodies together followed by both second antibodies, and theoretically this should always work. However, sometimes better results are obtained if the reagents are applied sequentially.

Enzyme labels are satisfactory if the two antigens are located on separate cells (e.g. when distinguishing different transplantation antigens in chimeric mice) or are found in separate cellular compartments (e.g. nucleus and plasma membrane). If not, then it is difficult to distinguish unequivocally a singly- from a doubly-labelled cell. In this case, fluorescent labels must be used together with a microscope which allows the operator easily to switch filter combinations (i.e. from those appropriate for fluorescein to those for rhodamine and back again) while observing a particular cell.

9. DNA PROBES FOR IN SITU HYBRIDIZATION

Specific sample protocols are given in *Tables 14−16*. First, however, some general statements on the principles of probe methods will be made.

9.1 **Principles of the method**

It is possible to replicate lengths of mammalian DNA in bacteria or yeasts. One method makes use of plasmids which are small circular molecules of double-stranded DNA. Plasmids replicate at the same time as their host. Because of their size, they can be easily separated from the DNA of the bacterium or yeast in which they exist. Using specialist enzymes, short lengths of foreign DNA can be inserted into a plasmid which are then amplified by growth in the host bacterium. The cultures are cloned and the colony carrying the desired length of DNA isolated. This colony can be grown on to give an inexhaustible supply of a specific piece of DNA.

If double-stranded DNA is heated, the two strands separate—this is called denaturation. On cooling, complementary strands will re-anneal or renature. Single-stranded DNA will also anneal to complementary mRNA. If the cloned DNA is suitably labelled, it may be used in this technique to detect a complementary sequence of DNA or mRNA. Probes for mRNA can be used to demonstrate the production of a specific protein which might be too labile or exported from the cell too rapidly to permit its immunohisto-chemical detection. Other probes may be used to detect viral DNA in a cell or to explore the genomic DNA.

DNA is often labelled by a process called nick translation. Breaks are introduced enzymatically into one strand of the DNA in the presence of a DNA polymerase and the four nucleotide triphosphates, one of which is labelled. The polymerase degrades one strand of the DNA from the nick and makes new DNA using the other strand as a template.

A detailed description of recombinant DNA technology can be found in ref. 11, or in three other books in this series (*DNA Cloning: A Practical Approach, Volumes I, II* and *III,* D.M.Glover, ed.).

For histochemistry, the DNA may be labelled with an isotope and visualized by autoradiography. For some applications, the localization achieved is not sufficiently high; autoradiography often gives a high background and development may take a week or more. The introduction of biotin into the DNA allows immunohistochemical methods to be used. These are faster and give better localization although, at present, the method is less sensitive than autoradiography.

9.2 **Experimental method**

The method falls into three parts—biotinylation of the DNA probe, hybridization of the probe to the section and visualization of the hybridized probe. Biotinylation is usually performed by nick translation in the presence of biotin-11-deoxyuridine triphosphate (which can be purchased from BRL−Gibco) which is utilized by the DNA polymerase in place of thymidine triphosphate. Particularly if genomic DNA is being probed, the amount of biotin bound to the section may be small. A sensitive method of detection

Table 14. Preparing sections for *in situ* hybridization.

A. *Frozen sections*

1.   Cut frozen sections onto prepared slides[a].
2.   Air-dry at 37°C for 1 h.
3.   Immerse in formol−calcium for 5 min at 4°C.
4.   Rinse in acetone.
5.   Immerse in chloroform−acetone for 5 min at −20°C.
6.   Rinse in acetone at −20°C.
7.   Wash thoroughly in 2 × SSC[b], 1 mM EDTA.
8.   Dehydrate in graded ethanols and air-dry.

B. *Paraffin sections from formalin-fixed tissue*

1.   Cut sections onto prepared slides[a], bring to water as usual, finally rinsing in distilled water.
2.   Digest in proteinase K solution (200 μg/ml of proteinase K, 2 mM $CaCl_2$, 0.02 M Tris−HCl, pH 7.4) at 37°C for ∼1 h. (Optimum time varies from one block of tissue to another and needs to be found by experiment.)
3.   Wash twice in distilled water by gentle agitation each time for 10 min at 4°C.
4.   Immerse for 5 min in 70% ethanol.
5.   Immerse for 5 min in 95% ethanol.
6.   Rinse in absolute ethanol.
7.   Air-dry.

[a]To prepare slides, wash in detergent, wash thoroughly in tap water, distilled water, and finally de-ionized water. Dry in oven. Place 20 μl of 1 mg/ml poly-L-lysine on a slide and spread. Mark the coated side of the slide and leave to dry in a dust-free container.
[b]To make 20 × SSC, dissolve 175.3 g of NaCl and 88.2 g of sodium citrate in 800 ml of water, adjust the pH to 7.0 with 10 N NaOH and make up to 1 litre.

has to be employed and this may involve several layers of antibody.

In this laboratory, we have successfully used probes for the human Y chromosome and for cytomegalovirus DNA. The method has been adapted by Mr C.D.Pallett (Institute of Cancer Research) from that described by Burns *et al.* (12). Examples of some other applications in which mRNA is detected are described in refs 13 and 14.

9.2.1 *Probe to the Y chromosome*

The DNA specific for the Y chromosome together with the kit for nick translation can be purchased from Amersham International and labelled according to their instructions. The probe is then purified by exclusion chromatography on a column of Sephadex G-50 (Pharmacia) equilibrated in 0.1 M NaCl, 1 mM EDTA, 0.1% (v/v) sodium dodecyl sulphate, 10 mM Tris−HCl, pH 7.4.

For every μg of probe DNA, 500 μg of salmon sperm DNA is added followed by 2 vol of ice-cold ethanol. After standing at −20°C overnight to precipitate the DNA, it is centrifuged, the supernatant aspirated and the remaining ethanol removed under vacuum. The DNA is redissolved in 1 mM EDTA, 10 mM Tris−HCl, pH 7.4, aliquoted and stored frozen.

The sections are prepared as described in *Table 14*. Cleanliness of the slides is important to remove any contaminating nucleases, to ensure good adhesion of the section and to prevent the probe sticking to the glass surface.

The DNA is hybridized to the section as described in *Table 15*. The incubation at

Table 15. *In situ* hybridization of DNA probe to Y chromosome.

1.	Dilute biotinylated probe to ~ 1 µg/ml in 50% (v/v) formamide, 10% (v/v) dextran sulphate, 2 × SSC, 0.1 mM EDTA, 0.05 mM Tris−HCl, pH 7.3.
2.	Apply 100 µl to the section.
3.	Place the section on a pre-heated metal tray in an oven set to 95°C[a] and leave for 8 min.
4.	Transfer to a moist chamber and leave overnight at 42°C.
5.	Incubate in 2 × SSC at room temperature for 30 min.
6.	Incubate in 0.1 × SSC at 42°C for 45 min changing the buffer after 30 min.

[a]As the slides take time to reach the temperature of the oven, their mean temperature during the 8 min will be less than this. The optimum temperature of the oven needs to be found by trial and error.

Table 16. Detection of biotinylated DNA probe.

1.	Wash the section twice.
2.	Incubate in avidin−alkaline phosphatase (Dako) diluted 1/100. Wash.
3.	Incubate in biotin at 1 mg/ml. Wash.
4.	Incubate in goat anti-biotin (Sigma) at 1/50. Wash.
5.	Incubate in alkaline phosphatase−rabbit anti-goat at 1/100 (Sigma). Wash.
6.	Incubate in alkaline phosphatase−goat anti-rabbit at 1/100 (Sigma). Wash.
7.	Repeat step 5.
8.	Develop the colour reaction for alkaline phosphatase using Fast Red TR (see Section 6.5).
9.	Counterstain lightly with Mayer's haemalum and mount in glycerine jelly.

All incubations and washings are in 0.1% BSA, PBS. All incubations are for 1 h.

the higher temperature for 8 min is to denature the DNA; that at the lower temperature is to allow the DNA to renature so that the probe hybridizes to the nuclear DNA.

The method for visualizing the biotin after hybridization is given in *Table 16*; the large number of steps is used to increase sensitivity. *Figure 8* shows Y chromosomes visualized as black dots in the nuclei of a male human lymph node. The tissue was fixed in formalin and embedded in paraffin wax.

9.2.2 *Cytomegalovirus*

This probe can be obtained biotinylated from Enzo Biochem. The method used is the same as that for the Y chromosome probe except that the hybridization buffer described in the manufacturer's work sheet is used (5 parts de-ionized 50% formamide, PBS:2 parts 50% dextran sulphate:1 part 20 × SSC:2 parts probe plus carrier DNA). The time for denaturation is 10 min and the hybridization is for 60 min at room temperature.

A section of human lung stained by this method is shown in *Figure 9*. The tissue was obtained *post mortem*, fixed in formalin and embedded in paraffin wax. The virus is visualized as a heavy black dot on the infected cell.

10. CYTOLOGICAL PREPARATIONS

Procedures for staining cytological preparations are the same as those for sections. The key to achieving good results lies in the method used to prepare the cells. On conventionally prepared smears, there is often protein or mucus overlying the cells and this can hinder good interaction between cellular antigens and antibodies. An

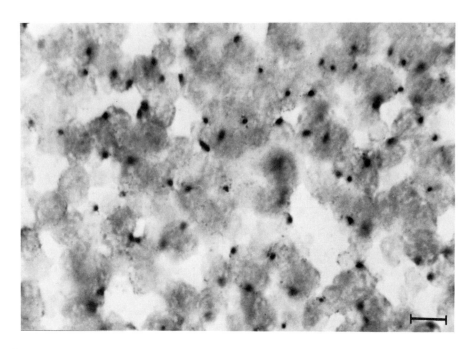

Figure 8. Male human lymph node stained for the Y chromosome by *in situ* hybridization. The bar represents 8 μm. Slide stained by Mr C.D.Pallett.

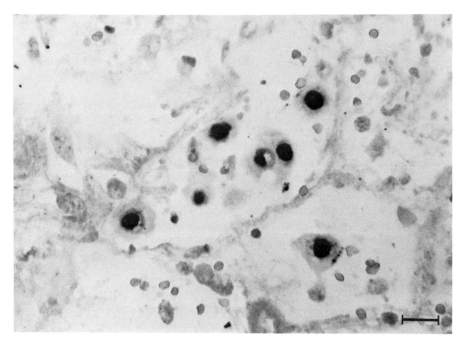

Figure 9. Human lung obtained *post mortem* stained for cytomegalovirus by *in situ* hybridization. The bar represents 20 μm. Slide stained by Mr C.D.Pallett.

Table 17. Preparation of smears from serous effusions.

A. *With little contamination from red blood cells*

1. Centrifuge the specimen at 300 *g* for 5 min.
2. Remove the supernatant and examine the deposit.
3. If the deposit is blood-stained, remove red cells (see below).
4. If the deposit is essentially free of blood, resuspend in 20 ml of PBS and re-centrifuge. Repeat this wash.
5. Remove the supernatant and resuspend the cells in as small a volume of PBS as possible.
6. Place a drop on a clean slide and smear with a second slide as for a blood film. Fix immediately in 95% alcohol and leave for a minimum of 1 h.
7. Store in alcohol or spray the smear with carbowax fixative and store at −20°C.

B. *From blood-stained effusions*

1. Wash the centrifuged deposit with 20 ml of PBS, and re-centrifuge. Remove the supernatant.
2. Mix the centrifuged deposit with 5 ml of PBS.
3. Underlay the cell suspension with 10 ml of Lymphoprep[a].
4. Centrifuge at 300 *g* for 20 min.
5. Remove the layer of nucleated cells at the top of the interface and transfer to a clean centrifuge tube.
6. Centrifuge at 300 *g* for 5 min and remove the supernatant.
7. Continue from step 5 as in A above.

[a]Lymphoprep is a mixture of sodium metrizoate solution and Ficoll and is manufactured by Nyegaard and Co.

Table 18. Preparation of smears from aspirates of bone marrow.

1. Aspirate 1−4 ml of marrow using a heparinized syringe and place in a 50 ml centrifuge tube containing 1000 units heparin plus 5 ml of tissue culture medium.
2. Mix the sample and then make it up to 35 ml with sterile medium.
3. Underlay with 15 ml of Lymphoprep (density 1.077) (see *Table 17*).
4. Centrifuge at 400 *g* for 20 min.
5. Aspirate the cell layer, transfer to a clean centrifuge tube and make up to 20 ml with PBS.
6. Centrifuge at 400 *g* for 15 min.
7. Aspirate down to 10 ml and make up to 20 ml with sterile PBS.
8. Centrifuge at 400 *g* for 5 min.
9. Aspirate down to ~0.6 ml, mix well and transfer to a 1 ml siliconized conical centrifuge tube.
10. Centrifuge.
11. Aspirate supernatant leaving volume approximately equal to pellet size.
12. Resuspend very gently and disaggregate the cells by taking up the suspenion into a 20 µl pipette, set at suitable volume, ~10 times.
13. Prepare thin smears and fix immediately in absolute ethanol. Leave for at least 30 min. Store at −20°C.

A similar method can be used to prepare nucleated cells from peripheral blood.

excessive number of red blood cells is also undesirable. It is important to wash the cells, and if necessary remove erythrocytes, before making a smear or centrifuging cells onto a glass slide.

Methods used in this laboratory to prepare smears of cells from serous effusions, bone marrow and cervical scrapes are given in *Tables 17−19*. The smears were subsequently stained with antibodies to epithelial antigens such as epithelial membrane antigen or cytokeratin. The methods might have to be modified for more labile antigens.

Figure 10 shows a photograph of a smear made from a cervical scrape and stained

133

Table 19. Preparation of smears from cervical scrapes.

1. Take a cervical scrape in a conventional manner using a wooden spatula.
2. Break the end from the spatula and drop into 10 ml Cellfix solution (1.0 g of DTT dissolved in 600 ml of PBS plus 400 ml of ethanol). Agitate violently. Remove the spatula. The cells can be stored in this form at 4°C.
3. Centrifuge the cells and wash in PBS.
4. Resuspend the cells in the smallest possible volume of PBS. Spread 10 μl on a clean microscope slide and allow to air-dry.
5. Store the smears at −20°C.

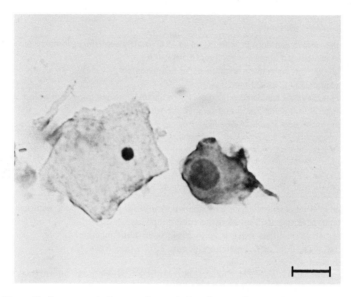

Figure 10. Two cells from a cervical scrape from a lesion diagnosed as cervical intraepithelial neoplasia, grade 2. The bar represents 10 μm.

by the indirect method. The primary antibody was rabbit anti-epithelial membrane antigen used at a dilution of 1/500; second antibody sheep anti-rabbit Ig conjugated to alkaline phosphatase and used at 1/100. Both reagents were produced in this laboratory. The colour was developed using Fast Red TR, counterstained in Mayer's haemalum, and the coverslip mounted in glycerin jelly. The normal squamous cell is negative while the neighbouring dyskaryotic cell has both its membrane and cytoplasm stained.

11. QUANTITATION

Immunohistochemical stains are not easily quantitated and consequently there have been few reports on this aspect. The best method presently available is to adapt the methods used for cytochemical assays. A scanning and integrating microdensitometer may be used to estimate in individual cells the average optical density at the isobestic wavelength for the stain used (15). The method is slow because each cell must be measured

Figure 11. Photograph of a tray suitable for immunohistochemical staining.

individually. It is probably easiest to perform on cytological preparations since the whole cell can be examined.

When undertaking quantitative measurements, all the variables, including temperature, should be carefully controlled.

12. EQUIPMENT

Apart from the normal equipment needed to cut tissue sections and to stain sections and cytological smears, the only special piece of equipment needed is a staining tray. This can be quite simple; it is used to keep slides horizontally in a humid atmosphere. We use trays, 36 cm × 36 cm, 5.0 cm deep with 2.5 cm high cross pieces and a lid, made of polymethylmethacrylate (Perspex, Lucite) (see *Figure 11*). The tray is levelled on the bench using a spirit level and by placing pieces of card appropriately under the edges. The slides are placed on the cross pieces and a little water is put on the bottom of the tray. With the lid in place, this ensures that the solutions of antibody do not evaporate during an incubation.

Some protocols require incubations at higher temperatures and it useful to have an incubator available. To fit in an incubator, a smaller version of the staining tray may be needed. This can easily be devised from a sandwich box sold in most hardware stores. For incubation at a lower temperature, a tray can be placed in a cold room.

The equipment needed and the methods used to prepare blocks of tissue and to cut sections are well described in standard books of reference (e.g. see ref. 16).

135

13. ACKNOWLEDGEMENTS

We thank Mr C.D.Pallett for the stained slides used to take the photographs for *Figures 8* and *9* and Drs J.P.Sloane and J.H.Westwood for their helpful comments. This work was supported by a joint programme grant from the Cancer Research Campaign and the Medical Research Council.

14. REFERENCES

1. Sternberger,S.S. and De Lellis,R.A. (eds) (1982) *Diagnostic Immunohistochemistry.* Masson Publishing USA Inc., New York.
2. Bullock,G.R. and Petrusz.P. (eds) (1982 and 1983) *Techniques in Immunocytochemistry.* Academic Press, London, Vols 1 and 2.
3. Cuello,A.C. (ed.) (1982) *Immunohistochemistry.* IBRO Handbook Series, John Wiley and Sons, Chichester, Vol. 3.
4. Polak,J.M. and van Noorden,S. (eds) (1983) *Immunocytochemistry.* John Wright and Sons, Bristol.
5. Polak,J.M. and van Noorden,S. (1987) *An Introduction to Immunocytochemistry: Current Techniques and Problems.* R.M.S. Handbook 11, Oxford University Press, Oxford, 2nd edition.
6. Johnstone,A. and Thorpe,R. (1982) *Immunochemistry in Practice.* Blackwell Scientific, Oxford.
7. Galfre,G. and Milstein,C. (1981) In *Methods in Enzymology,* Langone,J.J. and Vunakis,H.V. (eds), Vol. 73, p. 3.
8. Dearnaley,D.P., Ormerod,M.G., Sloane,J.P., Lumley,H., Imrie,S.F., Jones,M., Coombes,R.C. and Neville,A.M. (1981) *Br. J. Cancer,* **44**, 85.
9. Chui,K.Y. (1987) *Medical Laboratory Sciences,* **44**, 3.
10. Holgate,C.S., Jackson,P., Cowen,P.N. and Bird,C.C. (1983) *J. Histochem. Cytochem.,* **31**, 938.
11. Watson,J.D., Tooze,J. and Kurtz,D.T. (1983) *Recombinant DNA—A Short Course.* Scientific American Books, New York.
12. Burns,J., Chan,V.T.W., Jonasson,J.A., Fleming,K.A., Taylor,S. and McGee,J.O'D. (1986) *J. Clin. Pathol.,* **38**, 1085.
13. Liesi,P., Julien,J.-P., Vilja,P., Grosveld,F. and Rechard,L. (1986) *J. Histochem. Cytochem.,* **34**, 923.
14. Varndell,I.M., Polak,J.M., Sikri,K.L., Minth,C.D., Bloom,S.R. and Dixon,J.E. (1984) *Histochemistry,* **81**, 597.
15. Campbell,D.A., du Bois,R.M., Butcher,R.G. and Poulter,L.W. (1986) *Clin. Exp. Immunol.,* **65**, 165.
16. Bancroft,J.D. and Stevens,A. (eds) (1982) *Theory and Practice of Histological Techniques.* Churchill Livingstone, Edinburgh, 2nd edition.

CHAPTER 5

Histochemistry and the light microscope

RICHARD W.HOROBIN

1. INTRODUCTION

Light microscopy assisted by histochemical staining can contribute to the solution of a remarkable range of biological problems. The needs of pathologists, haematologists and other clinical investigators have perhaps had the most direct influence on the development of histochemical methods. However, biologists, from cytogeneticists to comparative zoologists also routinely use these procedures. The evolution of methodology has always been driven by the needs of such investigators, as with the recent development of microscope-based *in situ* nucleic acid hybridization methods. However, in addition to this biology-oriented core there are many further applications of histochemistry—food technologists, metallurgists, polymer scientists and many others use microscopic staining procedures.

Consequently, the experimental protocols cited in this chapter are illustrative only, merely sketching the range of available methods. This account offers useful advice and a clarifying overview for investigators from all backgrounds, but for practical details readers are directed to Chapters 4 and 6 and to the sources cited in the references.

1.1 Objectives of histochemical staining

1.1.1 *What is 'staining'?*

Typically, histochemical staining is carried out to facilitate the microscopic study of the content, geometry and function of biological materials. The process of staining may be regarded as the visual labelling of entities of interest. Labels for light microscopy are usually coloured—they may absorb or emit light. Very occasionally they are of some characteristic shape.

1.1.2 *Information obtained by staining*

Microscope-based histochemical staining can tell us what something is at the same time that we see where it is. Sometimes histochemical staining also provides information concerning how much of an entity is present, or perhaps how many of something.

Answers to the question 'what is it?' are not necessarily chemical, in spite of the generic name for the process. Histochemical staining is, of course, able to demonstrate chemical objects ranging from molecular fragments through particular molecules to classes of compounds. However, histochemistry is often also used to identify biological objects. Such entities are defined in terms of their biological significance, not their chemical composition. Thus, the presence of lysosomes in a cell may be important, whilst the particular enzymic content of these organelles is of no concern. Indeed, the

Table 1. Examples of histochemically demonstrable entities.

Type of entity	Specific examples
Chemical object	
Molecular fragment	Aldehyde group, disulphide bridge.
Individual compound	Glycogen, succinate dehydrogenase.
Class of compounds	Glycosaminoglycans (GAGs), neutral lipids.
Biological object	
Molecular	Antigen, hormone receptor.
Subcellular	Centriole, lysosome, mitochondrion.
Cellular	Helper T lymphocyte, mast cell.
Organism	Gram positive and negative bacteria.
Biological process	
Subcellular	Ciliary action, lysosomal digestion.
Cellular	Neuronal activity, phagocytosis.
Whole organism	Mobility, viability.
Morphology	
Connectivity	Neuronal wiring diagram.
Distribution	Lymphocyte types in germinal centres.
Surface form	Epithelial intracellular spaces.

entity to be demonstrated may not be an object at all and histochemical procedures exist to detect various biological processes, for example, the direct observation of neuronal activity and testing of viability. Finally, histochemical staining can be used to study morphology, another non-chemical property. Examples of many types of histochemical information gathering are given in *Table 1*.

1.2 Underlying physical chemistry of typical staining methods

For histochemical methods to be used critically, some understanding of their physical chemistry is necessary. Since practical accounts do not usually provide this, an introductory outline is provided here. More extended accounts are available from several monographs (1,2).

1.2.1 *Superficial diversity, fundamental unity*

The evolution of histochemical staining can be conceived as a series of technical enthusiasms, lightly integrated by physical chemical understanding. In the nineteenth century there were synthetic dyes and silver staining. Later came the wave of chromogenic histochemical methods, overlapping which was the development of enzyme histochemistry. Then came labelled antibody methods and currently there are the techniques of *in situ* hybridization of nucleic acids and the use of fluorochromes to study the properties of living cells. Each wave has added techniques to the general histochemical armamentorium, for examples think of the general oversight procedure haematoxylin and eosin, Cajal's gold chloride stain for astrocytes, the Feulgen nucleal procedure for DNA, or azo dye methods for hydrolytic enzymes.

However, many biologists use only one, or a few, histochemical procedures. Hence little appreciation exists of the mechanistic similarities of different types of histochemical technique, or indeed that certain problems and pitfalls are common to all such methods.

The protagonists of each new wave have re-invented many wheels.

In fact, whether considering smears, sections or cell suspensions, histochemistry is always concerned with two-phase systems. Silver impregnation or mordant dyeing are not merely examples of redox reactions and co-ordination chemistry. The periodic acid – Schiff procedure is not merely organic chemistry. Enzyme histochemistry is not just biochemistry. Immunocytochemistry is more than immunology. Like dyeing, such procedures depend on selective uptake of reagents from the solution phase into the solid specimen and selective losses of products and/or reagents from the specimen into solution. In routine practice these uptakes and losses depend on both equilibrium and rate factors.

1.2.2 *Equilibrium control and affinity effects*

When the distribution of staining reagents between solution and specimen favours the latter, the system is said to display a high stain – specimen affinity. The distribution equilibrium is influenced by all parts of the staining system: obviously reagent – specimen interactions are significant, but often so are those between reagent – reagent and solvent – solvent. Examples of common contributions to affinity will now be briefly discussed.

(i) *Reagent – specimen attractive forces.* The most widespread are short-range inter-molecular attractions such as dipole – dipole and dispersion forces; the so-called van der Waals attractions. These occur between all molecules, but are strongest when the interacting species are polarizable and have large dipoles. Many histochemical stains with large aromatic systems fit this specification, as do such components of biopolymers as aromatic amino acid residues and the heterocyclic bases of nucleic acids. Hence, variations in the sizes of the aromatic systems of staining reagents can markedly influence staining.

Coulombic forces are also important as many staining reagents and biopolymers are ionic. Since these electrical forces can be modulated by changes in pH or the amount of neutral electrolyte present, variations in such solution properties are used to control staining with ionic reagents.

Some staining systems involve reagent – specimen hydrogen bonding. In aqueous solutions hydrogen bonding with water will compete with this, so although these bonds are strong, their general contributions to affinity is uncertain.

However, the role of the strongest reagent – specimen attractions, namely covalent bonds, is clear. Some staining methods involve making and breaking covalent bonds to transform specimen sites into coloured derivatives; other techniques attach a coloured label onto a site. The character of the covalencies vary. Those involving metal ions are extremely polar and staining involving such bonds is sometimes termed 'mordanting' in the histochemical literature.

(ii) *Stain – stain attractions.* Staining affinity can arise from these, even if no stain – specimen interactions occur. Ionic crystals of lead sulphide, deposited at enzymic sites, are insoluble in water and alcohol due to strong coulombic interactions between lead cations and sulphide anions. The biological specimen merely provides a matrix to contain the crystals once formed. The metallic silver precipitated during metal impregnation provides another example. Metachromatic dyeing and azo dye enzyme histochemical

Table 2. Contributions to stain−specimen affinities.

Contribution	Examples of staining methods
Reagent−specimen attractions	
van der Waals attractions	Visualization of dehydrogenases using bis tetrazolium salts.
Coulombic forces	Basic dyeing of anionic GAGs.
Hydrogen bonding	Bests carmine stain of glycogen.
Covalent bonding	Chelation of calcium ions with Alizarin Red S[a].
Reagent−reagent attractions	
Any of the above	Precipitation of Thionin picrate in bone canaliculi.
Solvent−solvent attractions	
Hydrophobic bonding	Staining of xylem with hydrophobic basic dyes.

[a]For comments concerning dyestuff nomenclature, see Section 5.4.

procedures are similar, although they involve stain−tissue as well as stain−stain attractions.

(iii) *Entropy effects.* Another contribution to affinity not involving stain−specimen interactions is hydrophobic bonding. This entropy-driven process occurs only in aqueous solutions. It involves coupling the uptake of hydrophobic reagents onto hydrophobic biological structures with the disaggregation of hydrogen-bonded clusters of water molecules. As histochemical staining is a two-phase process, the distribution of staining reagents between solution and specimen is itself entropy driven. Examples of stains whose uptake is due to these various factors are given in *Table 2*.

This account should not be taken to imply that selective staining is always due to affinity effects. In practice the number of binding sites is as important as the reagent-site affinity. Moreover, staining is often carried out for short time periods, both for convenience and because biological specimens are often labile. Staining equilibrium is not always achieved and selective staining is often due to rate effects.

1.2.3 *Rate control of selectivity*

Rate effects can be exploited to generate selective stains, but they also occur unexpectedly, giving rise to staining artefacts.

(i) *Rates of diffusion.* Diffusion control may involve either the rates of entry of reagents into various compartments of an unstained specimen ('progressive staining'), or the rates of diffusion out of the various compartments of a stained specimen ('differentiation' or 'regressive staining').

Differential permeabilities of the various compartments of biological specimens are due to a wide variety of factors. Structures which are dense, or rich in highly crosslinked or poorly hydrated biopolymers, will usually be poorly permeable; and vice versa. Specimen preparation also influences diffusion rates. A protein-crosslinking fixative such as glutaraldehyde tends to reduce staining rates. Conversely, several specimen preparation steps can increase the state of dispersion of biological specimens. Freezing/thawing, coagulative fixatives, dehydrating agents such as alcohol and embedding in paraffin wax can all fracture cells and tissues. This increases surface area and hence permeability. The gross geometry of specimens also influences diffusion rates. Monolayers of flattened cells prepared for cytological and haematological diagnosis

Table 3. Influence of specimen permeability on rate controlled staining.

	Permeability factors	Staining examples
Poorly permeable structures		
A bands and Z lines of striated muscle	Proteins are close packed, so of high density.	Iron haematoxylin is retained by dense structures during differentiation.
Intact cell membranes of live or well fixed cells	Coherent, lipid-rich layers.	Exclusion of antibody from cells gives negative artefacts in immunostaining.
Very permeable structures		
GAGs, as in cartilage matrix and mucus	Hydration of ions and hydroxyl groups causes swelling.	Even large basic dyes, e.g. Alcian Blue, can enter hydrated GAGs.
Cytological smears	Cells form flattened monolayers, shattered by alcohol fixation.	Even large acid dyes of Papanicolaou stain enter cells rapidly.

stain rapidly as do thin sections cut from solid tissue blocks on the cryostat, or cut from blocks embedded in wax and stained after removal of the wax. On the other hand, plastic sections used with the embedding medium still present stain slowly, as do thick paraffin or cryostat sections cut for neuro-anatomical work. Examples of staining methods influenced by such factors may be seen in *Table 3*.

The molecular structures of reagents also influence diffusion rates. The simplest influence is size: the larger the stain, the slower it tends to diffuse. Macromolecular reagents such as labelled antibodies (cf. Chapter 4) are largely excluded from intact cells and plastic sections. Even the much smaller dyestuffs show dramatically slowed diffusion rates if their molecular or ionic weights are more than a few hundred daltons. Another significant factor is reagent—tissue affinity, since it is only low affinity reagents which diffuse freely. A synergistic enhancement of selectivity is possible by using two stains of differing sizes, applied to specimens containing structures of differing permeability. The classic staining systems exploiting this effect are the Trichromes which selectively stain highly permeable collagen fibres with large acid dyes, whilst staining relatively impermeable cell cytoplasms with small acid dyes.

All diffusion controlled methods are sensitive to the mode of fixation, staining time, temperature and other factors influencing diffusion.

(ii) *Rates of reaction.* In some metal impregnation methods the application of a silver salt to the specimen is followed by application of a reducing agent. The rate of reduction must not be too fast, or all structures will be covered with microcrystals of silver; nor must it be too slow, or no staining occurs. Another example is found in the enzyme histochemical demonstration of acid and alkaline phosphatases. Though these enzymes have different pH optima, they will both hydrolyse the same substrates. Prolonged incubation in such a substrate gives some staining at the sites of both enzymes, regardless of pH. As a final example consider the periodic acid—Schiff procedure. This involves an initial oxidation of tissue polysaccharides to yield aldehyde groups which are visualized using Schiff's reagent. Selectivity depends on the fast reaction of periodate with saccharides and its slower oxidation of other biopolymers.

1.2.4 *Catalysis and other less common influences on selectivity*

(i) *Catalytic staining.* The selectivity of some stains depends on certain specimen sites catalysing visualizing reactions. Such catalysts may be native to the specimen or may be chemically induced. Enzymes are obviously examples of the former type, and may be demonstrated using their specific catalytic abilities to convert substrates into coloured derivatives. Occasionally this can be achieved directly, by suitable choice of substrate. More commonly, enzyme demonstration requires incubation of the specimen with a suitable substrate, and conversion of a colourless intermediate into a coloured final reaction product using a visualizing reaction. Examples include the demonstration of dehydrogenases of the Krebs cycle. Tissue specimens are incubated with a suitable substrate and protons and electrons are formed at the sites of the dehydrogenase. These are trapped and visualized by their reaction with a tetrazolium salt, forming a coloured, insoluble formazan pigment. There are many variations of this strategy, allowing demonstration of a wide range of enzymes.

Sometimes catalytic sites are generated by chemical manipulation of the specimen. An example is a procedure used to demonstrate low intracellular concentrations of metal ions. These are first converted into their sulphides which, even if coloured, are present at a concentration too low to be seen. However, many metallic sulphide deposits catalyse the reduction of silver (I) to metallic silver, using a physical developer. Incubation with silver salt plus developer can be continued until sufficient metallic silver is deposited to allow visualization.

(ii) *Negative staining.* This is used to visualize surface profiles of specimens by depositing stain against them. Like certain other methods previously mentioned, no stain−specimen attractive forces occur. Negative staining enables the easy observation of intra- and intercellular channels and ducts, and the outlining of such entities as the cells of fungi or spirochaetes.

(iii) *Vital staining and fluorescent probes.* When live cells and tissues are exposed to dyes and fluorochromes, staining reflects in large part physiological and biochemical processes. A classic application was mapping the distribution of macrophages within the mammalian body, using dyes such as Evan's Blue as indicators of phagocytosis, since these dyes are retained in secondary lysosomes. Currently, fluorochromes are widely used for such purposes since the increases in sensitivity given by fluorescence allows lower reagent concentrations to be used, reducing toxicity problems. This approach is nowadays often termed 'use of fluorescent probes' and is applied to a wide range of problems. For instance, viability of cultured cells may be assessed, since intact cells exclude hydrophilic dyes but permit entry to hydrophobic species [see Section 3.2.2(i), *Table 18*]. Another complex function which can be probed is neuronal activity in the central nervous system, using potential-sensitive fluorochromes.

2. PREPARATION AND STORAGE OF SPECIMENS

2.1 Necessary characteristics of the specimen

For viewing in the light microscope specimens must be neither too thick nor too thin. If too thick, then an excess of detail will confuse analysis. If too thin, specimens may be insufficiently contrasty and the interrelation of components may be uncertain. Suitable layers for much biological work are a few micrometers thick—less than the diameter

Table 4. Preparing thin layers from biological specimens.

Biological material	Nature of thin layer	Procedure used
Fluids		
Cell rich suspensions (e.g. blood)	Cell monolayer on microscope slides.	Smear, or spin, across a slide.
Cell poor suspensions (e.g. urine and cerebrospinal fluid, CSF)	Cell monolayer on filtration layer.	Pass fluid through a filter.
Soft solids		
Bone marrow	Cell monolayer	Disperse cells in buffer and smear onto slide.
Spinal cord	Cell monolayer	Wipe across a slide.
Coherent solids		
Cervical epithelium	Cell monolayer	Wipe off superficial cells with swab, smear on slide.
Solid organ (e.g. liver)	Thin section	Freeze, section in cryostat.
Solid organ (e.g. kidney)	Thin section, plastic still present.	Embed in plastic, section on microtome.
Hard solids		
Mineralized bone	Thick section	Attach to slide, grind down.
Woody plant stem	Thin section or cell suspension	Soften, cut on microtome, or macerate and suspend.

of many mammalian cells, but more than that of many organelles. To obtain such thin layers requires radical transformation of the native forms of most specimens. For specialist purposes, thicker layers can be viewed and data overload avoided by selective staining, or by an optical technique such as stereoscopy.

Some biological specimens such as cellulose cell walls and mammalian hairs are remarkably robust. However, most living cells are mechanically fragile and obtaining thin layers from them directly would be totally disruptive of morphology. Hence, most specimens are observed after some stabilization procedure has been applied, to conserve both structure and chemical content. Such stabilization must not only protect against mechanical trauma but also prevent autolysis, attack by micro-organisms, and extraction and osmotic damage due to the solvents used in specimen processing. Such stabilizing procedures often kill cells or are applied to dead specimens. Occasionally cells are viewed whilst still living, in which case particular care must be taken not to damage them.

Of course the thin layers required for microscopy are fragile, needing careful manipulation before and after staining. The stained layers must also be treated in a way that limits fading and minimizes light scattering in the microscope.

2.2 Preparation and storage methods

2.2.1 *Preparing thin layers*

Procedures for preparing thin layers include smearing and spinning fluids, dabbing and smearing soft solids, cutting slices ('sectioning') of coherent solids on a microtome or cryostat, and grinding hard solids. Sectioning is occasionally carried out directly on the native specimen (more often after hardening the specimen by freezing) and most commonly by cutting a specimen embedded within a solid matrix. One common embedding procedure involves infiltrating the specimen with molten paraffin wax and allowing this to cool. In the more recently developed plastic embedding techniques, specimens are infiltrated with monomers which are then polymerized, yielding solid blocks.

The methods for obtaining samples from which such thin layers are made call for comment. Fluid specimens may be sampled with a pipette or syringe and if the cell content of the fluid is low, a concentration step utilizing centrifugation, filtration, or settling may be inserted here. A swab can be used to remove fragments, such as cells, from a wet surface and cells from a dry surface can be sampled by applying and removing sticky tape. Portions of coherent solids may be detached with a scalpel or other blade, or a thin cylindrical sample may be obtained using a needle. Sectioning hard solids is sometimes achieved by softening the specimen before microtomy.

Some general strategies for the preparation of thin layers of various biological specimens are given in *Table 4*, with some specific examples in *Tables 5* and *6*.

2.2.2 *Stabilizing specimens, both stained and unstained*

(i) *Fixation and its alternatives.* Traditionally, biological specimens are stabilized by fixation. This is the conversion of soluble native constituents of cells and tissues into insoluble derivatives. This can be achieved by a remarkable range of fixative agents, varying from heat to organic solvents or to reactive organic and inorganic compounds.

Table 5. Preparing thin layers of highly dispersed cells, namely those in urine samples using a filter-imprint procedure.

1.	Place a Millipore™ cellulose ester filter membrane with an 8 μm pore size into a suitable apparatus and moisten the membrane with physiological saline.
2.	Filter the urine[a] sample under negative pressure; apply 15−20 mm Hg using a water pump.
3.	Remove the membrane from the apparatus and place the membrane, cell side down, onto an albuminized slide. Apply an even pressure to the membrane, e.g. with an ink roller[b].
4.	Remove the membrane, and place the slide now carrying the cell imprint into a 95% ethanol fixative bath.

[a]If the urine is not filtered immediately after voiding a preservative agent should be added, e.g. a bactericide such as merthiolate or a fixative such as dilute formalin.
[b]A variant procedure is to transfer cells as above, but using a chilled slide just after its removal from a freezer.

Table 6. Preparing thin layers of a hard specimen, namely wood, using maceration.

1.	Cut woody specimens into thin shavings using a scalpel or razor blade.
2.	Macerate the shavings by soaking in a mixture containing equal volumes of 1 M nitric and chromic acids, until the ends of the fragments begin to fray[a].
3.	Wash away the macerating solution using several changes of water.
4.	Complete the separation of fibres by gentle teasing with a needle[b].

[a]Depending on the type of wood and size of fragment this may take up to 24 h, though the process may be accelerated by cautious warming of the solution.
[b]The resulting fibres may be stained as a suspension.

However, all fixative action involves one or more of four key processes.

(a)   The denaturation of proteins—if this is not achieved, most cells will disperse during specimen preparation or staining.

(b)   The crosslinking of constituents, such as proteins by glutaraldehyde, or unsaturated lipids by osmium tetroxide.

(c)   The formation of insoluble complexes, such as those between intracellular copper cations and sulphide anions.

(d)   Substances which are unchanged by fixation may be trapped within the molecular cages formed from proteins or lipids fixed by one of the other mechanisms.

Stabilizing the content of a specimen is thus achieved by selecting a fixative mixture which insolubilizes the components of interest. Stabilization of morphology is, to a first approximation, dependent upon stabilizing proteins. However, fixation does have marked limitations. Obviously there may be failures to retain substances in the specimen. For instance, retaining saturated lipids, or low molecular weight water-soluble substances such as amino acids, is difficult. Other problems arise from the very success of fixation. A clear example occurs when enzyme activity is to be demonstrated. Many proteins are not well retained in the cell unless they are denatured or crosslinked—but such modifications often reduce or destroy enzymic activity.

Two quite different strategies have been developed in response to these contradictory outcomes of fixation. The first is to reduce the unfortunate side effects, whilst keeping at least some of the benefits. This has resulted in compromise manoeuvres such as using fixatives for short periods of time or at low temperatures, deferring fixation until after

Table 7. Methods of stabilizing the contents of biological specimens.

Class of substance	*Method of stabilization*
GAGs	Precipitate with cetyl pyridinium salts.
Lipids	1. Section on cryostat, avoid organic solvents.
	2. Crosslink with osmium tetroxide or potassium dichromate fixative agents.
Nucleic acids (e.g. DNA, rRNA)	Trap by fixing associated protein; use weakly acid fixatives (e.g. acetic acid−methanol) to avoid solubilizing DNA.
Non-ionic polysaccharides (e.g. glycogen)	Trap by fixing surrounding proteins; use non-aqueous solutions (e.g. formal alcohol) to avoid extraction of polysaccharide.
Proteins	Use denaturing and/or crosslinking fixatives (e.g. alcohol or glutaraldehyde).
Protein antigens and catalytically active enzymes	1. Use unfixed cryostat sections.
	2. Fix in formaldehyde, then unmask with a protease (for antigens).
	3. Section on cryostat, use colloid protecting agent on semipermeable membrane (for enzymes).

staining, and persistent trials of new fixatives such a vapour phase fixation with benzoquinone. The second strategy is to avoid fixation altogether. Hence the use of unfixed cryostat sections for immunostaining or enzyme histochemistry. However, loss of content is usually substantial from such material and so even if the entity of concern is not lost, the morphology is degraded. Loss of macromolecules can be reduced or eliminated by coating cryostat sections with semipermeable membranes, or by adding colloid protecting agents to the staining solutions. However, stabilizing content and morphology simultaneously does usually involve compromise. To establish the best technical procedure for a particular specimen/question combination usually needs empirical study. However, that is not to say that certain general rules cannot be set out (see *Table 7*).

(ii) *Mounting stained preparations.* The stabilization of specimens after staining is another matter. It is usually achieved by mounting, that is by sealing a specimen into a medium of high refractive index. Such isolation from the environment prevents microbiological attack, and limits fading. Moreover, the high refractive index of the medium reduces light scattering at the specimen−air interface, facilitating microscopic observation. Mounting media are often applied as solutions, which evaporate leaving the specimen within a solid film, for example polystyrene applied in xylene and polyvinylalcohol applied in water. Others, such as gelatine and glycerol, are applied as non-drying solutions and immersion oil is sometimes used as a non-drying liquid mountant. Such a range of media reflects both the compromise of permanence with convenience and also the need to avoid mountants which will react with or extract the stain. Thus, specimens stained with hydrophilic reagents such as sulphonated acid dyes are mounted in hydrophobic media such as polystyrene, since aqueous gelatine would have an affinity for the dyes and hence result in their extraction.

2.2.3 *The need for problem-oriented specimen preparation*

Presented with biological material to be examined histochemically, the immediate problem is to select an appropriate specimen preparation method. Three issues must

Table 8. Some illustrations of the matching of technical methods to specimens and questions.

Specimen	Information sought	Possible technical procedures
Bone marrow	1. Identities and relative numbers of different cell types	Remove sample with syringe; smear; fix in alcohol.
	2. Identity and number of monocytes; spatial distribution in marrow	Take sample preserving spatial integrity; section in cryostat; stain enzymatic marker (e.g. esterase); fix in formaldehyde.
Small gut	Intracellular lysosomal localization in sub-mucosal macrophages	Remove tissue sample; fix briefly; embed in hydrophilic resin; demonstrate marker enzyme.

be considered in parallel:

(i)   the nature of the specimen;
(ii)   the available technical methods;
(iii) the type of information sought.

The need for such a three-pronged approach can be illustrated by stating a few of the typical requirements and dilemmas of specimen preparation. If the need for morphological excellence is paramount, then fixation and the use of sections cut from embedded material will be essential. On the other hand, if demonstration of enzymic activity is intended, then fixation should be avoided, as should use of embedding regimes using hot and hydrophobic media. However, enzymic activity and good localization may both be objectives. In this case, one possibility is brief fixation, followed by embedding in a hydrophilic resin such as glycol methacrylate, with resin curing being carried out at low temperature. However, this presupposes possession of a suitable microtome for cutting plastic sections and being able to afford the relatively expensive plastic media. Moreover, the nature of the original sample may not be controllable. A central histopathology laboratory, for instance, may receive biopsies from many peripheral hospitals. Such materials will usually be fixed in neutral formalin. Obtaining samples in any other form may be difficult or impossible; theatre staff have other clinical priorities and a patient's abdomen will hopefully be closed as soon as possible. However some generalizations concerning specimen preparation can be made (see *Table 8*).

3. WHAT CAN BE DEMONSTRATED? POSSIBLE EXAMPLES

3.1 Demonstrating chemical entities

3.1.1 Chemical fragments

A wide variety of chemical fragments can be demonstrated histochemically. These include organic radicles such as amino groups, carbon−carbon double bonds, and thiol groups, which can occur as substituents on biopolymers or as constituents of smaller molecules. Inorganic ions can also be detected, both anions such as phosphate and sulphate, and cations such as calcium or ferric iron. These may be present as substantially inorganic deposits (e.g. bone) or they may be found bound to organic ligands.

Sometimes the chemical fragments are themselves of biological interest, for example the calcium deposited during bone formation or as a consequence of pathological

Table 9. Method for demonstrating aldehyde groups.

1.	De-wax paraffin cut from specimens fixed in e.g. formalin, and bring them to water. Place cryostat or plastic sections in water.
2.	Place in Schiff's reagent, prepared as in *Table 10*, for 15 min[a].
3.	Wash in running tapwater for 5–10 min[a].
4.	Rinse in 100% ethanol, wash in xylene and mount in synthetic resin.

Results. Aldehyde-rich sites stain magenta.

Control. Stain a specimen known to be aldehyde-rich, as a positive control. Sections cut from glycogen-rich liver are well suited. Glycogen yields aldehydes by periodate oxidation; nuclear DNA yields aldehydes by acid hydrolysis.

Trouble shooting. If the control fails to stain, replace the Schiff's reagent. When experimental sections fail to stain, try increasing times of steps 2 and 3 before concluding aldehydes absent.

[a]Optimum times will vary with source of reagent, and modes of fixation and embedding.

Table 10. Preparation of Schiff's reagent.

1.	Dissolve 1 g of Basic Fuchsin (or Pararosanilin or New Fuchsin) in 200 ml of boiling water.
2.	Cool to 50°C, add and dissolve 2 g of potassium or sodium metabisulphate.
3.	Cool to room temperature, add 2 ml of concentrated HCl and suspend 2 g of activated charcoal in the resulting solution.
4.	Leave in the dark overnight, at room temperature.
5.	Filter off the charcoal, store in a refrigerator in a dark container.

Trouble shooting. If the solution is still pink after filtration, replace the metabisulphate. If the solution is orange and gives orange background staining, replace the Basic Fuchsin.

Table 11. Selective methods for generating aldehydes from biopolymers.

Hydrolytic depurination of DNA[a]
1.	De-wax paraffin sections and bring to water, hydrate cryostat and plastic sections[b].
2.	Place in aqueous 1 M HCl, preheated to 60°C, maintain at that temperature for 10–15 min[c,d].

Oxidation of 1,2-diol groups of polysaccharides[e]
1.	De-wax paraffin sections and bring to water, hydrate cryostat and plastic sections.
2.	Treat with 1% w/v periodic acid for 5 min (see footnote a, *Table 9*).
3.	Wash well with several changes of water.

[a]This procedure, added to the aldehyde visualizing steps described in *Table 9*, is termed the Feulgen nucleal method for demonstrating DNA.
[b]Most fixatives are suitable, unless acidic (e.g. Bouin's), since that pre-hydrolyses the DNA (see footnote d below), or if glutaraldehyde, which covers the tissue in aldehyde groups.
[c]As an alternative use 5 M HCl at room temperature.
[d]Short hydrolysis times may fail to generate enough aldehyde, long times may totally remove DNA from the specimen. Optimum times are fixative dependent, so if nuclei stain palely try varying the hydrolysis time.
[e]This procedure, added to the aldehyde visualizing steps described in *Table 9*, is termed the periodic acid–Schiff (PAS) method.

calcification of damaged tissues. On other occasions chemical fragments are used as histochemical markers for other chemical entities. Thus, tissue aldehyde groups are occasionally demonstrated, although they are not usually present in high concentrations. Hence, aldehydes may be generated from tissue components such as DNA and polysaccharides, and subsequent visualization of aldehydes serves to detect these other components.

(i) *Demonstration of aldehyde groups.* The method given achieves this using the traditional Schiff's reagent. This reagent reacts with aldehydes via its reactive primary arylamino substituents; the final magenta colour of aldehyde-rich sites is due to the formation of the triphenylmethane chromophore from the initial reaction products.

The procedure of demonstrating aldehydes is given in *Table 9*. The preparation of Schiff's reagent from the widely available dye Basic Fuchsin is described in *Table 10*. Although Schiff's reagent is commercially available many workers still prefer to prepare their own. Procedures for the selective generation of the aldehyde groups from various biopolymers are described in *Table 11*. Taken together, *Tables 9* and *11* provide a description of the standard histochemical methods for the detection of DNA and of polysaccharides as a class. The footnotes show that optimum staining may then require adjustment of staining conditions, depending on the precise mode of specimen preparation.

3.1.2 *Specific compounds*

A large number of compounds can be specifically stained, and hence, polymeric materials such as the polysaccharides callose and glycogen can be identified, and structural proteins such as collagen and haemoglobin and a wide variety of enzymes can be demonstrated. Histochemical methods can also demonstrate low molecular weight compounds, from bile pigments to cholesterol and various vitamins. In addition to such methods, the histochemical use of *in situ* hybridization permits the histochemical demonstration of many nucleic acid fragments, from the DNA sequences of genes to cytoplasmic mRNA, using fluorescent or isotopically labelled nucleic acid probes. Moreover, many biopolymers and indeed smaller species are antigenic and can, in principle be demonstrated immunocytochemically. The reader should consult Chapters 3, 6 and 9 for further material on DNA probes, fluorescence and histochemistry of chromosomes.

The mechanisms of the non-immunological, non-hybridization staining procedures with which we are concerned here are diverse. Some compounds are visualized by using a marker chemical fragment or combination of fragments—the bile pigments provide examples of this. The marker property can alternatively be a biophysical feature of the molecule, for example, charge, hydrophobic – hydrophilic character, planarity and so on. More complex chemical properties may be exploited to provide identification as is the case with enzyme histochemistry, in which a protein's specific catalytic properties are used to generate a stain. Selectivity may also be due to properties of molecular aggregates rather than individual molecules—collagen for instance may be selectively dyed by exploiting the gross swelling of collagen fibres in acidic solution.

(i) *Demonstration of succinate hydrogenase.* In this enzyme histochemical method the specimen is incubated in a solution containing succinate, the enzyme's natural substrate, plus a tetrazolium salt redox indicator. In sites containing the enzyme, succinate is catalytically dehydrogenated to yield fumarate plus protons and electrons. This enzymatic process is visualized by reduction of the tetrazolium salt by the electrons and protons, forming a blue – purple, insoluble formazan pigment.

A procedure for demonstrating succinate dehydrogenase is described in *Table 12* and the preparation of the necessary reagents in *Table 13*.

Table 12. Method for demonstrating succinate dehydrogenase.

1.	Cut $5-7$ μm cryostat sections and mount on coverslip[a]. Do not fix[b].
2.	Incubate sections at 37°C for $30-60$ min in the incubation medium specified in *Table 13*.
3.	Transfer sections into aqueous formal saline (as specified in *Table 13*) for 15 min[b].
4.	Wash in distilled water.
5.	If desired, counterstain cell nuclei with 2% w/v aqueous Methyl Green for 10 min.
6.	Wash with distilled water.
7.	Dehydrate through alcohols to xylene and mount in a synthetic resin.

Results. Blue-purple formazan is deposited at the sites of succinate dehydrogenase. If Methyl Green counterstain is used, the nuclei are green.

Control. Check that the stain is enzymatically generated by ommitting succinate from the incubation medium or by heating a dry section to 80°C for 1 h prior to incubation.

Trouble shooting. Traces of pink staining may be removed by an acetone rinse applied after step 4. If excessive pink staining occurs, replace the tetrazolium salt.

[a]To minimize amounts of expensive reagents required.
[b]The enzyme is inhibited by formaldehyde, hence the use of a post-staining fixation step to preserve morphology.

Table 13. Reagents for succinate dehydrogenase demonstration.

Stock solutions

Substrate solution

1.	Dissolve 6.75 g of disodium succinate in 10 ml of distilled water.
2.	If necessary adjust pH to $7.0-7.1$ using 1 M HCl.

Tetrazolium solution

1.	Dissolve 10 mg of Nitro Blue Tetrazolium (or Tetranitro Blue Tetrazolium) in 2.5 ml of distilled water.
2.	Add 2.5 ml of 0.2 M Tris buffer, pH 7.4, 1 ml of 0.05 M $MgCl_2$ solution and 3 ml of distilled water.

Working solutions

Formal saline fixative

1.	Mix 15 ml of commercial 40% w/v aqueous formaldehyde solution with 85 ml of water.
2.	Add 0.9 g of NaCl and dissolve.

Incubation medium

1.	Mix 1 ml of substrate stock solution with 9 ml of tetrazolium solution.
2.	Warm to 37°C.

3.1.3 *Classes of compounds*

On occasion the chemical information sought concerns general classes of compounds not individual chemical species. Thus, the distribution of lipid within the proximal tubules of the kidney may be under investigation but the nature of the particular lipids present is of no concern. Even if the chemical classification required is more precise, it may still fall far short of identifying single components. Thus, if a patient is suspected of suffering from the storage disease metachromatic leukodystrophy, the histochemical investigation of the occurence of sulphatide deposits in macrophages is a useful diagnostic aid. However, no finer chemical distinctions are called for.

Once again, the mechanisms of such methods vary enormously and include all types discussed previously.

(i) *Demonstration of lipids.* This procedure exploits the favourable partitioning of

Table 14. Standard method for demonstrating neutral lipids.

1. Cut cryostat sections from an unfixed kidney and fix for 1 h in formal calcium[a].
2. Mount the sections on slides and allow to dry.
3. Delipidize a section for a negative control—see *Control* below.
4. Rinse a test section plus negative control in 70% v/v aqueous ethanol.
5. Stain the test and control sections for 15 min in a saturated solution of Sudan Black B in 70% v/v aqueous ethanol. Filter prior to staining.
6. Differentiate the sections in 70% v/v aqueous ethanol until the delipidized control appears essentially colourless.
7. Wash in water and mount in glycerine jelly[b].

Results. Triglycerides and unsaturated cholesteryl esters stain blue—black; some phospholipids appear grey.

Control. Extract a section with 2:1 v/v chloroform:methanol for 2 h at room temperature to remove all free lipids, after which no staining should occur. Some lipoproteins may be delipidized by adding 1% v/v concentrated HCl to the chloroform:methanol.

Trouble shooting. If nuclei- or GAG-rich sites stain blue, replace the Sudan Black.

[a]Prepare the formal calcium by diluting 10 ml of commercial 40% w/v aqueous formaldehyde solution with 100 ml of water. Dissolve 1 g of $CaCl_2$ in the solution.
[b]Resinous media extract non-ionic dyes.

Table 15. Bromine—Sudan Black method for total lipids.

Take mounted sections from step 2 of *Table 14* and continue as follows:
1. Immerse the sections in 2.5% v/v aqueous bromine solution for 30 min at room temperature. IMPORTANT: use a fume hood.
2. Wash in water and then in 0.5% w/v aqueous sodium metabisulphite solution for 1 min.
3. Wash well in water.

Continue with step 3 in *Table 14*.

Table 16. Bromine—acetone—Sudan Black method for phospholipids.

Take brominated sections from step 3 in *Table 15* and continue as follows:
1. Dry the sections well[a].
2. Extract with anhydrous[a] acetone for 20 min at 4°C.
3. Remove from the solvent and let acetone evaporate.

Continue with step 3 in *Table 14.*

Results. Phospholipids stain blue—black.

Control. Remove total lipids (see *Control* in *Table 14*) after which no staining should occur.

[a]No water must be present, or phospho- as well as neutral lipids will be extracted.

hydrophobic non-ionic dyes between polar solvents and non-polar lipids. The enhancement of staining following bromination is due to conversion of unsaturated carbon—carbon double bonds to saturated bromo derivatives. These latter are less polar and typically of lower melting point than the unsaturated precursors—factors which increase dye—lipid affinity and staining rate, respectively.

The standard Sudan Black procedure for staining neutral lipids is given in *Table 14*, the bromine—Sudan Black method for total lipids in *Table 15*, and the bromine—acetone—Sudan Black procedure for phospholipids in *Table 16.*

Table 17. Demonstrating muscle striations.

---

1.   Cut longitudinally oriented 6 μm sections from a skeletal muscle embedded in paraffin wax. The tissue can have been fixed in any fixative.
2.   De-wax in xylene and hydrate through graded alcohols to water.
3.   Treat with ('mordant') 5% w/v aqueous ferric ammonium sulphate solution (i.e.iron alum) for 1 h[a].
4.   Rinse the sections in distilled water.
5.   Stain in haematoxylin solution[b] for 1 h[a].
6.   Wash in running tap water.
7.   To de-stain the section in a controlled manner, alternate a rinse in differentiating solution[c] with a rinse in tap water. Inspect the outcome microscopically after each cycle until banding of muscle is seen.
8.   Wash in running tap water for 10 min.
9.   Dehydrate through graded alcohols to xylene and mount in a synthetic resin.

Results. After staining in haematoxylin, all structures are black or grey. The diffferentiation in step 7 leaves stain in the least permeable structures; mitochondria, muscle striations, nuclear chromatin and chromosomes, myelin, red blood cells and keratin. This list is in increasing resistance to de-staining.

Trouble shooting. If staining is too pale after step 5, lengthen times in iron alum and haematoxylin solutions. If achieving desired degree of de-staining is hard to control, try the alternative differentiators described in footnote[c]. If staining fades on standing, increase post-staining washing time.

---

[a]Optimum times in iron alum and haematoxylin solutions vary with the fixative used. After use of formaldehyde solutions (also formal sublimate, and Susa and Bouin's and Carnoy's fluids) 1 h in each is usually satisfactory. Dichromate fixatives (e.g. Helly's and Zenker's often need 3 h and osmium tetroxide fixatives longer still, e.g. Flemming's solution requires up to 24 h.
[b]Prepare the staining solution by dissolving 0.5 g of haematoxylin in 10 ml of ethanol, adding 90 ml of distilled water and leaving the solution to air oxidize at room temperature for 4 weeks.
[c]The standard differentiation is a 5% w/v aqueous ferric ammonium sulphate solution. Alternative differentiators are prepared by diluting this solution with an equal volume of water, or by saturating ethanol with picric acid—both these solutions give greater control of the process.

3.2 **Demonstrating biological entities**

3.2.1 *Biological objects*

There is no limit to the range of histochemically demonstrable objects, nor indeed to the size range of these objects. Thus, biological objects may be in the molecular range; important examples are antigens and receptors. Larger objects include a variety of sub-cellular structures and organelles. Of course staining is also used to detect particular cell types and particular organisms, currently often using immunostaining. However, demonstrating biological objects is a traditional strategy, as shown by such terms as 'eosinophil' or 'Gram positive bacteria'.

Again, mechanisms of these methods are too varied for useful comment, other than pointing out that methods can use chemical markers of all the types discussed and illustrated above, or, as with the example given below, the selectivity may be based on physical differences.

(i) *Demonstration of muscle striations.* The traditional method for distinguishing A and I bands of muscle involves the *in situ* generation of an iron haematein co-ordination complex (or complexes?) in most tissue components. Selectivity is largely achieved by controlled destaining (differentiation).

The Heidenhain iron haematoxylin procedure is described in *Table 17.*

Table 18. Method for viability testing using fluorescein diacetate.

1.	If cells are monolayer cultures, detach by replacing the growth medium with a solution of trypsin−EDTA[a] in phosphate-buffered saline (PBS)[b]. Then incubate at room temperature for the minimum time needed to remove cells from slides, usually 5−15 min.
2.	Centrifuge the cells and resuspend in sufficient PBS to give an approximate concentration of 10^{6b} cells per ml.
3.	Mix a small volume of the cell suspension with 2.5 vol of the FDA working solution[c]—this should give a final concentration of ∼0.1 μg per ml.
4.	Mount this fluid on a slide under a coverslip and seal the edges with silicon grease.
5.	Examine under a fluorescence microscope or under dark-field conditions using a tungsten light and an interference filter.

Results. Viable cells immediately display a bright green fluorescence the intensity of which increases with time. Injured or dead cells fail to stain.

Control. Since some cell populations contain autofluorescent materials, examine a preparation made without FDA.

Trouble shooting. If no cells fluoresce, make another preparation taking great care to avoid mechanical abuse.

[a]Prepare trypsin−EDTA from 0.05% w/v trypsin and 0.003% w/v EDTA dissolved in PBS.
[b]Prepare PBS by dissolving 0.2 g KCl, 0.2 g of KH_2PO_4, 8 g of NaCl and 1.15 g of Na_2HPO_4 in water. Make up to 1 litre.
[c]Make a 0.1% w/v stock solution of FDA in acetone. Dilute 1 vol of this stock with 4000 vol of PBS to prepare the working solution. This must not be milky or contain precipitate.

3.2.2 *Biological processes*

This is concerned with the interaction of histochemical stains with living cells or organisms, so directly observing a variety of processes. Detection of chemical and biological entities, discussed previously, is also used to elucidate biological processes, albeit in a less direct way.

A wide variety of processes, both in terms of physiology and scale, may be visualized. Thus, ciliary and flagellar motions of protozoa are detected by adding coloured colloids to the liquid film surrounding the organisms, the changing pH of the lysosomes of cultured macrophages following phagocytic activity can be monitored following uptake of dyes or fluorochromes with pH indicator properties, and the electrical activity of neurones in the visual cortex can be observed directly after uptake of fluorochromes sensitive to electrical potential.

(i) *Demonstration of viability of cultured cells.* It is often important before working with cells grown in culture to know the proportion of live cells. One general strategy is to expose the cells to fluorescein diacetate (FDA). Due to its hydrophobic character this non-fluorescent compound readily penetrates the intact cell membranes of live cells. Lysosomal esterases then hydrolyse the compound, to yield the fluorescent dye fluorescein. This ionic, and hence hydrophilic, compound accumulates in the cells since it cannot cross hydrophobic lipid membranes. In dead or damaged cells, fluorescein is either not generated or is lost through the damaged membranes into the incubation medium. The procedure for carrying out such viability testing is outlined in *Table 18*.

3.3 **Demonstrating morphological entities**

The markers used for such demonstrations can again be either chemical or biological entities; the staining mechanisms are correspondingly varied and may involve both

Table 19. Method for visualizing the cavities within bone[a].

1.	Specimens fixed in any media other than those containing mercuric chloride are suitable.
2.	De-calcify in any standard solution[b].
3.	Cut cryostat sections or cut sections of specimen embedded in nitrocellulose (i.e. Celloidin).
4.	Hydrate the sections—ensure that all ethanol is removed from nitrocellulose sections.
5.	Stain in Thionin solution[c] for 5−20 min.
6.	Wash in distilled water[d].
7.	Stain in a saturated aqueous solution of picric acid (~1.2% w/v) for 30−60 sec.[d]
8.	Wash in distilled water[d].
9.	Differentiate in 70% v/v aqueous ethanol until blue−green clouds of dye cease to leave the sections; usually at least 5−10 min[d].
10.	Dehydrate rapidly through graded ethanols, clear in xylene and mount in a synthetic resin.

Results. Canaliculi and lacunae are black−dark brown, bone matrix is yellow−brownish yellow and cells are red.

Trouble shooting. If staining is weak, or if fine canaliculi are not visible, try:
(a) extending time in Thionin solution and
(b) reducing time in differentiating and dehydrating fluids.
  If larger cavities also stain, extend the washing time in step 7.

[a]This method is termed the Schmorl's Picro−Thionin procedure and it visualizes many cavities, e.g. the dentine tubules in teeth.
[b]De-calcify fixed tissue blocks by immersing in 7% v/v aqueous formic acid for 1−10 days, depending on the size of the specimen. Or fix and de-calcify simultaneously using aqueous EDTA−formalin solution (containing 5.5 g of EDTA and 10 ml of 40% w/v aqueous formaldehyde per 100 ml) for up to several weeks.
[c]Dissolve 0.125 g of Thionin in 100 ml of water. Filter and add one drop of concentrated aqueous ammonia solution immediately prior to use.
[d]Agitate the sections gently during this process.

physico-chemical and physiological factors. Current applications of such methods include

(a)   tracing neuronal pathways by exploiting the axonal transport of fluorochromes,
(b)   determining the intercellular spaces between the clefts within epithelial cells by
(c)   negative staining, and
        exploring the three-dimensional distribution of lymphocyte classes within germinal centres of lymph nodes.

(i) *Visualization of bone canaliculi and lacunae.* Osteocytes and their processes are situated inside cavities in the bone matrix. The first stage of this method is the non-specific staining of all tissue elements with Thionin, a cationic dye which binds avidly to proteins under alkaline conditions. In the subsequent short exposure to the anionic dye picric acid, the second dye has time to penetrate only the most permeable sites, namely the cavities. Here, a poorly soluble salt, Thionin picrate, is precipitated. The more superficial deposits are then removed by differentiating with aqueous ethanol.
  Practical details of this procedure are given in *Table 19*.

4. SELECTION OF METHODS

When a biologist thinks that histochemical staining may be an appropriate tool, the problems in hand are most likely one of the following.

(i)   The identification or characterization of an object seen in the microscope, either in chemical or biological terms.
(ii)   The detection, and sometimes quantification, of a specified entity, either chemical or biological.

Table 20. Sources of information on specimen preparation and staining.

Topic·	Source
Specimen preparation methods	
Fixation, embedding, mounting etc.	Refs (3−5) and the journal *Stain Technology*
Possible artefacts.	Ref. (6)
Type of specimen	
Bacteriological	Refs (3,4,7)
Botanical	Refs (7−9)
General	Refs (4,6,7,10) and the journals *Histochemistry, Histochemical Journal, Journal of Histochemistry and Cytochemistry* and *Stain Technology.*
Human histopathology	Refs (3,4,6,10)
Invertebrate	Ref. (7)
Mycological	Refs (3,4,7)
Neurological	Refs (3,4,11)
Protozoan	Refs (4,7)

This account assumes appropriate histochemical methods for achieving these ends may be selected by one of the procedures listed below.

(i)   Finding a method in the histochemical literature which precisely addresses the problem in hand.

(ii)   Finding a method in the histochemical literature which addresses a closely similar problem and hence is highly likely to prove appropriate.

(iii) If the literature contains no relevant procedure, re-defining the problem in terms of chemical or biological entities which can be demonstrated histochemically.

Cases (i) and (ii) are discussed and illustrated in the next section, followed by an example of case (iii). As this chapter is of an introductory character, problems requiring *de novo* design of histochemical methods are not discussed.

4.1 Selecting staining methods for familiar entities

As this chapter is addressed to benchworkers unfamiliar with the histochemical literature, it assumes that it is not possible to be over explicit; readers already well informed may skip a few paragraphs.

Case (i), above, is best discussed by formulating the problem in the following terms: Find a histochemical method to demonstrate a named *chemical or biological entity*, present in a named *type of specimen*, perhaps by using a named *type of staining method.*

It is the words in italics which must be sought within the contents pages and indices of relevant books and journals, or, if necessary, in abstracts and databases. An annotated listing of generally useful sources for histochemical questions is provided in *Table 20.*

A few problems should be mentioned at this point.

(i)   Relevant literature may not be readily available to the non-specialist; note that in addition to academic libraries such reference material may often be found on the laboratory bookshelves of hospital haematology, histopathology and microbiology departments.

(ii)   Indices vary greatly in quality—the best single histochemical source book may have a miniscule index.

(iii)   The lexicons and keyword systems of most abstracting and data retrieval systems are keyed to biological, not technical/methodological terms.

As an illustrative and naive case example consider a botanist wishing to demonstrate *lignified xylem vessels* in the *leaves of a woody plant*. Inspection of *Table 20* indicates that the book *Staining Procedures*, edited by Clark (7), may contain relevant information. The contents page of this book is reassuring, since there is a section titled 'Botanical sciences ... plant anatomy'. In line with this, the index has twelve entries for 'lignified vessels' in addition to one for 'xylem'.

Case (ii) is merely a more speculative version of case (i). Here, it is necessary to seek methods answering analogous questions to the one of immediate concern. For instance, if the problem is to visualize lysosomal acid phosphatase in some previously unstudied fungal species, then a starting point could be to see what methods have been successfully used to demonstrate this enzyme in other fungi.

4.2 **The histochemical marker approach to novel entities**

Here it is assumed that the procedure outlined above fails. For instance, suppose the biological question is posed as follows: To stain selectively the parasitic *protozoan* in the *livers* of infected *sheep*.

Assume no reference to the particular protozoan can be found in the histochemical literature and parasitological and veterinarian sources have no mention of histochemical investigations.

The problem must therefore be re-cast to bring it within the purview of histochemical methodology. This is achieved by considering the ways in which the protozoan may differ from the surrounding hepatocytes, either chemically, biologically or morpho-logically. Does the protozoan contain large numbers of lysosomes or mitochondria for instance? If so, possible histochemical markers are lysosomal enzymes such as esterase or acid phosphatase, or mitochondrial enzymes such as succinate dehydrogenase. Does the nucleus of the organism differ in shape and size from host hepatocytes and endothelial cells? Or does its nuclear−cytoplasmic area ratio differ? If any of these possibilities are true, then staining thick sections with a basic dye could be a useful diagnostic tool.

Note the use of thick sections, as discussed in Section 2.2.3: it is necessary to tailor specimen preparation methods to the questions asked as well as to the nature of the specimens.

5. ASSESSING THE OUTCOMES OF HISTOCHEMICAL STAINING

After use of a histochemical procedure, results must be evaluated in the light of the questions asked. As histochemical methods seek to determine what something is, where it is, and how much of it there is, tests of significance must cover all three issues. Moreover, the implications of both staining and non-staining must be considered.

Suppose that, after carrying out a procedure intended to demonstrate some particular chemical or biological entity, a specimen is found to be stained. Before concluding that the presence of the entity is proven, ask the following questions:

(i)   Was the staining actually due to the entity of interest? This is the issue of specificity/selectivity.

Table 21. Control methods for assessing the outcomes of staining.

Possible artefacts	Control procedures
When staining of specimen occurs	
1. Staining may be due to non-target entity.	Blockade or extract target entity.
2. Stain may not be deposited at *in vivo* site of target entity.	Compare staining patterns from different stabilization methods.
3. Stain intensity may not relate to amount of target entity.	Stain model or standard systems.
When staining of specimen does not occur	
1. Faulty or insensitive method.	Stain test objects, use different method, if target entity chemical, assay fresh specimen.
2. Target entity lost from specimen prior to staining.	Vary pre-staining procedure, e.g. stabilization; if target entity is chemical assay prior to staining (cf. above).
3. Stain lost before observation.	Vary post-staining procedures, observe during/immediately after staining.
In all cases	
Evaluate reagent purity	

(ii) Does the distribution of stain correspond to that of the target entity in the live cell or creature? This is the issue of localization.

(iii) Does the staining intensity reflect the amount of the entity present? This is the issue of quantitation.

Now suppose that after applying the method, the specimen did not stain. Before assuming the entity is absent, ask these questions:

(i) Does the lack of staining reflect a failure of the staining method rather than a lack of the target entity? This covers issues of low sensitivity and of technical malfunction.

(ii) Was the entity present in the live cell or creature, only to be lost during processing and/or staining of the specimen? This is the issue of stabilization of content.

(iii) Was the stain altered, or lost from the specimen, before observation? This involves such issues as fading and diffusion of stain.

Control procedures related to these issues are summarized in *Table 21*; discussion of some of these follows. Bench workers are advised to make routine use of relevant controls, rather than to use them only for trouble shooting puzzling staining outcomes.

5.1 Assessing selectivity

If staining is indeed due to the target entity, then removal of the entity from the specimen (extraction), or conversion of the entity to a chemically modified form (blockade), will be expected to inhibit staining.

An example of extraction was described in *Table 14*, namely the removal of total free lipids by organic solvents. Since such a solvent mixture will not remove other materials, any post-extraction staining occurring with Sudan Black may be assumed to be non-specific. Chemical and enzymatic hydrolyses are also used for selective

Table 22. Diastase extraction of glycogen.

1.	Take two positive control[a] and two test sections. If cut from paraffin blocks then de-wax. In all cases hydrate.
2.	Incubate one test section and one positive control in a 1% w/v aqueous diastase[b] solution, at 37°C for 1 h[c].
3.	Wash in running tap water for 5−10 min.
4.	Stain all sections by a method to demonstrate glycogen, e.g. the periodic acid−Schiff procedure described in *Tables 9* and *11*.

Results. Presence of glycogen is shown by the loss of staining after enzyme treatment.

Control. The positive control sections check that the enzyme is active and that the staining procedure is effective.

[a]Take sections from a glycogen-rich liver block.
[b]Dry diastase is a cheap and stable. Make a fresh solution each day.
[c]Tissues fixed in Gendre's fluid, or in aqueous glutaraldehyde, or osmium tetroxide solutions, resist enzymatic digestion. Therefore increase incubation times.

extraction of particular tissue components. The selectivity of such methods depends on the chemistry of the hydrolytic step, which reduces the biopolymers to fragments. These are then non-specifically extracted into the solvent.

(i) *Method for selective extraction of glycogen.* This exploits the rapid enzymatic hydrolysis of glycogen by amylase; no other common mammalian biopolymer is markedly affected. The practical details of this method are given in *Table 22*.

Blockades are varied in character. For instance, if a tissue section treated with an acid dye is stained intensely, this may be due to the presence of protein. If so, conversion of cationogenic amino substituents into non-ionic hydroxyls, by treatment with nitrous acid, will inhibit staining. Analogously, enzymes can be inhibited by destruction of their tertiary structure using heat or a reactive aldehyde (see *Table 12*), or by binding of an inhibitor. If staining still occurs after such treatment, it will be non-enzymatic artefact.

Control methods are not confined to blockades and extractions. One common tactic is to selectively omit reagents from a method and then to compare the staining outcomes of the complete and control procedures. Thus in the periodic acid−Schiff and the Feulgen nucleal stains (see *Tables 9* and *11*) the presence of native aldehydes is confusing. However, they are readily distinguished by omitting the oxidation and hydrolysis aldehyde-generating steps, respectively, from the standardized procedures, and looking to see what structures still stain.

5.2 Assessing localization

Components of specimens can migrate from their *in vivo* sites at every stage of processing and also during staining. Many stains exhibit mobility, even during storage of stained specimens. Numerous technical manoeuvres are adopted to minimize such artefacts; such as using perfusion, rather than immersion or fixation to reduce solvent flow through the tissue blocks. However, the actual assessment of localization is difficult. The only common tactic is to use more than one processing and staining schedule and to compare the localizations produced.

5.3 **Assessing sensitivity and quantitation**

For qualitative assessments of sensitivity, stain known specimens (both rich and poor in the entity of interest) along with the test specimens. As an example of this, in the context of glycogen staining, see *Table 22*.

For quantitative assessments several strategies can be adopted. The content of the entity present in various specimens may be assayed prior to staining and staining intensity correlated with content. This correlation can then be applied to further unknown specimens. In one important case, namely the quantitation of DNA, assays are not usually required, as comparison of diploid and haploid cell nuclei provides an internal standard. However, use of external standards, in the form of model systems of various kinds, is another possibility. For instance, known amounts of the entity of interest can be trapped within a polymeric matrix, which can then be cast or sectioned to form thin layers. Alternatively, the entity may be attached to the surface of particles of micrometer sized chromatographic support media. Such objects, containing known amounts of the entity, are then stained along with the biological test specimens.

5.4 **Assessing reagent impurity**

Commercial samples of staining reagents of all types—dyes, enzyme substrates and visualizing agents, chromogenic agents and labelled antibodies—may be impure. The reagent content may vary between batches. If the diluent is a salt or a surfactant it can influence the staining indirectly. There may also be impurities which themselves directly stain the specimen, but in a different manner to that intended. Indeed, the label on a bottle of stain is not always a guide to the contents—commercial quality control should not be innocently assumed. Moreover, it must be emphasized that histochemical methods display a more complete specimen than do separation-based biochemical assays. Hence, the consequences of reagent impurity can be even more deleterious for histochemistry than for biochemistry.

This being so, how should reagent purity be assessed? Chemical analysis of dyes and other reagents is a specialized skill, although should this prove necessary the analysis of such substances has been reviewed (12). If using standard histochemical methods, the best advice is to let others do the assays; purchase stains certified by the Biological Stain Commission. Whilst such stains are not guaranteed pure, their performance in the specified histochemical methods is assured. One other general recommendation is to keep samples of satisfactory stains for checking that a particular staining difficulty is not due to stain impurity.

However, in addition to the problems of impurity, use of incorrect staining reagents also arises due to confused nomenclature.

5.5 **Reagent nomenclature**

Histochemical staining reagents are usually complex organic molecules, whose correspondingly complex systematic terminology has encouraged the widespread use of trivial names, acronyms and initials. After all, would you rather say 'Hand me the 3,3'(4,4'-diortho-anisylene)-2,2'-di(para-nitrophenyl)-bis-(5-phenyl) ditetrazolium chloride' or 'Pass the Nitro BT'? In addition, many histochemical reagents had their

Table 23. The naming of dyes and related reagents used in this chapter.

Trivial name used	Synonyms or abbreviations	CI number (and generic name)
Alcian Blue 8G		74240 (Ingrain blue 1)
Alizarin Red S	Alizarin Carmine, Diamond Red	58005 (Mordant red 3)
Basic Fuchsin	Aniline Red, Rosanilin	None allocated
Evans Blue	Diazol pure Blue BF	23860 (Direct blue 53)
Fluorescein diacetate	FDA	None allocated
Iron haematoxylin	Heidenhains haematoxylin	None allocated
Methyl Green	Double Green SF, Light Green	42585 (Basic blue 20)
New Fuchsin	Isorubin, New Magenta	42520 (Basic red 9)
Nitro Blue Tetrazolium	Nitro BT, NBT	None allocated
Pararosanilin	Basic Rubin, Magenta O	42500 (Basic violet 2)
Picric acid	Trinitrophenol	10305 (none allocated)
Schiff's reagent		None allocated
Sudan Black B	Fat Black, Waxol Black	26150 (Solvent black 3)
Tetranitro Blue Tetrazolium	Tetra NBT, TNBT	None allocated
Thionin	Lauth's Violet	52000 (none allocated)

origins in the textile dyeing industry, where the same compound was often given different tradenames by different manufacturers. Such confusions can often be reduced by use of the Colour Index (CI) standard nomenclature; and indeed CI terms are used in parallel to the trivial names in many texts, journals and catalogues. A useful resource for resolving problems of nomenclature is the monograph *Conn's Biological Stains* (13).

Possible terminological ambiguities arising with the reagents cited in this chapter and a listing of relevant CI terms, are described in *Table 23*.

6. RESOURCES NEEDED FOR HISTOCHEMICAL WORK

6.1 Equipment and materials

Microtomy is the only stage of specimen processing demanding committed and complex (and thus relatively expensive) equipment. Moreover, separate devices are needed for cryotomy, and for specimens embedded in paraffin wax and in plastic. However, if a research programme involves large numbers of specimens, a processing/staining machine is a recommended labour saving device. The need for a good microscope goes without saying. However, though a modern automated research microscope is expensive, excellence came to microscopes long ago, and old machines are often optically satisfactory though they will demand more skill to operate. Many reagents used in specimen preparation and staining are inexpensive. Note that whilst embedding and staining kits are often more expensive, they often prove to be worth the money for the beginner, though only if they are of high quality. A number of standard histochemical reagents are volatile and hazardous, either being toxic, such as the fixative agent formaldehyde, or flammable, such as the clearing agent xylene. Hence an exhaust hood or equivalent device is required. All items discussed in this paragraph are widely available from commercial suppliers.

6.2 Skills, manual and cognitive

The practical skills most benchworkers find hardest to learn are those involved in microtomy and microscopy, and of course there is an inverse relationship between skills

Table 24. Histochemistry in perspective—comparisons between some information technologies.

	Advantages[a]			
	Histochemistry	Autoradiography	Biochemistry	Flow cytometry
Types of information obtained				
What is it?	**	*	***	*
Where is it?	***	**	*	
How much of it?	*	*	***	*
How many of them?	**	**		***
Problems involved in obtaining the information				
Technical flexibility	***	**	***	*
Minimum sample size	tiny	tiny	larger	small
Largest sample size	small	small	no upper limit	large
Cost	low	moderate	moderate	high
Type of specimen	insoluble or insolubilizable		soluble or solubilizable	particulates only

[a]Indicated by the number of asterisks—the more the better.

called for and the degree of automation (that is the cost) of the equipment purchased. If practical expertise is already available within your institution, consider yourself fortunate. If not, advice could be sought in such academic specialities as anatomy or zoology, or in such hospital departments as haematology or pathology. To acquire the understanding of the principles which must go hand in hand with the technical skills, first read extensively, starting with refs 1 and 2. Then talk to the well informed. If you lack in-house tuition think seriously of attending formal courses, such as those run by The Royal Microscopical Society. For information concerning these contact The Administrator, The Royal Microscopical Society, 37/38 St Clements, Oxford OX4 1AJ, UK.

As a tailpiece, remember that in the UK a difficult-to-obtain Home Office licence is required for performing perfusion fixation of live animals. Similar legislation exists, or will soon exist, elsewhere.

7. WHY USE HISTOCHEMICAL METHODS?

Biological problems calling for the localization of some entity immediately suggest microscopy as a tool. However, there are microscope-based methodologies other than histochemistry. Problems emphasizing identification are less obviously demanding of the microscope. As a guide to methodological choice, some advantages and limitations of several procedures for investigating the content and structure of cells and tissues are compared, rather simplistically, in *Table 24*. The summarized information emphasizes both the versatility of histochemistry and the complementary character of these different approaches.

8. ACKNOWLEDGEMENTS

This chapter was written whilst the author was Visiting Professor in the Department of Molecular and Cell Biology, at the University of Connecticut, CT, USA. I wish to thank members of the department, in particular Dr Emory Braswell, for their hospitality, and also the staff of the University's Wilbur Cross Library for their assistance. My thanks are also due to the University of Sheffield for granting me study leave.

9. REFERENCES AND FURTHER READING

Background theory

1. Baker,J.R. (1958) *Principles of Biological Microtechnique: A Study of Fixation and Dyeing*, Methuen, London.
2. Horobin,R.W. (1988) *Understanding Histochemistry: Selection, Evaluation and Design of Biological Stains*. Ellis Horwood, Chichester.

Sources of technical information—see *Table 20* for commentary.

3. Bancroft,J.D. and Stevens,A. (1989) *Theory and Practice of Histological Techniques*, Churchill Livingstone, Edinburgh, 3rd edn.
4. Lillie,R.D. and Fullmer,H.M. (1976) *Histopathologic Technic and Practical Histochemistry*, McGraw-Hill, New York, 4th edn.
5. Thompson,S.W. and Luna,L. (1978) *An Atlas of Artifacts Encountered in the Preparation of Microscopic Tissue Sections*. Thomas, Springfield, IL.
6. Pearse,A.G.E. (1968, 1972, 1988) *Histochemistry, Theoretical and Applied.* Vol. 1, Churchill, London, Vol. 2, Churchill Livingstone, Edinburgh, and Vol. 3, in press.
7. Clark,G. (1981) *Staining Procedures.* Williams and Wilkins, Baltimore, MD, 4th edn.
8. Gahan,P.B. (1984) *Plant Histochemistry and Cytochemistry: an Introduction.* Academic Press, New York.
9. Jensen,W.A. (1962) *Botanical Histochemistry, Principles and Practice.* Freeman, San Francisco, CA.
10. Thompson,S.W. (1966) *Selected Histochemical and Histopathological Methods.* Thomas, Springfield, IL.
11. Adams,C.W.M. (1965) *Neurohistochemistry.* Elsevier, Amsterdam.

Information on reagent purification and nomenclature

12. Horobin,R.W. (1969) *Histochem. J.,* **1**, 115.
13. Lillie,R.D. (1977) *Conn's Biological Stains.* Williams and Wilkins, Baltimore, MD, 9th edn.

CHAPTER 6

Fluorescence microscopy

JOHAN S.PLOEM

1. INTRODUCTION

Fluorescence microscopy is based on the property of some substances to absorb light in a certain wavelength range, the energy of which is subsequently partly emitted in the form of light. The emitted light differs from the absorbed light in regard to wavelength and intensity. According to Stokes' Law the wavelength of the emitted light is longer than that of the absorbed light and this difference in wavelength is the basis for the observation of fluorescence in fluorescence microscopy. Furthermore, the intensity of the emitted light is weaker, as the emitted energy is much smaller than the energy needed for excitation.

To observe fluorescence it is absolutely necessary to prevent the mixing of exciting radiation with the emitted fluorescence light in the microscope image. The quality of the image in fluorescence microscopy depends largely on the image contrast and the image brightness. The contrast is determined by the ratio of the fluorescence emission of the specifically stained structures to the light observed in the background. An optimal image contrast may however result in a dim fluorescent image due for example to the use of narrow-band excitation filters which often have a low transmission. For easy visual observation, for photography with reasonably short exposure times, and for measurement purposes a sufficiently bright image is required. Therefore, in most applications, a compromise between the intensity of the specific fluorescence and the level of non-specific fluorescence in the background must be accepted.

For reviews of fluorescence microscopy the reader is referred to the papers by Young (1), Price (2), Rost (3), Siegel (4) and Ploem and Tanke (5). Modern applications in the biomedical sciences are described in the book by Lansing Taylor *et al.* (6).

2. FLUOROCHROMES

Substances which show fluorescence upon radiation are called fluorophores or fluorochromes. Different fluorochromes are characterized by their absorption and emission spectra. The absorption or excitation spectrum is obtained by measuring the relative fluorescence intensity at a certain wavelength when the specimen is excited with varying wavelengths. The most intense fluorescence occurs when the specimen is irradiated with wavelengths close to the peak of the excitation curve. The emission curve results from excitation at a certain wavelength. An example of an absorption and an emission spectrum is given in *Figure 1*. Most excitation and emission curves overlap to a certain extent. Another characteristic of fluorochromes is their quantum efficiency (Q), which is determined by the ratio between emitted and absorbed energy.

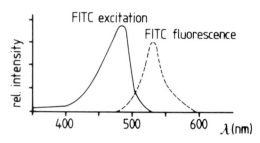

Figure 1. Excitation and emission spectra of fluorescein isothiocyanate (FITC). λ = wavelength of excitation (solid line) and emission (dotted line). Reproduced with permission from Oxford University Press.

A fluorochrome with a low Q will emit a low intensity of fluorescence in comparison to the absorption level needed and be termed a weak fluorochrome.

Decrease in fluorescence during irradiation with light is called fading. The degree of fading depends on the intensity of the excitation light and the exposure time (7). The reduction in fluorescence intensity can also be due to modification in the excited states of the fluorophore. These physico-chemical changes may be caused by the presence of other fluorophores, oxidizing agents, or salts of heavy metals. The phenomenon is termed quenching and is quite complex. To reduce this decrease in fluorescence, preparations to be studied by fluorescence are best stored in the dark at 4°C. Another way to reduce decrease in fluorescence is to add agents such as DABCO (1,4-diazo-bicyclo-2,2,2-octane), *N*-propylgallate and *p*-phenylenediamine to the mounting medium (8,9).

A 5% w/v solution of propylgallate in glycerol will for example, when applied to rhodamine-labelled specimens, lengthen the life of the fluorescence emission from 60−90 sec in the untreated to 15−25 min in the treated. This is particularly helpful in the case where photomicrographs are to be taken of the specimen. It is advisable, however, to observe or photograph the specimens directly after mounting when an anti-bleaching agent is used as storage may diminish fluorescence intensity due to quenching phenomena (8).

3. THE FLUORESCENCE MICROSCOPE

The fluorescence microscope must fulfil two main functions. In the first place the fluorophore in the preparation must be excited effectively, preferably with wavelengths close to the absorption peak of the fluorophore (10). Secondly, the microscope must collect as much as possible of the fluorescence light.

For the observation of fluorescence three main components are essential: the fluorophore, the light source, and the fluorescence detecting unit. Besides these essentials, filters are required to select the appropriate excitation (excitation filters) and emission (barrier filters) wavelengths and, in addition, lenses are needed. The fluorescence can be photographed, observed visually, or measured by a photomultiplier.

3.1 Types of illumination

There are two types of illumination for fluorescence microscopy: transmitted (1,11) and incident illumination (10,12).

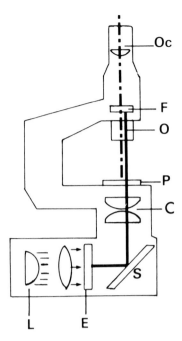

Figure 2. Schematic diagram of transmitted illumination. L, light source; E, excitation filter; S, mirror; C, condenser; P, preparation; O, objective; F, barrier filter; Oc, ocular. Solid line, excitation light; broken line, emitted light.

3.1.1 *Transmitted light illumination*

The illumination pathway is shown in *Figure 2*. A condenser focuses the exciting light onto a microscope field. The emitted fluorescence is collected by the objective and observed through the eyepiece. Essential to this configuration is that two different lenses are used, one to focus the exciting light (the condenser) and one to collect the emitted light (the objective). For optimal observation of fluorescing images these two lenses, which have two independent optical axes, must be perfectly aligned. This is not always easy to obtain and maintain in routine use. A disadvantage of transmitted light illumination by a bright-field condenser is that almost all the exciting light enters the objective. Since the intensity of exciting light is often several orders of magnitude greater than that of the emitted fluorescence light, filters of very high quality are required to separate exciting from fluorescence light. To avoid this problem, a dark-ground condenser may be used which illuminates the specimen at such an angle that no direct exciting light enters the objective. This facilitates separation of fluorescence light and unabsorbed exciting light by means of barrier filters. On the other hand, the transmission of dark-ground condensers for the exciting light is much less than for bright-field condensers. Transmitted illumination with high-power objectives requires the use of oil between the condenser and the slide, which is troublesome in routine applications. Another disadvantage of transmitted illumination is the difficulty in combining fluorescence with phase-contrast or absorption microscopy.

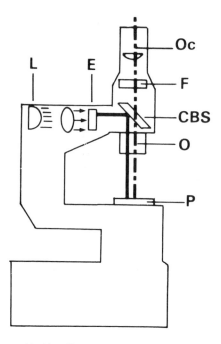

Figure 3. Schematic diagram of incident illumination. CBS, chromatic beam splitter. All other letters and lines as in *Figure 2*.

3.1.2 *Incident light illumination*

This type of fluorescence microscopy is shown in *Figure 3*. For focusing exciting light onto the specimen and collecting emitted light from the fluorescing specimen only one lens is used, namely the objective. In order to separate fluorescence emitted light from unabsorbed exciting light a special type of mirror, a chromatic beam splitter (CBS, sometimes known as a dichroic mirror), is positioned above the objective when incident fluorescence microscopy is used. These mirrors have a special interference coating which reflects light shorter than a certain wavelength and transmits light of longer wavelengths. Thus, these mirrors effectively reflect the shorter wavelengths of the exciting light onto the specimen and transmit the longer wavelengths of the emitted fluorescence towards the eyepieces. Furthermore, the exciting light reflected by the specimen or optical parts and directed towards the eyepieces is also effectively reflected by the CBS, thus blocking this light from reaching the eyepieces. In principle, the CBS acts as both excitation and barrier filter but in practice an additional barrier filter is usually needed to eliminate any residual exciting light. *Figure 4* gives an example of the separation of blue (460 nm) from green (530 nm) light. Similar CBSs exist for separation of other regions of the light spectrum, from the UV (300 nm) to the far red (700 nm).

   The advantage of epi-illumination is that the same lens system acts as objective and condenser. Focusing this lens onto the specimen consequently results in proper alignment of this part of the microscope. The illuminated field is the field of view. If an oil-immersion objective is used, oil on the specimen is sufficient, whereas transmitted light illumination would also require oil on high-aperture condensers for maximum

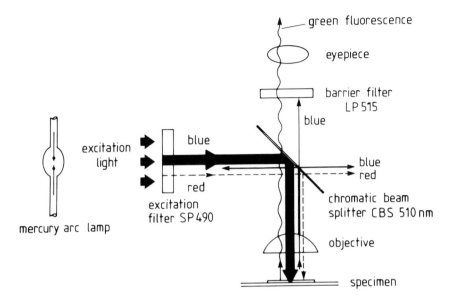

Figure 4. Separation of excitation and fluorescence light in incident illumination. The CBS is placed at a 45° angle to the light path. This CBS reflects the blue (wanted) excitation light onto the preparation and transmits the green fluorescence light towards the eyepieces. The blue (unwanted) excitation light is partly reflected backwards into the direction of the light source and only a small part is transmitted towards the eyepieces. (From ref. 5.)

illuminating intensity. Epi-illumination, moreover, permits an easy changeover between fluorescence microscopy and transmitted light microscopy since the substage illumination remains readily available. This is very useful for a number of applications, such as immunofluorescence in combination with phase-contrast microscopy.

3.2 Light sources

The choice of light source is determined by the excitation spectrum of the fluorochrome and its quantum efficiency. Weak fluorochromes with low Q require more exciting light for viewing than strong fluorochromes. Some light sources [mercury (Hg) lamps] have steep emission peaks in the region 300−700 nm. Other lamps such as tungsten and xenon (Xe) lamps show less distinct peaks in this wavelength region. Halogen lamps show an increase of emission towards the longer wavelength side and xenon lamps have distinct peaks in the deep red and IR region (see *Table 1*).

Tungsten halogen (12 V, 50 and 100 W) lamps are suitable and inexpensive light sources for routine investigations, provided that the specimen emits fluorescence of sufficiently high intensity. These lamps can be used for both transmitted and incident light illumination, and can be switched on and off easily and frequently without damage to the lamp (13).

A mercury lamp emits characteristic lines at, for example, 366, 405, 436, 546 and 578 nm, but also has a strong background continuum. In the blue region for instance, this continuum is usually much brighter than that given by a tungsten halogen lamp. If UV light or in general high-energy excitation light is required, mercury lamps are

Table 1. Lamps for fluorescence microscopy in their order of intensity.

Lamp (W)	Mean luminous density (cd/cm × cm)	Wavelength region (nm)	Lifetime (h)
Hg 100	170 000	Peaks 300 – 700	200
Hg 200	33 000		200
Hg 50	30 000		100
Xe 75	40 000	Continuum and peaks >800	400
Xe 450	35 000		2000
Xe 150	15 000		1200
Halogen		Increasing intensity towards longer wavelengths	50

recommended (14). Mercury lamps are available as 50, 100 and 200 W. It should be noted that the 100 W has a smaller arc than the 50 and 200 W lamps. Mercury lamps have a limited lifetime (about 200 burning hours) and are operated on AC current supply, although the HBO 100 W can be operated on DC supply for measurement purposes.

Xenon lamps emit a spectrum of rather constant intensity from UV to red (13). They are available as 75, 150 and 450 W with lifetimes of 400, 1200 and 2000 burning hours, respectively. Xenon lamps should be handled with care because even cold lamps are under pressure and safety glasses must therefore be used during removal and replacement. These lamps are operated on DC current supply.

Both mercury and xenon arcs are expensive. They should be burnt in under conditions of low mechanical vibrations, for example during the night, and with an AC voltage stabilizer to overcome large voltage fluctuations in the electric mains. This results in more stable burning points with greater stability of light output for fluorescence measurements, which should be performed with the use of DC current supplies.

Lasers offer monochromatic radiation of very high intensity and are therefore a potential source for special purpose fluorescence microscopy (15,16). Their relatively short lifetime and high cost limit their use in routine studies. Lasers can either provide a continuous output of energy or operate in a pulsed mode; with the use of short pulses of excitation energy (1 μsec to 1 nsec), fading of fluorescence can often be avoided. Lasers are used in fluorescence scanning microscopy (17), a new form of fluorescence microscopy, which will be discussed in Section 6.

3.3 **Lamp housings**

Lamps of the high pressure type should be mounted in strong protective housings, provided with cooling fins. Adjustment of the lamp in two directions and a focusing collector lens should be provided for. Lamp housings usually have filter holders for heat- and red-absorbing or -reflecting filters. Heat- and IR-reflecting mirrors are preferable to heat-absorbing filters since they crack less frequently. These filters should always be placed closer to the lamp than the coloured filters to prevent excessive heat absorption by the latter. In these instrumental configurations it is very important to follow the manufacturer's instructions to ensure correct illumination.

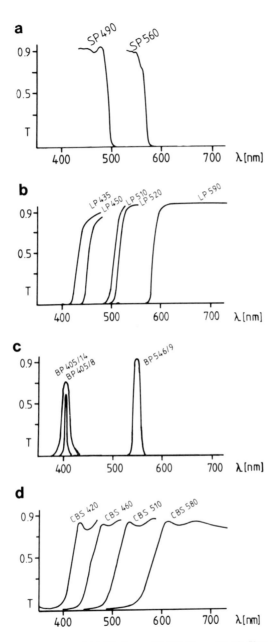

Figure 5. Transmission (T) curves of (**a**) SP filters, (**b**) LP filters, (**c**) BP filters and (**d**) CBSs.

3.4 **Filters**

Filters are very important components in the fluorescence microscope and they should be carefully chosen for specific applications. Filter choice depends on the spectral characteristics and Q of the fluorochromes used, as well as the light source involved. A filter is used to select a certain part of the excitation or emission spectrum. The effect is shown by its transmission curve.

Types of filters available include coloured glass filters, band-pass (BP) interference filters, short- and long-pass (SP and LP, respectively) interference filters and CBSs (*Figure 5*). Coloured glass filters more or less transmit light of their own colour and absorb complementary colours, whilst interference filters are glass substrates on which thin layers of metal salts are deposited. Filters are specified by symbols/letters and numbers. Symbols/letters indicate the type of filter according to their function or characteristics and numbers refer to the wavelength; in the case of BP filters the numbers correspond to the maximum transmission wavelength and in the case of LP or SP filters, to the wavelength at 50% of their maximum transmission. In addition, sometimes the band width (in nm) of a BP filter is given. Furthermore, filters can be characterized by their position in the microscope (excitation or emission side), consequently the terminology used by different manufacturers is quite confusing.

3.4.1 *Excitation filters*

Band-pass filters transmit one particular region (band) of the light spectrum. Glass filters with these characteristics still in use are the BG (blue glass) and UG (ultraviolet glass) filters. They have rather broad transmission characteristics. Band-interference filters are more selective. A disadvantage of these filters in the past was their low transmission value (30−60%). This problem has been overcome. At present, BP filters with high transmissions (90%) and very narrow-band characteristics can be produced for selective excitation of fluorochromes (line filters). Consequently, BP filters are mainly used as excitation filters, especially when mercury arc lamps are used as the source of excitation light. Depending on quality, they vary in price.

Short-pass filters transmit lower wavelengths and effectively block higher wavelengths (18−20). These properties make them suitable as excitation filters, especially when combined with LP filters. In such combinations, they are preferable to glass filters (for example BG type filters), since they generally do not leak unwanted excitation light. Short-pass filters, like all interference filters, are expensive and, in comparison with the older types of interference BP filters, they have a very high transmission.

Glass filters used for excitation have rather broad transmissions and are identified by the colour they transmit.

3.4.2 *Barrier filters*

The LP filter has a high transmission for longer wavelengths (90% or higher) and effectively blocks shorter wavelengths. In combination with a SP filter the LP serves as an effective emission selection filter. This is especially useful for sequential observation of two-colour fluorescence with minimal overlap. Although LP filters are predominantly used as barrier filters, they can also be applied as excitation filters in combination with a SP filter. Until recently all LP filters were glass type filters; nowadays interference type filters are being manufactured. Some glass LP filters may show autofluorescence at very high intensities; used as barrier filters in routine fluorescence microscopy this generally will not cause problems.

Table 2. Typical filter combinations.

Application	Excitation filter	CBS	Barrier filter
UV excitation (365 nm)	3 mm UG1	400 or 410	LP 430
Violet excitation (405 nm)	3 mm BG3 + SP 425 or BP 405	455 or 460	LP 470
Blue excitation (470 nm)	2× SP 490 + 2 mm LP 455	500 or 510	LP 515
Green excitation (546 nm)	BP 546 or SP 560 + LP 515	580	LP 580

3.4.3 *Chromatic beam splitters*

A CBS reflects light of wavelengths shorter than the specified wavelength and transmits light of longer wavelengths. It is placed at an angle of 45° to the optical axis and reflects exciting rays into the objective in epi-illumination, where the objective also serves as a condenser. Multi-wavelength vertical illuminators are available with sliding or revolving filter holders permitting epi-illumination in various wavelength bands. Various filter combinations for different wavelengths are given in *Table 2*.

3.5 **Objectives and eyepieces**

In epi-illumination the objective also serves as a condenser; the result obtained, therefore, strongly depends on the choice of the objective. Not all objectives are suited for fluorescence microscopy as they may be equipped with many extra lenses for correction of imaging errors. Suitable lenses to observe fluorescence are achromats and fluorites. The observed fluorescence intensity is proportional to the square of the numerical aperture (NA) of both condenser and objective in transmitted, and to the fourth power of the objective in incident illumination. The brightness is inversely related to the magnification of the objective. Therefore, fluorescence microscopy must preferably be carried out with objectives of high NA and with a moderate total magnification.

4. APPLICATION OF FLUORESCENT STAINS

4.1 **Nucleic acids**

A large number of staining reactions have been described for obtaining fluorescence from structures containing nucleic acids. Probably one of the best known fluorochromes for staining nucleic acids is acridine orange (21,22). The staining reaction is complex and consists of intercalation and electrostatic binding mechanisms. Acridine orange can be excited with blue excitation light and, generally stated, DNA then fluoresces green, and RNA red (23). Ethidium bromide and propidium iodide are frequently used dyes in flow cytometry. They both show red fluorescence upon excitation with blue or green (ethidium bromide) and green (propidium iodide) excitation light.

DAPI (4′,6′-diamidino-2-phenylindole) is another frequently used nuclear fluorochrome and is readily available (e.g. from Sigma). It is used at a concentration of 30 − 1000 ng/ml in buffer (0.1 M NaCl, 10 mM EDTA, 10 mM Tris, pH 7.0). When combined with DNA and excited with UV (see *Table 2*) at about 350 nm it gives rise to a blue emission at 450 nm. It can be used for staining the nuclear material both *in situ* and in biochemical fluorescence spectrophotometry (24,25).

171

Table 3. To vitally stain living freshwater plankton.

1.	Make a suspension of living freshwater plankton.
2.	To two drops of the above on a slide add an equal volume of a very dilute aqueous solution of acridine orange[a].
3.	Cover with a coverglass and examine.
4.	Preferably search with phase-contrast microscopy and then excite with blue-violet radiation in epi-fluorescence mode.
5.	With a yellow barrier filter in place the rotifers and other animals will show very considerable yellow-green fluorescence and much movement. The diatoms and other algae will show green nuclei and red chloroplasts and some can readily be seen to glide across the field of view.

[a]A stock solution of acridine orange can be kept indefinitely if sufficiently concentrated to, for example, give an orange-brown solution. Four drops of the stock added to 250 ml of water will give a working solution of a very pale yellow.

DAPI applied as a protein stain results in a blue-white fluorescence image using UV excitation. Although DAPI-stained cytoplasm shows a bright fluorescence, we observe a rapid decrease in intensity upon light exposure. SITS (stilbene isothiocyanate sulphonic acid)-stained cytoplasm, on the other hand, also gives a blue fluorescence with UV excitation but does not fade to the same degree as DAPI cytoplasmic fluorescence.

Feulgen-type staining procedures such as pararosanilin−Feulgen or acriflavin−Feulgen are mainly known as absorption stains. With the introduction of specific excitation wavelengths a sufficiently strong red fluorescence can be obtained from pararosanilin-stained nuclei (excitation with green light). Acriflavin-stained nuclei show a yellow fluorescence upon excitation with blue light.

4.1.1 *Vital staining with fluorochromes*

The very low concentrations of the dye required to produce contrast in fluorescence microscopy allows the use of such a dye as acridine orange to render the cells visible while keeping them in a living state. Some compounds also autofluoresce and, in combination with a dye, can give considerable information. An example is given in *Table 3*. The use of acridine orange as a vital stain for studying living trypanosomes was mentioned in Chapter 1.

4.2 **Immunofluorescence**

In immunofluorescence use is made of a specific reaction between an antibody and an antigen. Antibodies or immunoglobulins are produced by animals or humans as a defence mechanism against foreign invaders (antigens) such as bacteria, viruses, etc. When antibodies are labelled with a fluorochrome, they are called conjugates (see for instance Table 6 in Chapter 4). When conjugates are used in an antigen−antibody reaction, the site of the reaction can be seen in a fluorescence microscope by reason of the precipitated reaction product. The antigen−antibody reaction is very specific; the antigen will not react with antibodies that do not fit its structure. Antigens can also be fluorescently labelled to demonstrate antibodies. The most well known fluorochromes used for labelling are FITC and tetramethyl rhodamine isothiocyanate (TRITC). FITC is maximally excited in the blue wavelength region, which results in a green fluorescence (*Figure 1*). Green excitation light is used to excite TRITC, whereupon the reaction product shows a red fluorescence. The filter combinations for blue and green exciting light·using incident illumination are given in *Table 2*. Immunofluorescence is now a widely applied technique (Chapter 4 and refs 11,26). It is also applied in living cell investigations, as for example with the aid of membrane markers for the sorting of live cells using a flow cytometer with cell sorting facilities.

Immunofluorescence of small objects, such as for example banded chromosomes, can be very weak and the availability of strong blue laser light (15) can certainly contribute to an improved detection. A recent technology for the visualization of weakly fluorescing reaction products is laser scanning microscopy which is described under Section 7.

4.3 Neurotransmitter fluorescence

Biogenic amines such as catecholamines and 5-hydroxytryptamine can be converted to highly fluorescent products by treating dried tissues with formaldehyde vapour (27). Violet excitation light (see *Table 2*) results in a white-blue fluorescence of the catecholamines and a yellow fluorescence of the 5-hydroxytryptamines. When narrow-band excitation is used, the level of autofluorescence can be kept sufficiently low to provide a high-contrast image (28).

4.4 Double staining

The development of efficient filter systems for excitation with blue and green light has stimulated the use of two fluorochromes in one preparation. The use of double staining in turn has led to the development of vertical illuminators provided with an internal turret containing four CBSs and barrier filters and a revolver for excitation filters. For the observation of double-stained cells, special illuminators are available which permit an exchange of a complete filter set consisting of an excitation, barrier and fluorescence selection filter and a dichroic mirror in one movement. Examples of double staining are the acriflavin − Feulgen − SITS (AFS) staining for epithelial cells (29) and labelling with FITC and TRITC conjugates in one preparation.

5. MICROFLUORIMETRY

5.1 Introduction

Microfluorimetry is a method to measure the emitted light from a fluorescing object such as a cell or cellular organelle. The intensity of the fluorescence is directly proportional to the absorbance of light, provided that the local absorbances of the compound are low (30). Thus, emitted light is proportional to concentration under conditions of low local absorbances. Two phenomena cause reduced fluorescence at higher local absorbance values. The first is the inner filter effect, which is a reduction of excitation energy in the layers furthest from the light source. As the layers near to the light source absorb a large portion of the exciting energy—as will occur with high local absorbance values—the deeper layers cannot be as adequately illuminated. The second consists of the reabsorption phenomenon, which means that the emitted fluorescence is partly reabsorbed by surrounding layers (30,31). This may occur due to the fact that most absorption and fluorescence spectra overlap to a certain extent.

The result is a reduction in fluorescence yield. To avoid this phenomenon, fluorescence measurements should be done outside the overlap region of the spectrum, at the higher wavelength range of the emission spectrum.

To measure an object in fluorescence microscopy, the fluorescence value can be assessed by recording the fluorescence from the entire object. Since the total fluorescence equals the sum of the local intensities, it is not necessary to perform scanning measurements as is mandatory in absorption measurements. However, fluorescence intensity measurements are directly related to the intensity of the exciting light, since the reading from an empty spot, being composed of unwanted excitation light and autofluorescence from optical parts and the preparation, cannot be used as a reference signal. The intensity of the exciting light must therefore be kept constant to allow fluorescence readings on different days and from different slides. In practice this means that standards have to be used to compare different measurements.

5.2 Fluorescence standards

An example of an instrumentation standard is uranyl glass, which gives a strong yellow-green fluorescence (32). The procedure is as follows.

(i)   Start the epi-illumination light burner (HBO 100) and wait for 20 min before beginning measurements.
(ii) Place a 25 × 50 mm slide of uranyl glass (GG 17 or 21, Schott, FRG) with a thickness of 2 mm on the stage.
(iii) Lower the oil-immersion objective into a droplet of non-fluorescing oil on the slide. Take care that each time the same oil-immersion objective is used for calibration. Also, filters and other optical parts should remain identical.
(iv) Use a diaphragm size of 2 mm diameter.
(v)   Use the maximal reading with the objective just touching the glass.
(vi) Adjust the current supply to the photomultiplier before each immunofluorescence measurement in order to obtain the same reading with the uranyl glass every time.

Other instrumentation standards include microcapillaries and beads of known volume or dimension with known concentrations of the fluorochrome (33,34).

Furthermore, standards are needed to control the fluorescent staining reaction. This would be a preparation containing cells with a known amount of substance stained with a defined staining procedure. An example is the use of diploid cells as a reference for DNA measurements.

5.3 Instrumentation

A microfluorimeter essentially consists of a fluorescence microscope with a light-sensing device, a photomultiplier, attached (*Figure 6*). Microfluorimetry instruments preferably have to be equipped with two types of illumination: substage transmitted and incident illumination. Incident illumination is most suited for fluorescence intensity measurements, as functions of illumination and observation are combined in the objective. When exact focusing of the objective is maintained, excitation conditions will be reproducible during movement of the slide on the microscope stage. As soon as a new part of the specimen is focused, it is automatically placed in the optimal position

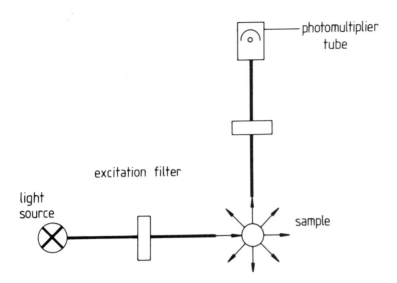

Figure 6. The schematic beam path in microfluorimetry.

for measurement. Alignment of the substage condenser together with alignment of the objective for reproducible illumination and measurement conditions is more complicated than alignment of the objective alone. Moreover, and more important, the substage illumination pathway can now be used to search for the objects to be measured.

A typical microfluorimeter is schematically drawn in *Figure 7*. Substage low-intensity illumination with a tungsten lamp or by a mercury high pressure arc with thick excitation filters, enables searching of fluorescent objects without premature bleaching. Under conditions of very effective light collecting power, obtained by using optics with low magnification and high NA, the intensity of the substage illumination can be kept very low. This also allows full opening of the field diaphragm to visualize the whole field, facilitating the searching of cells to be measured. A HBO 100 mercury high pressure lamp, operated on highly stabilized DC power supply, provides the epi-illumination for measurement purposes. Specific excitation and barrier filters and CBSs are required to reduce unwanted excitation light. Just before the photomultiplier a second fluorescence selection filter is placed, which only transmits the fluorescence outside the overlap region of excitation and emission spectra. The objective is preferably of moderate power and high NA to ensure a high light-collecting power. The field diaphragm of the epi-illumination pathway is kept only slightly larger than the variable measuring diaphragm to prevent unnecessary fading of the remaining microscope field by the strong light used for measuring the fluorescence. The variable measuring diaphragm can be adjusted to just surround the image of interest to avoid disturbance from the background light. In the MPV II (Leitz) and the MP 03 (Zeiss), for example, the measuring field is visible as an illuminated spot superimposed on the microscopic image. In automated measuring microscopes, the light paths can be alternately switched on and off by means of electronically operated shutters (35,36). During the period of searching the substage electronical shutter is open. The measuring operation is started by pressing a button

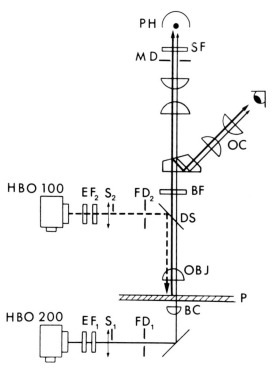

Figure 7. The schematic diagram of a microscope equipped with transmitted illumination for observation of fluorescence and incident illumination for measurement of fluorescence. EF, excitation filter; S, shutter; FD, field diaphragm; BC, bright-field condenser; P, preparation; OBJ, objective; DS, dichroic mirror (CBS); BF, barrier filter; OC, ocular; MD, measuring diaphragm; SF, selection filter; PH, photomultiplier.

that initiates the electronic circuit. The shutter in the transmitted light path is now closed and the shutter in the incident light path opened for a very short period between 0.001 and 1 sec. Connection of the microfluorimeter to signal analysing and recording units provides automated data output.

5.3.1 *Inverted microfluorimeters*

In inverted microscopes the objective is placed below the stage, so that cells at the bottom of wells of test trays can be observed. These microscopes can also be equipped with both transmitted and incident illumination. The light source for incident illumination is, in this type of instrument, placed below the stage. Immunological tests in microtitre plates (96 wells) or in Terasaki-type trays (60 wells) are examples of studies carried out with inverted microscopes. The microfluorimeter can be equipped with computer-controlled stepping motors for moving the stage along rows and lines of wells to be measured.

5.3.2 *Scanning microfluorimeters*

Scanning fluorimetry can contribute to a better interpretation if objects of interest cannot be sufficiently isolated from the background and centered in the field of a measuring diaphragm. Scanning microfluorimeters can be divided into object scanners, such as stage scanners and flying spot scanners, and image plane scanners, such as TV scanners (see Section 6). In stage scanning the stage is moved by stepping motors operated under computer control. A conventional microfluorimeter can be further adapted to such options.

Microfluorimetric scanning has proved to be very useful in the localization of neuronal and extraneuronal fluorescence. Mathematical analysis of the histograms of the scan areas (50 × 50 μm) resulted in detailed information on neuronal and extraneuronal intensities of formaldehyde-induced fluorescence or immunologically labelled neuronal substances (37).

5.4 DNA measurements

As an illustration of the potentialities of microfluorimetry, a study with cervical cells— carried out in our laboratory—is described (38). Microfluorimetry in this study was used to measure the DNA of cervical cell nuclei. The goal was to investigate whether pre-cancerous and cancerous lesions of the cervix all contained cells with more than 5C DNA and whether these cells were absent in specimens without these types of lesions. The study was performed with a MPV II (Leitz, FRG) equipped with transmitted illumination, for searching and finding the cells to be measured, and with epi-illumination for measurement. Any other make of microfluorimeter can be used, as long as the microscope is equipped for transmitted as well as epi-illumination. For this particular application the following instrumentation and methods were used.

5.4.1 *Equipment*

(i)   A microfluorimeter microscope with two mercury lamps: a HBO 200 W for transmitted illumination with AC current supply and a HBO 100 W for incident illumination for measurement purposes with DC current supply (see *Figure 7*).

(ii)   Transmitted illumination is performed with violet light (425 nm) for the observation of both nucleus and cytoplasm after AFS staining (Section 5.4.3). The excitation filters following the heat absorbing filter are: SP 425 + 3 mm BG 3. The fluorescence beam then passes a dichroic mirror (470 nm) necessary for the epi-illumination. As a barrier filter, a LP 460 is used.

(iii)   Epi-illumination is used for measurement purposes. The exciting wavelength is 485 nm (blue light) for excitation of the acriflavin-stained nucleus. The following filter combination is placed after the heat-absorbing filter: LP 475 + 2× SP 490 + 4 mm BG 38. Epi-illumination demands a dichroic mirror in the light path, which, in this case, is DM 470. It has to be noted that this dichroic mirror is not optimal for blue light excitation, but is a compromise between the view mode at 405 nm excitation and the measurement mode at 485 nm. The barrier filter is a LP 590.

(iv)   Oil-immersion objectives are used: for screening, a ×25 objective (NA 0.75) and for measurement, a ×63 Planapo (NA 1.4).

(v)   The fluorescence obtained by transmitted illumination is viewed through oculars (×6.3) and the fluorescence from incident illumination is directed towards the photomultiplier. After conversion of the photomultiplier signal in an A/D converter, the values can be displayed via a microcomputer.

(vi) A variable measuring diaphragm can be fitted around the object to avoid disturbance from background light. The field diaphragm for epi-illumination is set such that the illuminated area is only slightly larger than the measured area to prevent premature bleaching of the rest of the microscope field.

(vii) Electronically operated shutters take care of the switching from transmitted illumination to epi-illumination. They are activated simultaneously with the automated measuring procedure by pushing a simple button.

5.4.2 *Preparation of material*

(i) Scrape cervical cells from the cervix with an acrylic-cotton swab. Place the swab in a preservative solution [phosphate-buffered saline (PBS) or Polyionic with 25% ethanol].

(ii) After vortexing to dislocate the cells from the swab, wash the cells by centrifugation.

(iii) Resuspend the pellet in about 0.5 ml of Esposti fixative (acetic acid: methanol:distilled water, 7:33:30).

(iv) For cell dissociation syringe the suspension 50 times through 21-gauge needles, then add lymphocytes (extracted from fresh blood after centrifugation).

(v) Make the preparation by putting one or two droplets of the suspension on a glass slide.

(vi) Dry in paraformaldehyde vapour at 50°C for 30 min.

(vii) Just before staining fix the preparations further in Carnoy solution (15 min).

5.4.3 *Staining procedure*

The steps in the AFS staining procedure are:

(i) Rinse with water twice, each for 3 min.

(ii) Hydrolyse for 30 min in 5 M HCl (28°C) and rinse with water, again for 3 min.

(iii) Stain with 0.01% acriflavine (Chroma, FRG) for 15 min (28°C) followed by 1% HCl in 70% ethanol for 10 min.

(iv) Rinse with water for 5 min.

(v) Add 0.01% SITS (Polyscience, UK) in phosphate buffer, pH 7.6, for 15 min.

(vi) Rinse for 3 min with water followed by 100% ethanol (6 × 5 min), ethanol: xylene (1:1) for 5 min and finally with xylene (2 × 5 min).

(vii) Mount with Fluormount (Gurr, UK) under a coverglass.

Using this staining procedure, the nucleus can be visualized without interference from the cytoplasm, whereas the information of the cytoplasm is still available, either separately or in a total cell image. These preparation and staining procedures (29) have recently been modified and adapted to centrifugation buckets (39) but have only been extensively evaluated for absorption. Another cervical smear procedure is given in Table 20 in Chapter 4.

5.4.4 *Measurement procedure*

(i) Switch on light sources and computer and wait for about 20 min to allow stabilization of the lamps. In the meantime set the field diaphragm in the incident light path slightly larger than the size of the measuring diaphragm for a large nucleus, and the field diaphragm below the stage fully open. Apply Köhler illumination (see Chapter 1) using a test slide.

(ii) Put the slide on the stage and search for lymphocytes using transmitted illumination: both nuclei and cytoplasm will show weak fluorescence. Take care that only nuclei well separated from each other are chosen for measurement.

(iii) Set dark current value at zero.

(iv) Change to ×63 objective and adjust the measurement diaphragm to just surround the nucleus. For measurement use the option for non-automated continuous measurement and adjust the read-out value to 200 arbitrary units. Repeat this procedure with different lymphocytes until values are around 200 units.

(v) From this moment on the read-out values are no longer adjusted. Switch to automated measurement mode with print-out of values.

(vi) Search for and measure 10 lymphocytes and 10 intermediate epithelial cells throughout the preparation. Before each measurement the ×63 objective is switched into the light path manually and the measurement diaphragm is each time manually adjusted. Then a button is pushed and the fluorescence value obtained with incident illumination is registered.

(vii) Search for epithelial cell nuclei which are visually expected to have a high DNA content. Every time such a nucleus is found, its fluorescence is registered using the measurement procedure described under (vi).

(viii) Screen the whole slide for cells mentioned under (vii).

(ix) After termination of the measurement the values will be printed.

(x) Although lymphocytes and normal epithelial cells are both diploid cells, i.e. their DNA content is 2C, their fluorescence values are not similar using the staining method described above. The local absorbances, especially in lymphocytes, are higher than the 0.1−0.3 required for linearity. The fluorescence values for normal epithelial cells were found to be 1.3 times higher than lymphocyte values. To calculate the ploidy of the selected cells, the fluorescence values are divided by half the mean value of the lymphocytes multiplied by 1.3. Subsequently, a histogram relative to diploid values can be obtained.

We investigated about 800 cervical samples and could conclude that all cervical samples with a severe pre-cancerous lesion, or cancer, contained cells with more than 5C DNA. In 10% of the samples with no abnormality or only inflammatory changes, these cells were also present. Morphological codes given to the measured cells revealed that most high DNA-containing cells in normal specimens were enlarged, regularly shaped pale nuclei, whereas the nuclei in the abnormal specimens were often irregularly shaped and very hyperchromatic. The results of this study were the basis for the development of automated image analysis in cervical cytology using a TV analysis system called LEYTAS (see Section 6).
Nucleic acid quantitation by microfluorimetry has also been performed by many other laboratories (30,31,40).

6. IMAGE ANALYSIS IN FLUORESCENCE

Instead of viewing the microscope images by means of eyepieces, the image can also

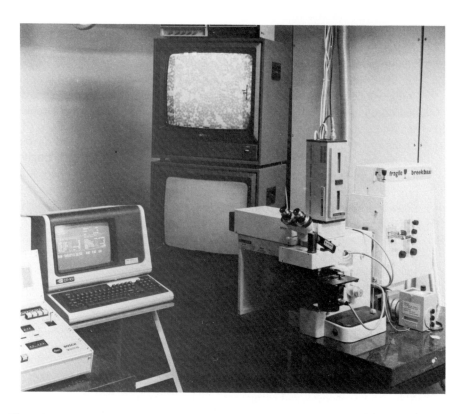

Figure 8. The image analysis system, LEYTAS, for analysis of fluorescent specimens. The microscope is connected to two TV cameras, one for absorption (on top of the microscope) and one for fluorescence (left side of the microscope). Microscope functions are performed via computer control. The cables seen in front of the figure connect the microscope to the computer. The screening can be followed on a TV monitor and the communication with the system is done via the terminal. The screen of the terminal displays the quantitative data of the screening.

be fed into a TV camera connected to the microscope. Image analysis computers coupled to such microscopes can then analyse these TV images after conversion to digital images. An example of such an instrument is the Leyden television analysis system (LEYTAS). This system consists of a microscope with automated functions, a TV camera, the image analysis computer (Leitz, FRG) and a buffer memory (*Figure 8*). Experiments in the analysis of fluorescence images with this set-up have shown that, provided the fluorescence is sufficiently bright, the signal-to-noise ratio is adequate. For an example it can be mentioned that the TV images of the antinuclear factor test showed sufficient brightness and contrast. With special LEYTAS procedures it is possible to automatically select nuclei above certain pre-selected thresholds, while artefacts are eliminated (41). This option is highly suited for the search of cells occurring in such low frequencies that a visual procedure is far too time consuming. An application of this type of investigation is described under the next heading.

6.1 **TV analysis in the location of rare mutant cells**

In the scope of environmental research, it is considered highly desirable to develop a test that can demonstrate the eventual toxicity of chemicals or radiation directly in the human. Based on the fact that point mutations in red blood cells result in deviating haemoglobins, such as haemoglobin-S (HbS), this medium was chosen for further investigation as a mutagenicity test. Haemoglobin-S cells, which can be made visible by a FITC-labelled antibody against HbS (42), occur in extremely low frequencies in 'normal' non-exposed human beings. A visual counting pointed to a frequency of 1 in 10 million blood cells. Such a visual counting procedure could not of course be carried out routinely. A machine such as an image analysis system would be much more suited for this purpose: the screening speed would be higher and the screening process not subject to fatigue or dips in concentration. Flow cytometry cannot be used to detect these rare mutants. Artefact rejection procedures in the latter method are not as sophisticated as in image cytometry and therefore it would not be possible to detect the few true mutants in the huge number of artefacts. However, even if the number of artefacts could be kept low, flow cytometry does not allow visual relocation of positive cells and thereby is unsuited for rare event detection.

For this investigation large microscope slides of 8 × 8 cm are used onto which 50 million cells are centrifuged. Strong polyclonal antibodies against HbS can be raised that result in measured high FITC fluorescence intensities in red cells containing HbS and low fluorescence readings of normal haemoglobin-A (HbA)-containing red cells. The contrast in the microscope image is quite sufficient for automated image cytometry.

A newly constructed Leitz multiparameter microscope can perform simultaneous image analyses on the haemoglobin absorption images of red cells and of fluorescence images of FITC-marked HbS cells, enabling millions of cells to be examined. Connected to the image analysis computer, the system (LEYTAS) first focuses and counts all erythrocytes along one row of microscope fields, by using dia-illumination with violet light (415 nm). After one 8 cm scan line, illumination is changed under computer control from dia-illumination to epi-illumination and all the fields of the scan line are re-examined, following a fitted curve through the previously defined focus positions to permit detection of the fluorescing objects. From each detected object both the fluorescence and the absorption image are stored in a grey value memory. This memory is displayed on a TV screen, thereby enabling visual evaluation of the selected objects (*Figure 9*). From each alarm the fluorescence image and two absorbance images are stored. Visual comparison of the high magnification absorption image with the fluorescence image renders information on the type of alarm. In this way artefacts can be distinguished from possibly true alarms. The confirmation that a mutant is an alarm is done in the microscope. As all co-ordinates are stored, relocation is automatically performed. Most alarms are artefacts, only frame number 79 in *Figure 9* is probably a mutant cell (43), a finding that can be verified by using the microscope.

Other antibodies against mutant erythrocytes have been used by Langlois *et al.* (44), who detected these mutants by flow cytometry. The frequency of the mutants detected by these workers is presumed to be much higher than the HbS frequency.

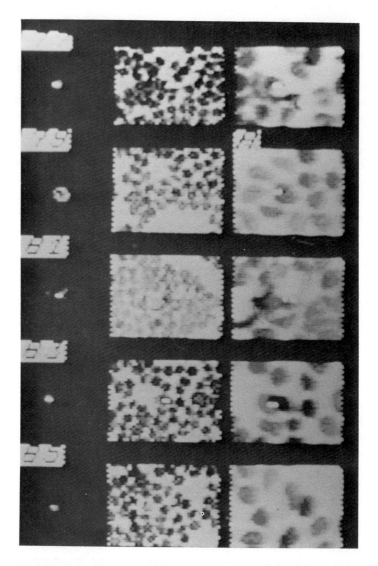

Figure 9. LEYTAS detects alarms in a peripheral blood sample stained with a FITC-labelled HbS antiserum. The alarms are displayed on a TV monitor. From left to right: alarm number, fluorescence image of the alarm, absorption image of the alarm, absorption image of same alarm but at larger magnification.

Besides the detection of rare mutants, image analysis can also be applied to detection of rare cancer cells in residual disease and to the detection of virus infected cells in an early state of the disease.

7. LASER SCANNING MICROSCOPY

Traditionally in microscopy, images of very small structures are obtained by illuminating the object with a lamp, whereafter the image is magnified by an objective. The addition of TV cameras did not change this principle as the image is still created by the objective

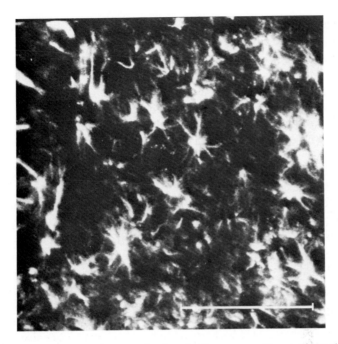

Figure 10. Fluorescence images by laser scanning from the hippocampus of the rat. The sections have been treated with fluorescently labelled anti-guanine monophosphate. Magnification: objective, ×40; zoom, ×10. Scale bar = 100 μm.

and only afterwards scanned by the TV camera. All scanning techniques have in fact been mainly applied to the limited field of microscope photometry to measure parameters in absorption, fluorescence and reflection contrast. Recently, scanning techniques have also been used in the build-up of high-quality images, by exploiting the high intensity and good collimation of laser beams. In optical laser scanning microscopy the object is not illuminated in total, but scanned step by step (17,45−47). From each illuminated point the transmitted, reflected or emitted light is measured. The image is built up by storing these point by point measurements, after analogue to digital conversion, as a matrix in a computer image memory.

In the instrument used in our laboratory (Zeiss, FRG) the scanning is performed by servo-controlled galvanometers. The scanning time is relatively short (2 sec/field of 512 × 512 pixels). A microprocessor controls the scanning process. A photomultiplier is used as a photodetector. Its signals pass an analogue signal processor, which controls brightness and contrast. After digitization, this signal is led to a frame memory, the read-out of which is in video frequency. Consequently the image on the monitor is stationary provided there is no movement of the stage. A continuously variable magnification can be obtained by a zoom unit. The laser scan microscope can be used with a conventional as well as with a laser light source. Both incident illumination (for reflectance and fluorescence) and transmitted illumination (for absorbance, phase- and differential interference contrast) can be obtained with the laser source. As a conventional source only a tungsten lamp guided by a glass fibre can be used. To provide better focusing and searching possibilities in the preparation and to allow investigation of double-stained probes, we have added a conventional epi-illumination device to the laser scanner. The principle advantages of laser scanning are:

(i)   The low level of autofluorescence of optical parts owing to the pointwise illumination.

(ii)   High sensitivity owing to the strong laser light that is focused pointwise. Weakly fluorescing DNA adducts can still be observed.

(iii)   The use of low-power optics. The fluorescence intensity is sufficiently strong to allow observation with ×2.5 objective. This is of special advantage in brain research.

(iv)   Low level of fading, as the illumination time per spot is extremely short.

(v)   Multiparameter analysis.

(vi)   Consecutive scanning of several layers in confocal scanning microscopy.

Examples of applications include studies where there is a weak fluorescence such as that which occurs with hybridization techniques. Brain research studies also benefit from laser scanning by the availability of low-power optics, which are necessary for evaluation of connectivity of nervous networks. In *Figure 10*, nerve cells from the hippocampus of the rat are displayed. The cells have been treated to demonstrate specific molecules. In comparison to normal fluorescence microscopy, the contrast between cells and background is much higher in images obtained by laser scanning: there is a very low level of unwanted background light. This reaction can also be quantified by laser scanning, as this technique allows thresholding to improve the contrast between cell and background in preparations with very low levels of the reaction product.

Laser scanning further facilitates the study of thin sections cut for electron microscopy by means of light microscopy techniques without the use of additional staining (48,49).

8. REFERENCES

1. Young,M.R. (1961) *Quart. J. Microsc. Sci.*, **102**, 419.
2. Price,Z.H. (1965) *Am. J. Med. Technol.*, **31**, 45.
3. Rost,F.W. (1972) In *Histochemistry, Theoretical and Applied.* Everson Pearse,A.G. (ed.), Churchill Livingstone, Edinburgh, Vol. 2, p. 1171.
4. Siegel,J.I. (1982) *Int. Lab.*, **12**, 46.
5. Ploem,J.S. and Tanke,H.J. (1987) *Introduction to Fluorescence Microscopy.* Royal Microscopical Society, University Press, Oxford.
6. Lansing Taylor,D., Waggoner,A.S., Murphy,R.F., Lanni,F. and Birge,R.R. (1986) *Applications of Fluorescence in the Biomedical Sciences.* Alan Liss, New York.
7. Patzelt,W. (1972) *Leitz-Mitt. Wiss. und Techn.*, Bd V, Nr 7, 226.
8. Giloh,H. and Sedat,J.W. (1982) *Science*, **217**, 1252.
9. Johnson,G.D. and de C.Nogueira Araujo,G.M. (1981) *J. Immunol. Methods*, **43**, 349.
10. Ploem,J.S. (1967) *Zeitschrift Wiss. Mikroskopie*, **68**, 129.
11. Nairn,R.C. (1976) *Fluorescent Protein Tracing.* Livingstone, Edinburgh.
12. Kraft,W. (1973) *Leitz Techn. Inform.*, **2**, 97.
13. Tomlinson,A.H. (1971) *Proc. Royal Microsc. Soc.*, **7**, 27.
14. Thomson,L.A. and Hageage,G.J. (1975) *Appl. Microbiol.*, **30**, 616.
15. Bergquist,N.R. and Nilsson,P. (1975) *Ann. N.Y. Acad. Sci.*, **254**, 157.
16. Wick,G., Schauenstein,K., Herzog,F. and Steinbatz,A. (1975) *Ann. N.Y. Acad. Sci.*, **254**, 172.
17. Wilke,V. (1982) *Proc. Royal Microsc. Soc.*, **17**, 21.
18. Rygaard,J. and Olson,W. (1969) *Acta Path. Microbiol. Scand.*, **76**, 146.
19. Ploem,J.S. (1971) *Ann. N.Y. Acad. Sci.*, **177**, 414.
20. Lea,D.J. and Ward,D.J. (1974) *J. Immunol. Methods*, **5**, 213.
21. Boehm,N. (1972) In *Techniques of Biochemical and Biophysical Morphology.* Glick,D. and Rosenbaum,R.M. (eds), Wiley-Interscience, New York, p. 89.

22. Pearse,A.G.E. (1972) *Histochemistry, Theoretical and Applied.* Churchill Livingstone, Edinburgh.
23. Rigler,R. (1966) *Acta Physiol. Scand.*, (Suppl.), **267**, 1.
24. Baer,G.R., Meyers,S.P., Molin,W.T. and Schrader,L.E. (1982) *Plant Physiol.*, **70**, 999.
25. Stoehr,M. and Goerttler,K. (1979) *J. Histochem. Cytochem.*, **27**, 564.
26. Wick,G., Baudner,S. and Herzog,F. (1978) *Immunofluorescence.* Medizinische Verlaggesellschaft, Marburg.
27. Falck,B., Hillarp,N.A., Thieme,G. and Torp,A. (1962) *J. Histochem. Cytochem.*, **10**, 348.
28. Ploem,J.S. (1971) In *Progress in Brain Research.* Eranko,O. (ed.), Elsevier, Amsterdam, Vol. 34, p. 27.
29. Cornelisse,C.J. and Ploem,J.S. (1976) *J. Histochem. Cytochem.*, **24**, 72.
30. Prenna,G., Leiva,S. and Mazzini,G. (1974) *Histochem. J.*, **6**, 467.
31. Boehm,N. and Sprenger,E. (1968) *Histochemie*, **16**, 100.
32. Ploem,J.S. (1970) In *Standardization in Immunofluorescence.* Holborow,E.J. (ed.), Blackwell Scientific Publications, Oxford, p. 137.
33. Sernetz,M. and Thaer,A.A. (1970) *J. Microsc.*, **91**, 43.
34. Haaijman,J.J. and van Dalen,J.P.R. (1974) *J. Immunol. Methods*, **5**, 359.
35. Ruch,F. and Leeman,U. (1973) In *Molecular Biology, Biochemistry and Biophysics. Vol. 14, Micromethods in Molecular Biology.* Neuhof,V. (ed.), Springer Verlag, Berlin, p. 329.
36. Wasmund,H. (1976) *Leitz-Mitt. Wiss. und Techn.*, **6**, 217.
37. Schipper,J., Tilders,F.J.H. and Ploem,J.S. (1978) *J. Histochem. Cytochem.*, **26**, 1057.
38. Ploem-Zaaijer,J.J., Beyer-Boon,M.E., Leyte-Veldstra,L. and Ploem,J.S. (1979) In *The Automation of Cancer Cytology and Cell Image Analysis.* Pressman,N.J. and Wied,G.L. (eds), Tutorials of Cytology, Chicago, p. 225.
39. Driel-Kulker,A.M.J. van, Ploem-Zaaijer,J.J., van der Zwan-van der Zwan,M. and Tanke,H.J. (1980) *Anal. Quant. Cytol.*, **2**, 243.
40. Ruch,F. (1970) In *Introduction to Quantitative Cytochemistry.* Wied,G.L. and Bahr,G.F. (eds), Academic Press, New York, Vol. 2, p. 432.
41. Ploem,J.S., van Driel-Kulker,A.M.J., Goyarts-Veldstra,L., Ploem-Zaaijer,J.J., Verwoerd,N.P. and van der Zwan,M. (1986) *Histochemistry*, **84**, 549.
42. Bernini,L.F., Ploem,J.S., Tates,A.D., Adayapalam,T. and Natarajan,A.T. (1986) *7th International Congress of Human Genetics.* Heidelberg, FRG.
43. Verwoerd,N.P., Bernini,L.F., Bonnet,J., Tanke,H.J., Natarajan,A.T., Tates,A.D., Sobels,F.H. and Ploem,J.S. (1987) In *Clinical Cytometry and Histometry.* Buerger,G., Ploem,J.S. and Goerttler,K. (eds), Academic Press, London, p. 465.
44. Langlois,R.G., Bigbee,W.L. and Jensen,R.H. (1986) *Human Genet.*, **469**, 1.
45. Bartels,P.H., Buchroeder,R.A., Hillman,D.W., Jonas,J.A., Kessler,D., Shack,R.V. and Vukobratovich, D. (1981) *Anal. Quant. Cytol.*, **3**, 55.
46. Deinet,W., Linke,M., Mueller,R. and Sander,I. (1983) *Microsc. Acta*, **87**, 129.
47. Brakenhoff,G.J., Blom,P. and Barends,P.J. (1979) *J. Microsc.*, **117**, 219.
48. Ploem,J.S. (1986) In *Applications of Fluorescence in the Biomedical Sciences.* Lansing Taylor,D., Waggoner,A.S., Murphy,R.F., Lanni,F. and Birge,R.R. (eds), Alan Liss, New York, p. 289.
49. Ploem,J.S. (1987) *Applied Optics*, **26**, 3226.

CHAPTER 7

Micrometry and image analysis

S.J.BRADBURY

1. INTRODUCTION

The need to measure objects seen with the microscope has been recognized for a very long time; the human eye is very good at pattern recognition, separating shapes from each other even though they may be of low contrast, touching or even overlapping. The eye, however, is very poor at estimating sizes, areas or numbers, and for this purpose some form of instrument is always required. Length estimations have been used in microscopy since the eighteenth century, using simple measuring devices which were the forerunners of those available today. With the perfection of the compound microscope in the nineteenth century, the accuracy of these basic measuring methods was improved so that they could be used at the limits of resolution and magnification allowed by the improved microscopes. It is only in the last 30 years, however, with the advent of electronics and, more recently, microcomputers, that measurement techniques have become precise and sophisticated. Now it is possible to measure much more than the mere size of an object; its area, perimeter, shape, optical density and many other parameters may be obtained. The data so acquired may be manipulated, stored and subjected to statistical analyses to determine degree of correlation and significance levels, and to show hitherto unsuspected relationships between multiple variables.

2. SIMPLE MICROMETRY

Although measurements are nowadays often obtained by the use of sophisticated electronic equipment attached to the microscope, it is probable that the average microscopist will generally use simpler methods for measuring objects. Such methods are, however, limited to determination of length, angles and number; by the use of stereological techniques it is possible to obtain accurate estimates of other parameters such as area, surface density, volume fraction, shape factors and grain size.

2.1 Linear measurements

Most commonly we need to know either the length of an object or, if it is circular, its diameter. Every measurement of length is made by comparison of the object with a scale. In everyday life it is simple to place a rule alongside the object and read off its length. It is not so easy if the object is seen under a microscope. Although it is possible to produce microscope slides which bear accurate calibrations, it is not usually practicable to mount the object to be measured directly on such a slide. Instead the microscopist superimposes an image of a transfer scale with arbitrary divisions, which serves as a ruler, onto the image of the object. After suitable calibration against the image of an accurately calibrated measuring scale, the transfer scale (which is, in fact, a graticule placed in the eyepiece) can then be used for measurements. Other methods will be considered in Sections 2.1.3 and 2.1.4.

2.1.1 *Eyepiece graticules*

The simplest form of transfer scale is a series of divisions engraved on a circle of glass called an eyepiece graticule. This is mounted in the microscope in a position where the scale will appear in sharp focus and superimposed on the object. The graticule normally lies on the eyepiece diaphragm (i.e. in its focal plane, which is coincident with the primary image plane of the objective). As the plane of the eyepiece diaphragm is conjugate with the object plane, the graticule is, therefore, automatically located so that it will appear in focus along with the image of the object. The size of the graticule disc required is determined by the internal diameter of the eyepiece tube; graticules are normally available on 16, 19 or 21 mm discs, although other sizes may be available to special order. In the more expensive graticules the pattern is produced in a layer which is sandwiched between two glass discs. This means that any surface scratches or dust on the graticule disc are usually out of focus when the scale itself is in sharp focus. A cheaper alternative is the surface type of graticule which bears the pattern in a layer bonded to the underside of the glass disc and reads correctly when viewed through the glass.

If the image of the scale is not absolutely sharp when the disc is inserted into the eyepiece, it may usually be brought into focus by slightly unscrewing the top lens of the eyepiece. A more satisfactory solution is to mount the graticule in a focusable eyepiece (one which is provided with extensive movement of the eyelens), so that sharp imaging of the graticule may be achieved with certainty. Such an eyepiece is commonly called a 'micrometer eyepiece' (*Figure 1*).

The scale engraved on the eyepiece graticule may take many forms (*Figure 2*), the most basic of which is a line (which may be orientated either horizontally or vertically in the field of view) marked off into numbered divisions. Cross-line graticules and net graticules are also available and are used for special purposes such as marking a reference point in a field of view or for dividing the field of view up into small areas for the easy counting of particles. A typical eyepiece graticule would have a line 10 mm in length divided into 10 major subdivisions of 1 mm, each of which is further subdivided into 10 parts. It is obvious that the actual values of the divisions of the eyepiece graticule will not correspond to the real length of objects in the field of view. The calibration of each division of the eyepiece graticule will vary according to the objective lens in use, to the tubelength of the microscope and to variations in the interocular separation of the binocular eyepiece tubes. For this reason it is essential to calibrate the eyepiece graticule for each objective lens on the microscope. Once these calibrations have been established with the help of a calibration scale (stage micrometer), they will be maintained as long as the same lenses are used and there is no change in the microscope tubelength or interocular distance. A calibration scale or stage micrometer as it is often called, is a glass slide of standard size bearing an accurately engraved scale. These scales are available in differing lengths, but the most common is 1 mm, with 100 divisions each 10 μm long. Alternative versions for use with lower powers are 2, 5 or even 10 mm

Figure 1. A micrometer eyepiece. This eyepiece contains a graticule which is focused by moving the sliding part of the tube on the left which contains the eye lens. The line scribed around the tube indicates its normal setting. Note the knurled locking ring for the sliding tube.

long, with correspondingly larger subdivisions. Stage micrometers are available for use with both transmitted and reflected light.

The accuracy of measurements made with an eyepiece graticule is not very high, being limited by the accuracy of the scale and, more importantly, by the observer's ability to read the coincidence of the scale with the edges of the object being measured. The procedure, however, requires very little equipment, is relatively easy to carry out and in most cases provides measurements of sufficient accuracy. The method of calibration and use of the eyepiece micrometer is given in Section 2.1.5.

2.1.2 *Filar or screw-micrometers*

The filar micrometer eyepiece, sometimes called a screw-micrometer eyepiece, permits measurements to be carried out more rapidly than with an eyepiece graticule or simple micrometer eyepiece. It is also easier to use than either of these. The filar micrometer (*Figure 3A*) has a moveable thread (often called a 'fiducial line') in its focal plane. Rotation of a screw thread driven by an external drum on the outside of the eyepiece traverses this line across the field of view. The drum bears a scale so that the amount of rotation may be accurately measured. In some filar eyepieces there is a fixed line which serves as the reference point to which one end of the object to be measured is aligned. Alternatively, if only the fiducial line is present, then this is used on its own, being aligned first with one end of the object and then with the other. The micrometer

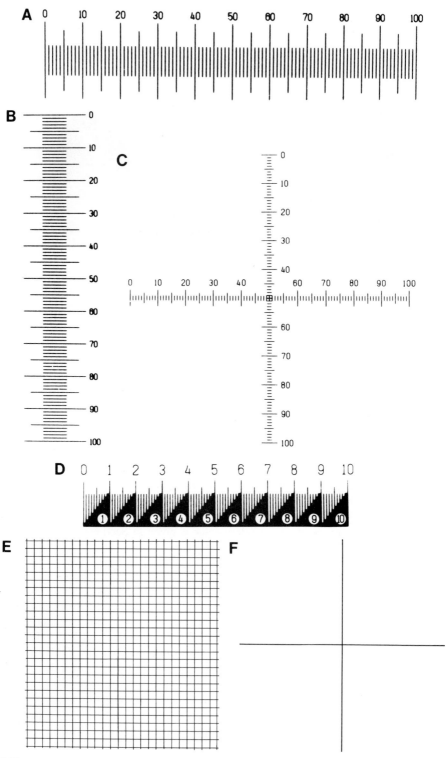

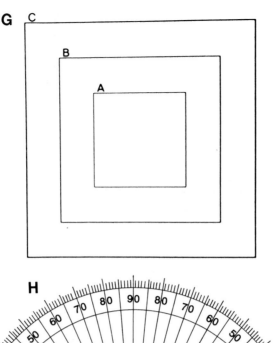

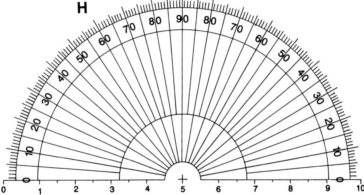

Figure 2. Diagrams of some of the rulings available as eyepiece graticules. (**A**) and (**B**) horizontal and vertical micrometer scales; (**C**) crossed micrometers; (**D**) step micrometer; (**E**) squared grid; (**F**) crosslines; (**G**) squares; (**H**) half protractor, graduated in degrees.

drum is read at each position and the difference between the two readings gives, when the eyepiece is calibrated with a stage micrometer, the apparent distance traversed by the line and hence the dimensions of the object. In another form of the filar eyepiece (known as a 'curtain' eyepiece), claimed by some workers to be much easier to use, the fine fiducial line is absent. In its place there is an area on one side of the field in which the image appears darker. As the drum is rotated a second dark area (or 'curtain') appears on the opposite diameter of the field and approaches the stationary curtain. The edges of the curtains are used as the markers for the spanning of the object dimensions just as with fiducial lines.

2.1.3 *Image shearing eyepiece*

This eyepiece (*Figure 3B*), originally produced by Watsons and later by Vickers, represented a means of measuring microscopic objects with an accuracy about ten times

191

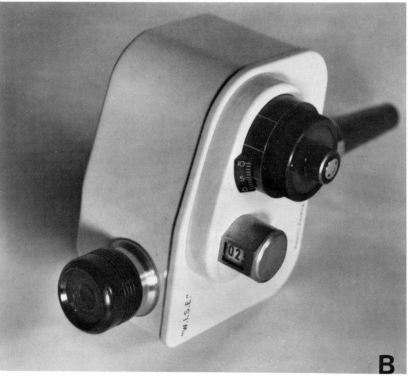

better than was possible with a standard eyepiece micrometer. Although it is now no longer in production, many image shearing eyepieces may still be available in laboratories. The principle of operation involves splitting the microscopic image into two identical images which may be separated or sheared by a variable and measured amount (*Figure 4*). This shear is obtained inside the eyepiece either by rotation of prisms on a vertical axis midway along their length or by rotation of two mirrors mounted on a carrier. The two sheared images are coloured, one in red and one in green to provide contrast and to allow very precise setting of the degree of shear. At the point of coincidence a normal image will be seen but as the setting control is rotated the image will shear into two components, one red and one green. When the two images contact one another, any overlap will result in the appearance of a dark line of very high contrast (*Figure 4*). This system allows very precise edge to edge settings to be made (hence the accuracy of the measurements) and the object does not have to be moved into coincidence with any reference or fiducial line. In addition, everything in the field of view is sheared by the same amount so that if the setting is made for one object then other objects may immediately be detected as being of larger or smaller size. The data from an image shearing eyepiece is usually obtained from a digital counter, although sometimes a vernier is fitted to the control knob to allow a more precise estimation of the amount of shear. As with all eyepiece micrometers, the image shearing eyepiece must be calibrated against a stage micrometer. This is done by choosing a suitable number of divisions (e.g. ten) which are sheared in one direction, their images are thus superimposed so that line number one is on top of the image of line number ten. The reading of the counter is then noted. The control is now rotated to shear the images in the *opposite* direction until there is again coincidence of an equal number of lines when a second reading of the counter is taken. The difference between the first and second counter readings, divided by two, represents the amount of shear for that given number of stage micrometer divisions. The actual measurement corresponding to the shear introduced when the counter is advanced by one unit may thus be calculated. A similar method is used to take measurements of objects; counter readings are taken when the object is sheared completely in one direction and then completely in the opposite sense. The difference between these two readings, divided by two and multiplied by the calibration factor for one unit of the counter, gives the required dimension. Alternatively counter readings may be taken when the image is first set so there is absolute coincidence (zero shear) and then for complete shear in one direction only. In this case the simple difference between the two readings, multiplied by the unit calibration factor, gives the size of the object. With any given shearing eyepiece, the operator will soon establish with accuracy the reading for zero shear and so this second method which requires only one setting of the eyepiece in the position for complete shear is much quicker, although less accurate than the double shear technique.

Figure 3. (**A**) Filar micrometer eyepiece. The knurled ring around the upper part of the tube serves to focus the fiducial lines. The graduated micrometer drum which controls the movement of one of the lines is seen on the right. (**B**) The image shearing eyepiece. The eye lens is seen at the bottom, whilst the digital counter and the operating drum are visible on the right hand side of the housing. The counter advances one unit for the movement of each division of the operating drum past zero on its scale. The graduations on this allow the estimation of fractions of a unit.

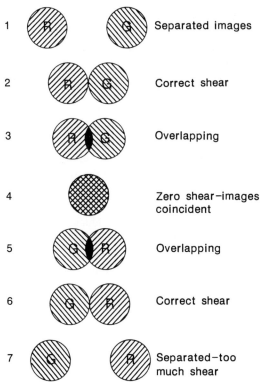

1		Separated images
2		Correct shear
3		Overlapping
4		Zero shear—images coincident
5		Overlapping
6		Correct shear
7		Separated—too much shear

Figure 4. A series of diagrams illustrating the operation of the image shearing eyepiece. (**4**) shows the appearance when the images are coincident. (**1**−**3**) show the separation in one direction, whilst (**5**−**7**) show the same process in the opposite sense. (**2**) and (**6**) show the images at the point of exact shear; these represent the settings at which the readings would be taken.

2.1.4 *Other methods of obtaining linear measurements*

Although the methods of measurement outlined in Sections 2.1.1 to 2.1.3 above are most often used, occasionally the need arises for a quick and rather crude size measurement. If the object is large (and hence a low power of the microscope is in use) then it is possible to obtain its size by traversing the object with the mechanical stage control across the intersection of the lines of a cross graticule (*Figure 2F*). The size is obtained by reading the scale and vernier of the stage at the beginning and end of the traverse. Such a method suffers from the drawback that it is often not easy to read the scales of a mechanical stage with any accuracy and so the measurement is likely to be in error by as much as 100 μm.

It is also possible to obtain measurements from a projected image. This may be either onto a specially designed viewing head fitted with a ground glass screen or onto a sheet of white card fixed to a wall or the bench. The image of the object is thus enlarged sufficiently for it to be measured with an ordinary ruler, the divisions of which have been previously calibrated by the measurement of the image of a stage micrometer projected with the same optics and projection distances as the specimen. This type of measurement technique has the advantage that the image of the object is much larger than that in an eyepiece micrometer so the actual measurement is much easier.

Several electronic television-based systems are now available which attach to a microscope in order to allow accurate linear measurements. In most instances a scale is superimposed onto the image of the specimen appearing on the TV monitor. The size of the scale is then adjusted electronically to match the dimensions of the object of interest, whose size is then directly displayed either in a digital read out or on the screen itself. As with all measurement systems, it is essential to use a standard scale for the initial calibration of the apparatus. In some of the more sophisticated systems, there is provision for displaying a grey level profile of the objects so that the measuring rectangle or cursors may be more accurately positioned with respect to the edges of the object of interest by setting them at the 50% level between white and black levels. Other instruments allow the use of a measuring rectangle (or circle) with inner and outer tolerance levels set by the user indicated; they may also permit the measurement of angles and allow simple cell counts. These systems (for example the Crystal, the RMC-D3 and the Portascan—see Section 6 for sources) are excellent when many routine measurements have to be made and when their price (which may be several thousand pounds) is acceptable.

2.1.5 *Calibration and use of the eyepiece graticule or filar micrometer*

The following procedure is recommended.

(i)   Set up the microscope for correct Köhler illumination, using the required objective and a contrasty specimen.

(ii) Replace the slide by a stage micrometer and place the image of the scale across the field of view.

(iii) Point a micrometer eyepiece towards a bright surface and focus its scale by altering the focus slide of the eyepiece. Care must be taken to avoid accommodation of the eye.

(iv) Replace the eyepiece of the microscope by the micrometer eyepiece and rotate it until its scale lies parallel to the image of the scale of the stage micrometer.

(v)   Align the scale in the eyepiece with that of the stage micrometer so that there is a point of coincidence at one end (*Figure 5*) where the 0 of the eyepiece graticule coincides with the 0 of the stage micrometer.

(vi) Count the number of divisions of the eyepiece graticule between this coincidence and the point of the next suitable coincidence between graticule and stage micrometer.

(vii) Calculate the value of the eyepiece graticule divisions in terms of micrometers.

In the example illustrated in *Figure 5*, when the zeros of both eyepiece graticule and stage micrometers coincide, the 10 mark of the eyepiece graticule (i.e. 100 divisions) corresponds to 1270 μm on the stage micrometer. Thus, for this combination of objective, tubelength and eyepiece in our example, each division of the eyepiece graticule equals 1270/100, or 12.7 μm.

It is now possible to measure objects by simply spanning their image with that of the eyepiece graticule, counting the number of divisions which they occupy and multiplying this number by the calibration factor (12.7 in our example) in order to obtain the size in μm.

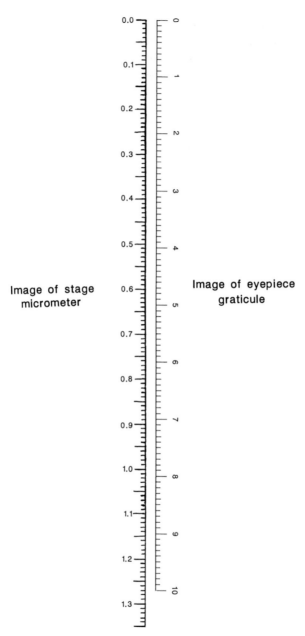

Figure 5. A diagram to show the appearance in the microscope of (**left**) the image of the stage micrometer and (**right**) the image of the eyepiece graticule. They have been brought into co-incidence at the zero points. It is seen that 10 divisions of the eyepiece scale are equivalent to a real distance, as indicated on the stage micrometer of 1270 μm.

If the microscope is an older monocular model which is fitted with a draw tube the setting of this should be adjusted to 160 mm before commencing calibration. Care should be taken not to alter this value after calibration and during measurement, as this would

196

alter the magnification of the final image and thus invalidate any measurements. Although the objectives are corrected to give their best image when used with a draw tube setting of 160 mm, it is permissible, during the calibration of an eyepiece micrometer, to alter the extension of the draw tube slightly if this is necessary to bring the calibration to an exact round number when a given number of eyepiece divisions are superimposed on a known number of stage micrometer divisions. After such a procedure it is important to note the exact setting of the draw tube and to reproduce it in any future measurement sessions.

The calibration procedure is exactly the same with a filar micrometer, except that the fiducial line is set to a given mark on the stage micrometer and the drum reading noted. The moveable line in the eyepiece is then moved across the image of the stage micrometer by rotating the micrometer drum until the image of the line coincides with a suitable mark further along the stage micrometer. This should be a considerable distance from the first reference point, since the greater the distance traversed, the more accurate will be the calibration. When the fiducial line is correctly registered on the second mark, the drum is read again and the amount of rotation calculated by subtracting the two readings. From this figure the absolute value of the amount of line traverse introduced by a given amount of drum movement is calculated in the same manner as for an eyepiece graticule.

2.2 **Angular measurements**

It is sometimes necessary to measure angles under the microscope, especially when studying crystals and minerals. Formerly, special goniometer eyepieces were available for this purpose. In a goniometer eyepiece, the upper part (which was fitted with a cross wire) was capable of rotation on the lower part. The rotating portion carried a pointer which moved around a scale graduated in degrees mounted on the lower, stationary part of the eyepiece. To measure an angle between two faces of a crystal, for example, one of the cross wires was aligned so that it was parallel with respect to one of the faces and the scale reading of the pointer noted. The upper part of the eyepiece was then rotated until the same cross wire was aligned with the second face and another scale reading taken, moving the object if necessary in order to achieve accurate alignment. The difference between the two readings gave a direct measure of the number of degrees of rotation and hence the required angle. Such eyepieces are now unavailable but two alternative methods are possible. The first involves using a protractor graticule which is in the form of a circle or half circle graduated in degrees (*Figure 2H*). The graticule is fitted into a micrometer eyepiece in place of the conventional scale and its image focused and superimposed on that of the image. It is then possible to estimate angles by lining up one side of the object with the radial graduations and reading off directly the value which is coincident with the two sides which include the angle of interest. The accuracy of this method which depends very much upon the observer's ability to position the object with respect to a reference line and to read off the value in degrees is usually sufficient for all practical purposes.

The second method of estimating angles is to use a microscope fitted with a circular rotating stage which is graduated in degrees. If a cross wire graticule is used in the eyepiece as a fiducial reference, both sides of the object containing the angle to be measured may be lined up with this in turn; the difference in the stage readings gives the required included angle. Before beginning, the rotating stage should be carefully centred to the optical axis of the microscope. As many of these rotating stages are provided with a vernier scale, the accuracy of the estimation (approaching 0.1°) is much better than that possible with a simple eyepiece protractor.

If television-based image measuring equipment (e.g. the Portascan) is available, such angular measurements are usually very simple and are also extremely accurate; the manual appropriate to the individual system should be consulted for details of procedure.

2.3 **Measurement of depth**

There are many possible applications for measurements of depth (i.e. dimensions in the z axis) in microscopical preparations. For example, it may be necessary to determine the thickness of a microscopical section of tissue or the height of a layer in an integrated circuit. The standard method for obtaining depth measurements (see Section 2.3.1) relies on the fact that high aperture objectives have a small depth of focus (of the order of 1 μm or less). Galbraith (1) has shown that the real depth as measured by the method of Section 2.3.1 is not accurate. He suggested a correction, which may be obtained by multiplying the apparent depth by n_2/n_1, where n_2 is the refractive index of the object and n_1 is the refractive index of the immersion medium of the objective lens. He demonstrated that the refractive index of the *mountant* is unimportant in the calculation. Other complications may arise if the object is not plane-sided or approximates to a vertical line (e.g. a flagellum); for further details the original paper should be consulted. The method of depth measurement given in Section 2.3.1 is not of great accuracy (even with the correction), so Galbraith suggests that wherever possible it is preferable to make linear measurements in the x or y direction, using the same object (if it is a sphere) or another specimen lying in a different orientation if the object is elongated.

2.3.1 *Using the calibrations on the fine-focus to measure depth*

The following procedure is recommended.

(i)   Set up the microscope for correct bright-field Köhler illumination. Use an objective of fairly high power and numerical aperture (NA) (e.g. ×40, 0.65 or ×100, oil 1.3).

(ii)   Place the fine-focus adjustment knob in the middle of its range of travel. This will minimize errors due to lack of linearity in the screw focus mechanism.

(iii)   Focus carefully on the upper surface of the object.

(iv)   Read the value indicated on the calibration scale of the fine-focus knob and note it (e.g. 53 units).

(v)   Use the same focus knob to lower the focal plane of the objective so as to bring the plane of the bottom of the object into focus. In practice this plane is taken as that of the surface of the support slide, just to one side of the object itself.

(vi)   Note the new reading of the calibration scale on the fine-focus control (e.g. 49 units).

(vii)   Calculate the distance moved by the objective by subtracting the two readings

(53 − 49 = 4 scale units, in our example). This is the 'apparent depth' of the object.

(viii) Calculate the real depth of the object by multiplying the apparent depth (determined as outlined above) by the calibration factor for each scale unit. This is often engraved on the microscope stand, close to the fine-focus knob. It is often 0.5 μm. In our example, therefore, the real depth of the object would be (0.5 × 4) = 2 μm.

If there is no calibration on the microscope stand then this may be determined by repeating steps (i) to (vii) above but using a specimen of known depth or thickness. A good calibration specimen for transmitted-light microscopy would be latex microspheres (obtainable from E.F.Fullam, see Section 6) which are available as an aqueous suspension in a wide range of diameters from 0.5 to 25 μm. A good calibration standard for incident-light microscopy would be a glass micro-balloon of known size. These are available (mounted on a scanning electron microscope specimen holder) in various sizes between 50 and 230/240 μm from the same supplier.

2.4 Counting chambers

Estimation of the number of objects (e.g. erythrocytes or yeast cells) in suspension in a fluid is easily done by using a counting chamber. This is a specialized microscope slide that is designed either to contain a known volume of liquid or if of a definite depth have a series of accurately etched squares on its floor which demarcate units of a specified volume. Several different patterns of ruling are available designed for specific purposes or to facilitate the counting of blood cells. Slides especially intended for this latter purpose are often described as haemocytometers. Most counting chambers are designed with surfaces which are optically flat and highly polished, to be used with a specially thick coverslip (so that it will resist deformation which might cause considerable errors in the contained volume) and with a 'moat' around the counting area to catch surplus fluid when the chamber is filled to overflowing, as it must be in order to ensure the correct volume of suspension is included. When this is done, the volume enclosed by each set of rulings is given by the area covered by the ruling multiplied by the depth of the chamber. The simplest type of counting cell (which is now commercially available moulded in plastic or made with much greater accuracy in glass) is the Sedgewick Rafter (microlitre) cell (*Figure 6c*). This has a central area 1 mm deep and 20 mm × 50 mm in size. The base of this area is ruled into squares with sides of 1 mm. When filled with liquid, the grid demarcates 1 μl aliquots of fluid so that the average number of particles counted/square represents the number/μl of fluid. Errors may arise in counts made with this chamber if the particles are not distributed in a random fashion over the squares of the chamber. This possibility must be borne in mind; it may, however, be corrected for by counting a sufficiently large number of squares taken at random from each sample.

Typically, one would count, say, 20 squares taken at random from the whole area of the plate. From these figures, the mean, SD and SEM are determined using the accepted formulae. If the sample is truly random and from a normally distributed population, it is then easy to estimate the minimum number of particles which must be counted in order to estimate the population SEM with any given set of confidence limits. The SEM is given by the expression

$$\text{SEM} = \frac{\text{SD}}{\sqrt{N}}$$

where SEM is the standard error of the mean, SD the standard deviation and N is the number sampled.

$$N = \left(\frac{\text{SD}}{\text{SEM}} \right)^2$$

If SEM is used in this equation, then the answer gives the minimum number of particles to be counted for a probability level of 0.682. If a probability level of 0.954 (set by ± 2SEM) is desired then the figure for SEM used in the denominator of the expression must accordingly be divided by two.

Specialized types of counting chamber are available for counting blood cells; these chambers are called haemocytometers. In one such, (the improved Neubauer haemocytometer) the depth of the cell is 0.1 mm. There is a central square with sides of 3 mm, subdivided into 9 squares each with sides of 1 mm. Each of the outer squares is subdivided into 16 squares each with sides of 0.25 mm whilst the centre square is further

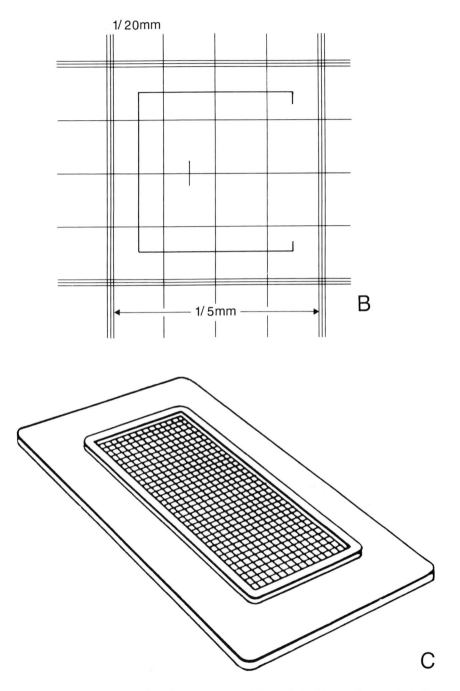

Figure 6. The rulings of the Neubauer haemocytometer. (A) The whole of the counting area. Note the central group of subdivided squares indicated by the triple rulings. The five special counting squares are indicated by the letter 'E' engraved into them. The four large squares with the engraved letter 'L' are for the special purpose of counting leucocytes in a blood sample. (B) An enlarged view of one of the small central squares to show the triple ruling which delineates it. (C) Sedgewick Rafter cell.

subdivided into 400 small squares with sides of 0.05 mm in length (*Figure 6A*). By means of a distinctive triple ruling, these small squares are grouped together into 25 blocks each containing 16 squares (*Figure 6B*). As the chamber is 0.1 mm deep, the volume of fluid held above each of these smallest squares is thus 2.5×10^{-4} mm^3. Detailed instructions for using a haemocytometer are given in Section 2.4.2.

It is often necessary to dilute the suspension of particles before counting; concentrated suspensions result in too many particles lying in each square which makes counting difficult and liable to error. A dilute suspension with only a few particles in each square is quick and easy to count. If working with particles it is advisable to make the suspensions in whole numbers of units of volume, say 10 ml or 100 ml; if still further dilutions are needed because the particles are too numerous to count easily, then this should be done in convenient steps of, for example, 10:1, 50:1 or 100:1 etc. When the actual counts are made, it is important to count as many squares as possible and to establish a convention for counting particles which appear to lie on the etched lines forming the sides of the squares. Usually one counts particles lying on the top and left hand side and excludes those intersecting the bottom and right hand margins. In all counts involving these chambers, some kind of hand tally counter is of great value in keeping the total number of counts.

2.4.1 *Technique for using a standard counting chamber*

The procedure is as follows.

(i)   Set up the microscope correctly, using either bright-field Köhler or a suitable contrast technique. Use sufficient magnification to allow easy visualization of the particles.

(ii)   Prepare a suitable dilution (e.g. 1:100) of the liquid sample containing the particles.

(iii)   Fill the Sedgwick chamber with the dilute suspension using a clean Pasteur pipette. Place the cover in place and press down firmly to displace excess fluid into the moat.

(iv)   Place on the microscope stage and allow the particles to settle.

(v)   Count the particles in, for example, 20 squares, each of which demarcates 1 μl of sample. It is customary to include in the count those particles which overlap the margins on the top and left hand side and exclude those intersecting the bottom and right-hand margins.

(vi)   Calculate the mean, SD and SEM for the number of particles/μl suspension.

(vii)   Use these figures to calculate the number of particles which must be counted in order to attain the required statistical degree of accuracy in your estimate of particle number.

(viii) Count further squares of the preparation as necessary and recalculate the mean and its SD and SEM.

2.4.2 *Technique for using a haemocytometer*

(i)   Dilute the sample of particles with water or saline or other suitable fluid, choosing a suitable value, for example 1 in 100 or 1 in 200.

(ii)   Place the coverslip in position on the counting chamber and press firmly into position. Usually a series of interference fringes will be seen between the coverglass and the sides of the chamber itself. This ensures that the chamber holds the correct quantity of fluid and that the coverslip is firmly adherent to the walls of the chamber.

(iii) Apply the tip of the pipette containing the diluted sample to the grooved edge of the coverslip and let sufficient fluid flow into and just fill the counting chamber by capillary attraction.

(iv) Allow the particles to settle.

(v) Place the counting chamber carefully on the microscope stage and focus with the ×10 objective. Switch to the ×40 objective.

(vi) Count the particles in at least five groups of the 16 smallest squares of the chamber. One group should be taken from each corner and one from the centre of the group of squares delineated by the triple ruling (these are indicated by the letter E in *Figure 6A*). When counting, the middle line of the triple ruling is used in each case. Particles touching the delineator line at the top and on the left hand side of the group of squares should be included; those touching the bottom and the right hand side are to be excluded. The counting should be started at the top left hand square of the block of 16 and proceed downwards through the four squares, then upwards through the adjacent four squares, downwards again through the next and finally upwards through the four squares on the right hand side of the block.

(vii) Calculate the number of particles/mm^3 from the following expression:

$$\frac{\text{number of particles counted} \times \text{dilution (D)} \times 4000}{\text{number of the smallest (0.05 mm) squares counted}}$$

As an example, let us assume that the total number of particles counted was 525, the number of smallest squares was the standard 5×16 ($= 80$) and D was 1 in 200. The total number of particles/mm^3 is then

$$\frac{525 \times 200 \times 4000}{80} = 5\ 250\ 000$$

A quick form of calculation is available if the standard five groups, each containing 16 of the smallest squares are counted, at a dilution of 1 in 200 and a chamber depth of 0.1 mm (1/10 mm^2). In this case, with an area of count of 0.2 mm^2 (1/5 mm^2), assume the number of particles to be P. The number of particles/mm^3 is then given by $P \times 5 \times 10 \times D$. For a dilution of 1 in 200 this becomes $P \times 10\ 000$ (i.e. P with the addition of four zeros). For a dilution of 1 in 100 the total number/mm^3 becomes $P \times 5000$.

3. AREA, SURFACE AND OTHER ESTIMATIONS

Quantitative microscopy has expanded greatly in the last decade or so to include the measurement of areas, surface densities, shapes, curvatures, measurements of preferred orientations and other topological descriptors. These techniques may be used for the quantitative assessment of features in two-dimensions (even though the objects themselves are three-dimensional). Alternatively, similar methods may take the two-dimensional image and use it to derive information on the three-dimensional structure. These latter techniques form the basis of stereology. Quantifying a structure in two-dimensions is often called morphometry. Some authors, however, use this word in a broader sense, as an inclusive term to include stereology as well. It is not possible to cover morphometry or stereology in detail in this chapter but several comprehensive works are available for the interested worker. The texts by Aherne and Dunnill (2), Elias and Hyde (3), Russ (4), Weibel (5,6) and Williams (7) listed in the references may be consulted for full details.

The ideal image for quantitative analysis would be from an infinitely thin section. This is not possible to obtain in practice but opaque surfaces imaged with incident light or projected images of well-separated particles approach the ideal for many types of measurement. If the image is that of a thick section then there may be problems due to structures overlapping or touching; if the thickness is reduced then portions of features are more likely to be cut off or be of low contrast and hence not available for measurement. Techniques are available for compensation for such errors and these are fully described in the references cited above. Other problems arise if the components of a tissue or the separate objects in a matrix which are to be measured are not arranged in a truly random or isotropic fashion. Such non-randomness (or anisotropy) again require special techniques which are beyond the scope of this introduction to the subject.

In morphometry and stereology, we must consider points which are dimensionless, lines which have lengths and hence have one dimension, areas which have two dimensions and volumes of three dimensions. If a thin section is taken through an object of (n) dimensions it is clear that the result must be a $(n-1)$ dimensional profile. Thus, a fibre (which is effectively a line) appears as a point, a two-dimensional surface as a line and a three-dimensional solid as an area. The principles of stereology are simple and straightforward, although as two (at least) separate systems of symbolism are in use, this may not be apparent at first sight. The numerical data are usually expressed in terms of a reference volume which may be in unitary terms (i.e. per mm^3), or may refer to some other volume, such as the volume of the organ in question or the volume of a specific component of an organ.

Most methods of morphometry and stereology produces estimates rather than exact answers; by repeating the number of counts or trials the estimates may be made to any desired degree of accuracy. The methods involve counting either the number of points or intersections of reference lines which fall upon the objects of interest when the reference set of points or lines is superimposed several times at random to the image. This superimposition may be achieved by the use of an eyepiece graticule bearing a suitable set of test points or lines. Alternatively an overlay drawn out on transparent plastic may be placed on photographic prints or the image may be projected onto a drawing of the test pattern. The first method is usually employed with the optical microscope, whilst the use of a transparent overlay is the method of choice for electron micrographs. Several types of test pattern are in current use, three of which are shown in *Figure 7*. Simple patterns are the easiest to use and for most purposes a grid of squares (*Figure 7A*) is quite suitable. A widely used alternative (*Figure 7B*) due to Weibel *et al.* (8) is based on equilateral triangles, this uses the ends of the lines as points for area and volume fraction estimations, whilst the lines themselves serve as test probes intersecting membrane profiles for the determination of surface density. The curved pattern

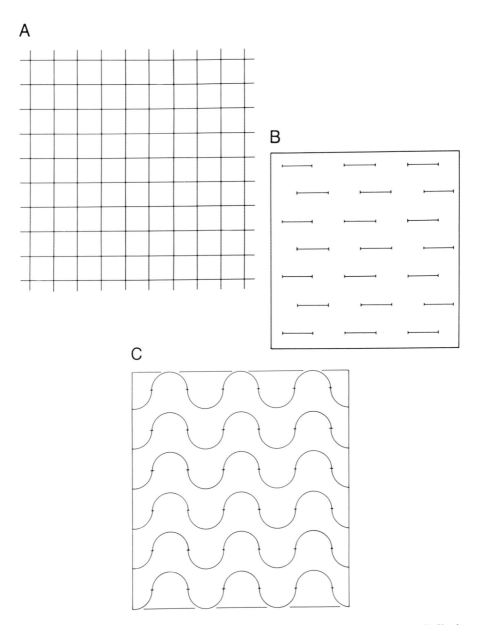

Figure 7. A diagram of three different graticules which can be used for morphometry or stereology. (**A**) Simple net ruling. (**B**) Weibel graticule with lines, the ends of which serve as the point counting indicators. (**C**) A Mertz graticule intended to compensate for anisotropy in the specimen.

(*Figure 7C*) was devised by Mertz (9) and is intended for minimizing the effects of anisotropy in the specimen.

In use the pattern is placed at random on the image and the number of points falling on each phase or object of interest recorded. The pattern is then displaced, again entirely at random, over the image and the test repeated several times. From the data the required parameters may be calculated. In practice, in optical microscopy it is more satisfactory in order to avoid any bias to move the slide with respect to the eyepiece graticule. This is most easily done by the use of a mechanical stepping stage which prevents any selection (conscious or unconscious) by the observer of which fields to measure. With electron micrographs it has become customary to avoid such bias by taking micrographs for analysis in a systematic but random manner by sampling only from the upper left hand corner of the grid squares.

The most frequently measured parameters are probably area fractions and volume fractions. Surface areas again are easily determined and it is now possible with the rapidly developing techniques of stereology to measure size classes, assess the shape of particles, the thickness of laminae, the degree of curvature of surfaces, degree of anisotropy and many other parameters. Full details of these are given in the references (2−7).

3.1 Principles of the measurement of area and volume fraction

It is very easy to estimate the fraction of a plane transect which is occupied by profiles of a given structure. This is done by superimposing on the image a grid of discrete points, such as those in *Figure 7*, and counting the number of points which fall on the features of interest as well as the total number of points which fall on the whole object. This exercise is repeated a number of times, the counts are averaged and the fraction, known as the 'area fraction' (usually symbolized A_A), of the whole occupied by the component of interest, is then calculated. It has been known for over a hundred years that a simple equivalence relationship exists between the A_A of a component and the fraction which the same component occupies in the bulk structure, that is the volume fraction (V_V). This relationship holds, irrespective of the manner in which the phase is shaped or dispersed within its matrix provided that the sampling plane is truly random with respect to the structure of the material. This equivalence was originally suggested by the French geologist Delesse (10) who made a study of minerals exposed on the polished surface of a rock section. Originally the A_A was determined by drawing the outlines of the phase in question, carefully cutting them out and determining the ratio of their weight to that of the picture as a whole. Later it was shown by Rosiwal (11) that the A_A used to determine V_V could be substituted by another parameter called the fractional linear intercept (L_L). An intercept is the name given to those parts of a series of lines superimposed on the image which fall on the profiles of interest. If we add the combined length of all the intercepts and divide by the total length of all the lines on the image we arrive at the L_L. This again is equivalent to the V_V. The L_L method is easier to perform than weighing cut-out areas and it has recently proved valuable as the basis by which many automatic image scanners determine A_A and V_V. The knowledge of the equivalence of the A_A and the L_L with V_V was further extended about fifty years ago. At that time Glagolev (12) showed that all of these were determinable by counting the number of points of a test graticule image superimposed on the section which coincided with the phase of interest. This point count (P_P) was subsequently rediscovered by Chalkley (13) and several authors have now demonstrated mathematically the truth of the equivalences

$$V_V = A_A = L_L = P_P$$

For manual determination of A_A or V_V the use of point counting is the method of choice as it is easy, relatively quick and can give any desired degree of accuracy.

3.1.1 *The determination of A_A in a two-phase object*

The procedure is as follows.

(i) Prepare the sample and examine under the microscope using a suitable magnification and contrast technique.

(ii) Place a Weibel graticule (*Figure 7B*) in the eyepiece focal plane so that it appears sharply superimposed on the image of the specimen (*Figure 8*).

(iii) Count the number of test points (in this case the ends of each line) which fall
(a) upon the background (= 7 in the example)
(b) upon the principal phase A (= 30)
(c) upon the second phase B (= 5)
The P_P (and also the A_A and the V_V), using phase A as the reference area, is thus for this single trial $5/30 = 0.166$.

(iv) Repeat stage (iii) for more trials (e.g. nine) in order to obtain more points. Use new microscope fields for each addition trial.

(v) Calculate an interim value of V_V from the data obtained from these measurements.

(vi) Calculate the relative standard error (RSE), a parameter derived by Hally (14), from the equation

$$RSE = \frac{\sqrt{1 - V_V}}{\sqrt{n}}$$

so that

$$\sqrt{n} = \frac{\sqrt{1 - V_V}}{RSE}$$

and

$$n = \left(\frac{\sqrt{1 - V_V}}{RSE} \right)^2$$

The interim value of V_V calculated in (v) above is used in this equation.

(vii) Repeat the sampling until sufficient points have been counted to achieve the desired accuracy of the estimate.

Stage (vi) above may be illustrated as follows. To obtain a final RSE of 5%, then using the value of V_V postulated in our example above of 0.166, we would have

$$n = \left(\frac{\sqrt{1 - 0.166}}{0.05} \right)^2$$

$$= \left(\frac{\sqrt{0.84}}{0.05} \right)^2$$

$$= 336 \text{ points.}$$

As V_V, however, only occupies 16% of the whole tissue we must count *at least* $336/16 \times 100 = 2100$ points, *falling on both phases of the tissue*, in order to be sure that our estimate of V_V remains within the 5% figure for phase B. The smaller the value of V_V then the larger is the value of n for any given RSE. If the V_V of the phase of interest had been of the order of 45%, then the same RSE would be obtained by counting about 500 points whilst for a V_V of about 85% then only 100 counts would be needed.

3.2 **Principles of the measurement of surface area**

A very important measure of correlation between structure and function in many biological systems is the determination of surface area. There are obvious relationships between the surface area of the alveolar walls of the lungs and the ability to exchange gases such as oxygen and carbon dioxide between the tissues and the air. Also, the ability of the small intestine to absorb digested nutrients depends to a large degree on the surface area of the villi lining the intestine, as well as on the well-being of the individual absorptive cells forming these villi.

The techniques of point counting enable these, as well as many other surface areas, to be measured in microscopical sections. In any section of a two-phase material, the objects forming the included phase will present as two-dimensional profiles in an enclosing matrix. The A_A and V_V of this phase may be determined as outlined in Section 3.1.1 above. If we now superimpose a line (or series of lines) of total length L on the image of the profiles it will intersect the surface of the profiles several times. The mean distance between all these intercepts is called the 'mean linear intercept' and is symbolized by $\bar{L}$; it can be shown that the total surface (S) of a phase enclosed in a given volume of tissue (V) is inversely proportional to its mean linear intercept, that is: $S = 2V/\bar{L}$.

The mean linear intercept is calculated quite easily if we superimpose a series of test lines on several randomly orientated sections in turn, and total up the number of intersections (or intercepts) made by the lines with the surface of the phase of interest. If this figure is called I, the total length of the line in our graticule is L and the number of superimpositions is n, then the mean linear intercept ($\bar{L}$) is given by $\bar{L} = (n \times L)/I$.

In practice a graticule with a series of parallel lines may be used or even a cross graticule as shown in *Figure 2F*. If we wish to combine the determination of surface with a measurement of the A_A or V_V, then the Weibel graticule shown in *Figure 7B* may be used. The ends of the lines serve as markers for the volume or area estimation, whilst the lines themselves serve as test probes for the intercepts.

3.2.1 *Determination of volume and surface ratio*

This procedure determines the volume and surface ratio by the single step method of Chalkley *et al.* (15).

(i)   Place a Weibel graticule (*Figure 7B*) in the eyepiece focal plane so that it appears sharply superimposed on the image of the specimen.

(ii)   Use a stage micrometer to measure the length of the lines in the graticule. This must be done using the objective which will be used for the actual measurements. Record the total line length (L) which might be say 15 μm.

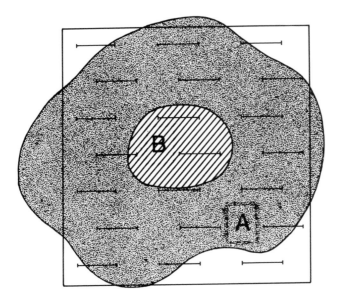

Figure 8. A diagram of the appearance of a two-phase object (**A** and **B**) with the image of a Weibel graticule superimposed on it for the purpose of determining A_A. In this example seven 'hits' are obtained on the background, 30 on phase A and five on phase B.

(iii) Replace the stage micrometer with the sample and examine using the same magnification and a suitable contrast technique.

(iv) Count the number of line ends falling on the phase of interest; record this number (p). In the example (*Figure 8*) $p = 5$.

(v) Count the number of times the lines intersect the surface of the same component; record this number (h). In the example $h = 1$.

(vi) Calculate the surface/volume ratio from the general formula

$$\frac{S}{V} = \frac{4 \times p}{L \times h}$$

In the example, for the single cell shown,

$$\frac{4 \times 5}{15 \times 1} = \frac{20}{15} = 1.33$$

(vii) Repeat the measurements of p and h for a sufficiently large number of cells to obtain meaningful data.

Such repetitions are very important in morphometric work as each field of the microscope represents a very small area of the whole organ. Most of the references give an excellent coverage of the statistical tests necessary in order to establish whether or not sufficient points have been counted to obtain any specified degree of accuracy in the results.

Point counting methods for obtaining various types of numerical data from sections are now well established and are of great sophistication. The above paragraphs are only intended to serve as an introduction and any microscopist intending to use such techniques should consult some or all of the listed references before commencing work.

4. MEASUREMENT WITH DIGITIZER TABLETS

With the rapid development of the microcomputer in recent years, the techniques of data acquisition and measurement from the microscope have been revolutionized. Microcomputers with large amounts of internal memory and with access to external storage of 30 megabytes or more in the form of a Winchester disc unit are now easily available and there are commercially available programs (software) which allow data to be gathered and processed.

One method of data entry when a microscope or electron microscope image forms the source is the digitizer pad (*Figure 9A*). The image is projected onto the active surface of the pad (alternatively, a drawing or photograph may be taped to the surface) and a cursor or pen connected to the electronics of the pad is moved by the operator around the periphery of the object of interest (*Figure 9B*). The movements of the cursor are translated electronically into a stream of co-ordinates which are transmitted to the microcomputer which then works out the areas, perimeters, linear dimensions, form factors etc, which are required by the microscopist. Up to a year or so ago, the digitizer pad was the most satisfactory method of entering data into a microcomputer for obtaining measurements. Now, however, the image is usually acquired by means of a TV camera and the digitizer pad, if fitted to the system, is solely used for interactive editing of the image which is stored in the computer memory.

Most digitizer pads rely on the principle of magnetostriction to obtain the position of the cursor. The active surface of the tablet carries a regular grid of magnetized wires embedded in it. The wires carry a series of very high frequency impulses generated by coils surrounding their ends; as these impulses pass along the wires they lead to minute changes in their size. The receiver in the cursor or tracing pen is another coil which senses these changes and the microprocessor in the electronics of the tablet itself measures the time delay in the x and y directions and so establishes the geographical location of the cursor. This is determined very accurately (usually better than 0.1 mm) and is dependent on the frequency of the impulses in the wires rather than on their physical spacing in the tablet.

The use of a digitizer pad for data acquisition allows measurements of several parameters to be made at one time with a single tracing of an object profile. More importantly, however, this type of system allows the microscopist to select the relevant parts of the image for measurement and to disregard those details which are irrelevant or artefactual. The use of digitizer pads in this way has been criticized on the grounds of operator tedium when large numbers of profiles have to be measured. In addition, there is the possibility of error arising due to operator inaccuracies in the tracing of the profiles. If the actual area of the profiles is kept above a certain value (usually ~ 16 mm^2 for a circular profile) then errors from this source will be of the order of 6%, a figure acceptable for the majority of biological applications; for details see ref. 16. This figure is for a simple shape; a very complex outline will naturally lead to an increase in tracing error which may be compensated for by increasing the magnification at which the image is presented on the digitizer tablet.

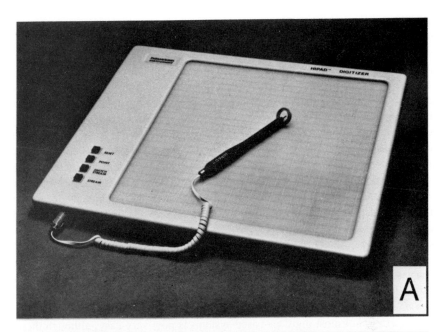

Figure 9. (**A**) A typical digitizer pad used for acquiring numerical data with a microcomputer. Note the tracing pen, with its sensing coil, lying on the active area of the pad. (**B**) The pad in use with structures on a photograph being outlined with the pen.

Digitizers may be used to obtain spatial co-ordinates, point-to-point distances, areas, perimeters, maximum chord lengths, and Feret diameters, to name some of the most common measurements provided by commonly-available software. Many systems now allow the storage of the measured data, along with secondary (calculated) parameters such as shape factors and equivalent circle diameters. This latter uses the measured area of an object to calculate the diameter of a circle which would have the same area; this is a very useful parameter as it allows size comparisons to be made between two sets of objects irrespective of their orientation and shape. For example, in order to obtain the size spectrum of a group of nerve fibres, cross sections would have traditionally been used to take a series of measurements of fibre diameters using an eyepiece graticule. Such measurements would give good results only if the objects were perfectly circular. Any irregularity in the profiles (or tendency to ellipticity due to section obliquity) posed problems in deciding which of several possible diameters should be used. By the use of a digitizer and microcomputer to measure areas and then calculate the equivalent circle diameter, such difficulties may be avoided.

A further advantage of measurements from a digitizer linked to a microcomputer is the ease with which the raw data may be stored on magnetic disc. This facility allows easy subsequent recall for the addition of more data or the performance of different types of statistical analysis.

The operation of most digitizer-based systems is now very easy, although there are variations according to the type of microcomputer which is used and the type and construction of the software modules. It is not possible, therefore, in a short account to give the steps in detail; users should consult their system manual. Most systems require first that the program be loaded from the computer disc memory and that the operator select from some form of displayed menu the options which are required, for example area, perimeter, length, shape factor etc. The image is then projected onto the active area of the digitizer and the objects of interest traced around with the pen or active cursor. If the image is projected onto the digitizer by means of a prism or mirror then the cursor or pen is simply moved around the outlines of interest. It is also possible to work with a drawing tube on the microscope; this allows the active area of the digitizer to be seen at the same time as the microscope image. In this case the cursor or pen is fitted with a miniature light-emitting diode which appears in the field of view and indicates the position of the cursor. Movements of the cursor on the tablet then move the spot of light around the image seen through the microscope so that again, the objects of interest may be outlined. If measurements are required in real terms rather than in machine units then it is of course necessary to calibrate the system with a stage micrometer before beginning any measurements. At the end of all the measuring procedures, the required parameters are usually calculated and displayed on the monitor or sent to the printer linked to the system.

5. MEASUREMENTS WITH TV-BASED IMAGE ANALYSERS

In most current image analysers (e.g. the new Magiscan, *Figure 10*) the microscope image is projected onto the photocathode of a vidicon tube or onto the elements of a photosensitive solid state array. This image produces a pattern of conductivity which varies according to the brightness of the different regions; when a fine beam of electrons

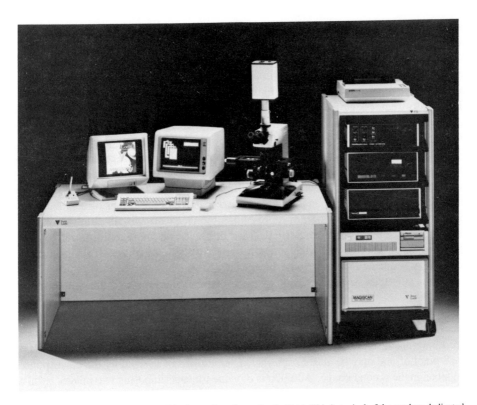

Figure 10. A photograph of the new Magiscan, from Joyce-Loebel Ltd. This is typical of the modern dedicated image analysers. The display monitors, microscope and TV camera which serves to input the image are seen on the desk, with the computing and storage modules in the rack on the right. Photograph by courtesy of Joyce Loebel Ltd.

is scanned in a regular raster fashion over the rear surface of the photocathode, charge neutralization produces a potential difference in a collector which is proportional to the original input brightness pattern. If the value of this signal is sampled at regular time intervals, then we have a means of obtaining a numerical representation of the image. Each number represents an individual picture point or pixel. In many image analysers, the image would be digitized into 512×512 pixels, at each of which the optical density or 'grey level' is represented by a number from 0 (which represents black) to 255 (which represents white). Any image digitized in this manner may then be stored in the computer memory for subsequent processing and measurement. The number of digitized images which may be stored depends on the memory available in the microcomputer and on the number of grey levels into which the picture is separated; many image analysers which have their own dedicated computers have large memories (of the order of 8 megabytes or more) which allows the storage of 12 or more images digitized at an 8-bit level, that is into 256 grey levels.

5.1 Image-processing and discrimination ('segmentation')

Before measurements may be made on a digitized image it is necessary to ensure that

213

the computer only operates on those parts of the image which are significant. This often requires processing of the image in order to allow the elimination of irrelevant detail. When the data is taken from a digitizer tablet this simplification and recognition of structure is done by the operator who outlines only the relevant parts. When a TV-based system is used the processing must be done as far as possible by the computer; this usually requires extensive image manipulations in order to make the image into a form suitable for analysis. Good surveys of digital image processing techniques will be found in the technical manual (17) and at a more advanced level, in the book by Gonzalez and Wintz (12). Because each pixel is represented in the computer memory by a number, it is relatively easy to write software routines which will alter the values of some or all of these pixels according to pre-established rules. In the simplest cases, each pixel may be operated on individually, irrespective of the values of neighbouring pixels. For example, a histogram representing the frequency distribution of the grey levels of all the pixels may be calculated and displayed on the monitor screen to give the operator an impression of the spread of grey levels in the image. If the histogram has a very limited range (as would be obtained from an image of low contrast), the computer may then be instructed to spread the pixel values out over the whole range ('stretching' the histogram) in order to enhance the contrast. Another example of individual or 'point' processing is when the value of each pixel is set to its comple-ment, an operation which results in the inversion of the image, all the white pixels becoming black and vice-versa. Yet again, if we have two images in memory, then each pixel of one may be added, subtracted, multiplied or divided by the correspond-ing pixel in the second image. Addition is of value for superimposing two images, whilst subtraction gives the ability to remove constant background patterns or to detect movement of an object.

More usually, image processing before measurement requires the use of group processing, whereby the final value of a pixel is determined by the value of the adjacent pixels. Such a process is often called 'spatial convolution'. A digitized image serves as the input and the value (i.e. the brightness) of the output pixel depends upon the values of a *group* of pixels called the 'kernel' which surrounds the input pixel. The kernel may be of any size, but is often a matrix of 3×3 or 5×5 pixels. This latter size allows a considerable degree of freedom in the image manipulations which are possible but minimizes the computation time involved. The process may be illustrated by considering a simple averaging process. Here the value of a pixel is calculated by summing the values of the pixels which form the kernel of the input image and then dividing by the number of pixels in the kernel. If we have a 3×3 kernel with the values

$$\begin{array}{ccc} 35 & 38 & 105 \\ 20 & 200 & 180 \\ 180 & 210 & 200 \end{array}$$

then the value of the central pixel becomes 130 after averaging. It is possible to produce a weighted average by attaching a specified factor to each term used in the average; such a factor is often called a 'convolution coefficient' and the matrix of convolution coefficients

$$\begin{array}{ccc} A & B & C \\ D & E & F \\ G & H & I \end{array}$$

used on any given kernel is termed a 'convolution mask'. Thus for a 3×3 kernel, there will be nine separate weighting coefficients $(A - I)$ which is applied to every pixel in the input image in turn. Suppose the convolution mask contains the following values in each cell

$$
\begin{array}{ccc}
-1 & -1 & -1 \\
-1 & 9 & -1 \\
-1 & -1 & -1
\end{array}
$$

and we apply it to the central pixel of the following input series

$$
\begin{array}{ccc}
10 & 15 & 15 \\
10 & 38 & 10 \\
10 & 10 & 15
\end{array}
$$

The final value of the central pixel of the output image is obtained by multiplying input pixel no. 1 by the value in cell A of the convolution mask, the result of which is added to the value of input pixel 2 multiplied by the value in cell B of the mask and so on. The value of the central pixel thus becomes

$$(-10) + (-15) + (-15) + (-10) + (342) + (-10) + (-10) + (-10) + (-15) = 247$$

which replaces the original value of 38. The operation is subsequently repeated on each individual pixel in the original image, so it is obvious that a considerable number of multiplications and additions must be performed to produce a modified image from a given input image. This explains the need for considerable computing power for carrying out image processing.

A convolution mask such as the one used in the example above is termed a 'high pass' filter. The coefficients sum to 1 and there is a large central value surrounded by smaller negative values. This means that the central pixel in the group of input pixels being processed will carry a large weighting, whereas the surrounding pixels act to oppose it. If the central pixel is much brighter than its neighbours (i.e the numerical value of the pixel is higher) then the effect of the surrounding pixels becomes negligible and the output pixel becomes a brighter version of the original, thus emphasizing gradients or sharp transitions. If there is little difference between the central pixel and its surrounding neighbours then the result is an averaging of all the pixels involved.

The converse type of convolution mask is called a 'low-pass filter'. This has the effect of blurring details and hence serves to suppress noise in the image which typically is represented by occasional pixels with very high numerical values. A typical 3×3 low-pass convolution mask or filter would have the composition

$$
\begin{array}{ccc}
1/9 & 1/9 & 1/9 \\
1/9 & 1/9 & 1/9 \\
1/9 & 1/9 & 1/9
\end{array}
$$

from which it is apparent that all the components are positive numbers which sum to 1. When a low-pass filter is implemented, areas of relatively little change in grey levels are not much affected whilst areas of rapid change are averaged to give only the slow changing aspects in the output image. This has the effect of blurring detail in the picture. There are many other convolution masks which may be used. Some, usually termed 'gradient filters', serve to enhance edges in a specific direction whilst others (such as the Laplacian filters) are omni-directional edge enhancers. Different convolution masks may be applied in sequence to an image and by their use it is often possible to process an image in such a way that the relevant details may be recognized by the computer.

Many of the modern image processors also include elaborate software routines for the geometrical transformation of images. These operations may include the facility to shift images either laterally or vertically with respect to a reference image, to rotate images and to distort an image according to a defined set of rules.

A further group of image operations performed on a binary image, that is, one in which all the pixels are either 0 (black) or 255 (white), is that of erosion and dilatation or a combination of these (*Figure 11*). In dilatation, a structured element (often a square or octagonal array of pixels) is stepped automatically over each pixel of the binary image in turn. The centre point of the array will sometimes fall onto background or sometimes onto an image feature. When this latter occurs, all of the pixels of the structured dilating element that do not overlap any pixel of the image are added to the image. The effect is to add to the boundaries of the image and to fill in holes, as illustrated diagrammatically in *Figure 11*. The inverse of this operation is erosion, where the central pixel of a structured array is again stepped over the detected image. If all of the pixels in the structured element fall onto image points then the central pixel of the image is retained. If, however, one or more of the surrounding pixels of the structured element fall onto background when the central pixel of the element is on the image then the central pixel of the original image is set to white, that is, it is 'eroded'. This process is repeated for all of the pixels which comprise the binary image of the object. The result is to remove a layer from the boundary and to enlarge holes and separate objects (*Figure 11*). It is possible to perform erosion and dilatation several times in sequence or to follow a dilatation with an erosion or vice-versa. Dilatation followed by an erosion of the same amount is technically called 'closing' and it is very useful for filling in holes and cracks in the images of objects. Care must, however, be exercised in its use as objects which are close but not in contact with each other may become fused so that separate measurement of their features is then impossible. The converse process, erosion followed by dilatation, is called 'opening' and is very valuable for cleaning up an image by, for example, the removal of protrusions or whiskers (*Figure 11*). It is especially useful when measurements involving perimeters are involved as the opening removes small surface irregularities which may seriously affect the perimeter measurement.

Once the image has been amended the relevant parts for measurement must be indicated to the computer. This process is called 'segmentation' or 'discrimination'. Often if the pre-processing has been adequate it is simply a matter of setting a threshold, that is, using the grey level values to indicate the relevant sections of the image; if the image is very complex it may be necessary to use further steps to ensure that only relevant structures are measured. For example, constraints of either size or shape may be used so that only objects which satisfy the requirements will be accepted. In extreme cases of difficulty it may be necessary to resort to an interactive approach in which the operator edits the image with a light pen or other device. It is worth stressing that it is always easier to perform computer-based measurements on a 'good' image, that is, one of high contrast and containing discrete rather than contiguous objects. Effort spent in the preparation of the specimen and in its imaging is well worthwhile, as it eases significantly the subsequent image processing, segmentation and measurement.

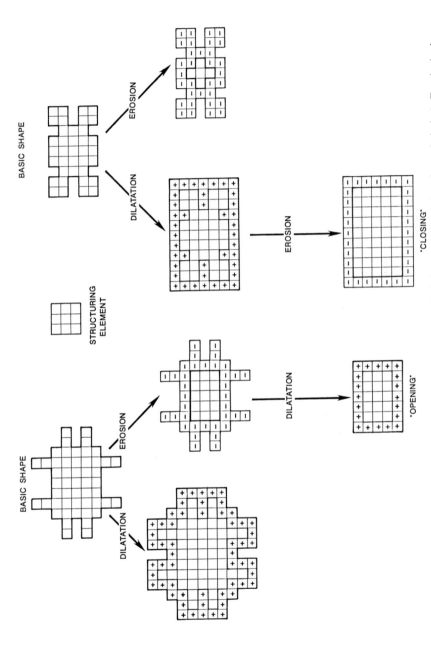

Figure 11. A diagram to illustrate the principles of erosion and dilatation, together with their sequential use (opening and closing). Two basic shapes are shown, that on the left representing an object with protrusions whilst that on the right has fissures. The structuring element used to erode or dilate is a kernel of 3 × 3 pixels. Note that when opening is carried out on the image with protrusions these are removed, leaving the basic shape whilst the converse operation closing has to be performed on the fissured image to achieve the same effect.

5.2 Computer-derived measurements

Once the image has been segmented many different types of measurements are possible; these are often pre-programmed into the instrument and both single or multiple measurements are executed with great speed. Data so obtained may be either field-specific or object-specific. Field-specific data is derived from *all* the objects present in the field, whilst in the object-specific mode each individually segmented object gives rise to a set of measurements. Which mode is used depends on the purpose of the investigation. Four major field-specific measurements (total area, perimeter, intercept number and count) are usually available for each detected phase. The first two of these, as with all dimensioned measurements, may be presented either in machine units (pixels) or, if the input peripheral devices have been calibrated, in absolute units such as mm, μm, nm etc. It should be remembered that perimeter is very much dependent upon the resolution and magnification at which the image is presented to the instrument. In general the greater the resolution, the larger will be the value obtained for perimeter. It is essential, therefore, for the operator to consider the optimum resolution and magnification to use if this particular measurement is desired. The intercept is a measure of the number of transitions which the scan lines make with a change of detected phase from, for example, black to white. It is now seldom used, being a relic of the days when computer-based image analysers had very little memory, as the number of times a line crossed a boundary could be stored very economically. Intercept is also very dependent upon the orientation of the objects with respect to the direction of the scan and hence to their orientation in the field of view.

Object-specific measurements are now more general and again area and perimeter are the basic parameters available from all instruments. Many image analysers also allow the rapid determination of the equivalent circle diameter (ECD). This is the diameter of a circle which has the same area as the object and is easily calculated from the formula

$$\mathrm{ECD} = 2 \times \sqrt{(\pi/A)}$$

where A represents the area of the object. Most current instruments also allow the determination of Feret diameters. These represent the distance between the tangents bounding an object in any given direction and simulate the measurement which is obtained when an object is measured by spanning it with a measuring caliper. Feret diameters are orientation dependent and it is clear that a whole series may be obtained at all possible orientations of the object with respect to the scan direction. If possible such a series of Feret diameters at different orientations should be made. This then allows the calculation of a shape descriptor known as the 'convex perimeter' which is, in effect, the perimeter of the polygon formed by the tangents defining all the Feret diameters. The convex perimeter is calculated from the formula

$$\mathrm{convex\ perimeter} = 2\tan(\pi/2n)\Sigma f$$

where Σf is the sum of n Feret diameters.

It is now common to find that image analysers provide, for each object, x and y co-ordinates which specify its location in the measuring field and also its centre of mass. Other measurable parameters which are often implemented in the software provided by the instrument designers are a series of shape factors, information on the nearest neighbours of an object, optical densities and integrated optical density for an area.

The simplest shape factors are derived by calculating the ratio of two measurements, say the Feret diameters at right angles. This ratio is a useful measure of elongation and is often supplemented by a measurement of ellipticity obtained by assuming the object to be an exact ellipse and calculating the ratio of the major and minor axes. Measures of departure from circularity are always implemented and the commonest is provided from area (A) and perimeter (P) by calculating the expression $4\pi A/P^2$ which, for a true circle has the value of one. There are many other more specialized shape descriptors which are considered in the considerable literature on image analysis. Extensive routines for reconstructing grain boundaries in metal specimens and deriving valid measures of the size and shape distributions of the grains are very common. A further measure of the surface roughness now available uses the so-called 'Fractal dimension' of a surface. This concept, originally due to Mandelbrot, has been extensively developed in image analysis terms by Flook (19−21) who realized that it could easily be derived from the closing operation described above. It is not possible to consider this in detail in this chapter; interested readers are referred to his original papers.

As the image analysers available commercially differ so much in detail, instructions for their operation and for processing, segmenting and measuring images with them should be sought in the literature supplied by their makers.

6. INSTRUMENTATION AND SOURCES OF SUPPLY

This list presents some of the principle sources of supply of instruments known to the author and available at the time of writing; it does not pretend to be a complete list of what is available. In such a rapidly-moving field as image analysis, this is almost impossible. For addresses please see Appendix.

Graticules, counting chambers, micrometers

Micro Instruments
Gallenkamp Group Service Organisation
Graticules Ltd
E.F.Fullam Inc., (latex microscopheres and glass microballoons) UK agents Graticules Ltd

Electronic measuring attachments

RMC-D3 Digital Video Measurement System, Brian Reece Scientific Instruments
Portascan, Scan Systems
Crystal, Quantel

Digitizer-based image analysers, TV-based measurement systems and image analysers using microcomputers

Video Image Analyser, Brian Reece Scientific Instruments
QA 1000 System, Emscope Computers Div., Emscope Laboratories Ltd
VIDS II, III & IV (Apple IIe and IBM PC and compatibles), Measuremouse (Amstrad 1512), AMS
Quantimet 520, Cambridge Instruments Ltd
Cue 2 Image Analysis System, Olympus Optical Co. (UK) Ltd

Imagan 2 (Leitz/GIS), E.Leitz (Instruments) Ltd
Image Manager, Sight Systems
Nachet 1500 (for use with Apple GS), Nachet Vision, Emscope Laboratories Ltd
Software package DIGIT for Summagraphics Bit Pad digitizer tablet and BBC
Microcomputer, Institute of Opthalmology

TV-based image analysers using dedicated computers

Autoscope P and MIAMED, E.Leitz (Instruments) Ltd
Quantimet 970, Cambridge Instruments Ltd
New Magiscan: Joyce-Loebel
Tracor TN5700, Tracor Europa
Optomax V, AMS
I3000 system, Seescan Ltd
Context vision GOP 300, ISI Europe
IS100, Kenda Electronic Systems
Micro-Semper Image processing System, Synoptics Ltd
Goodfellow Image Analyser, Goodfellow Metals Ltd
Omnimet II, Buehler UK Ltd

7. REFERENCES

1. Galbraith,W. (1955) *Quartz. J. Microsc. Sci.*, **96**, 285.
2. Aherne,W.A. and Dunnill,M.S. (1982) *Morphometry*. Edward Arnold, London.
3. Elias,H. and Hyde,D.M. (1983) *A Guide to Practical Stereology*. Karger, Basel.
4. Russ,J.C. (1986) *Practical Stereology*. Plenum Press, NY.
5. Weibel,E.R. (1979) *Stereological Methods*. Vol. 1, *Practical Methods for Biological Morphometry*. Academic Press, London.
6. Weibel,E.R. (1980) *Stereological Methods*. Vol. 2, *Theoretical Foundations*. Academic Press, London.
7. Williams,M.A. (1977) In Glauert,A.M. (ed.), *Quantitative Methods in Biology*. Vol. 6, *Quantitative Methods in Electron Microscopy*. North-Holland, Amsterdam.
8. Weibel,E.R., Kistler,G.S. and Scherle,W.F. (1966) *J. Cell Biol.*, **30**, 23.
9. Mertz,W.A. (1967) *Mikroskopie*, **22**, 132.
10. Delesse,M.A. (1847) *C.R. Acad. Sci. (Paris)*, **25**, 544.
11. Rosiwal,A. (1898) *Verh. K.K. Geol. Reichsanst.* (Wien), **5/6**, 143.
12. Glagolev,A.A. (1933) *Trans. Int. Econ. Min. (Moscow)*, **59**, 1.
13. Chalkley,H.W. (1943) *J. Natl. Cancer Inst.*, **4**, 47.
14. Hally,A.D. (1964) *Quart. J. Microsc. Sci.*, **105**, 503.
15. Chalkley,H.W., Cornfield,J. and Park,H. (1949) *Science*, **110**, 295.
16. Bradbury,S. (1984) *Proc. R. Microsc. Soc.*, **19**, 139.
17. *Image Analysis Principles and Practice* (1985) A technical handbook published by Joyce Loebel Ltd, Gateshead, England.
18. Gonzalez,R.C. and Wintz,P. (1987) *Digital Image Processing*. Addison-Wesley, 2nd edition.
19. Flook,A.G. (1978) *Powder Technology*, **21**, 295.
20. Flook,A.G. (1979) *Proceedings of PARTEC* (Second European Symposium on Particle Characterisation, Nuremburg), p. 591.
21. Flook,A.G. (1982) *Acta Stereologica*, **1**, 79.

CHAPTER 8

Video microscopy

DIETER G.WEISS, WILLI MAILE and ROBERT A.WICK

1. VIDEO MICROSCOPY AND THE EQUIPMENT REQUIRED

1.1 Introduction

The technique now known as video microscopy has produced a revolution in light microscopy when applied to biological systems equivalent to that of the development of the immunofluorescence technique. It has once more made the traditional light microscope a powerful tool for those who are working on dynamic aspects of small biological systems, for example biochemists, molecular and cell biologists. It has given further resolving power to the light microscope enabling the observation of particles which bridge the size range between those normally studied by electron microscopy and those which are already well known to light microscopists as a whole, with the added advantage in that specimens can be examined alive. As well as allowing small particles to be resolved, the technique has the capacity to clean up the image, so allowing greater visibility. Also, changes of such parameters as amounts, concentrations, transport or metabolism of specific molecules in both time and space can be quantitatively determined.

The improvement in resolution is achieved as the video-enhanced microscope is able to detect differences in intensity which are smaller than those detectable by the human eye. For the same reason, weakly self-luminous objects can be found and characterized as images.

The greatest improvements in the microscope images are only possible with very precise optical microscopy arising from a thorough knowledge of the principles of light microscopy, for example of Köhler illumination and conjugate optical planes. However, it should also be said that, on occasion, the less skilled microscopist might be able to generate good images when using video microscopy. A major advantage of the technique is that it provides easy recording of information on video tape or disc for later quantitative analysis. Certainly the ability to store and clean up images will enable all microscopists to re-examine their previously obtained results for comparison with more recently produced images.

Video microscopy, as dealt with here, involves the generation or improvement of microscopic images in three basic ways.

1.1.1 Video enhancement

Video enhancement is the procedure of increasing contrast electronically in low contrast or 'flat' images. This process not only clarifies images containing details visible to the eye, but renders visible structures 5–20 times smaller than could be detected by vision alone or in photomicrographs (*Figure 1*).

221

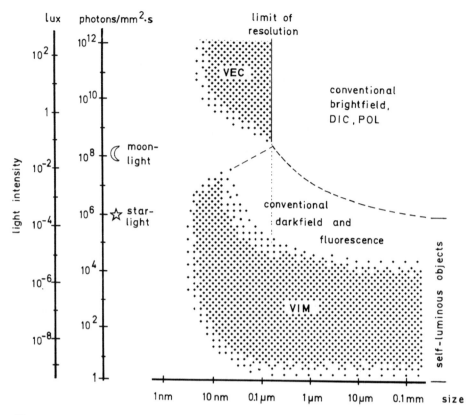

Figure 1. Video microscopy works beyond the former limits of light microscopy. New applications are opened (dotted areas) both in low light situations (VIM) and when working with very small objects (VEC microscopy). All borders are meant to be approximate.

Video-enhanced contrast (VEC) microscopy was developed in the laboratories of S.Inoué (1,2) and R.D.Allen (3−5). Allen *et al.* (3,4) noted that the use of video microscopy with polarized light methods allowed the introduction of additional bias retardation, which, after offset adjustment and analogue enhancement, permitted much better visualization of minute objects. AVEC (Allen video-enhanced contrast) microscopy is the term used to describe this technique.

1.1.2 *Video intensification*

Video intensification is the procedure for making visible low light level images and images generated by too few photons to be seen by the naked eye (*Figure 1*). Video-intensified microscopy (VIM) amplifies low light-images so that very weak fluorescence and luminescence can be visualized (6,7). This is especially important in biology because living specimens benefit from the sparing application of potentially hazardous vital dyes or excessive illumination.

1.1.3 *Digital image processing*

Microscopic images that have been picked up by a video camera can be converted to a digital signal allowing digital image processing to be performed. Image processing can be used to reduce noise in the image by digital filtering or averaging, to subtract undesired background patterns, to further enhance contrast digitally, or to perform measurements in the image (e.g. intensity, size, speed, or form of objects). It is only since the development of procedures for noise reduction and contrast enhancement *in real time*, that is at video frequency, that the microscopist has been able to generate electronically optimized pictures *while* working at the microscope.

1.2 **General strategies of electronic image improvement**

Strategies when planning to use video microscopy might be as follows.

1.2.1 *In photon-limited situations*

In photon-limited situations, especially with fluorescent, luminescent and most dark-field specimens, video intensification is required. VIM techniques require the use of low light level cameras which, unfortunately, are not high resolution cameras. It may be possible, however, to visualize highly fluorescent objects considerably below the limit of resolution if they are well separated from one another (e.g. a diluted suspension of fluorescent actin filaments). The procedure in low light level microscopy would usually require the full range of video microscopic techniques, that is the use of a video intensification camera, analogue enhancement, and a variety of digital image processing steps such as real time background subtraction (due to the uneven sensitivity of many VIM cameras), averaging to reduce noise, and digital enhancement (see Section 3). VIM images with their smooth transitions (low spatial frequencies) are usually well suited for digital image analysis of intensities and intensity changes (see Section 4.2.).

1.2.2 *For high image fidelity and detail*

If high image fidelity and detail are desired and the smallest objects of interest are larger than the limit of resolution of the microscope (~ 200 nm), then high resolution cameras are appropriate. They require a fair amount of light but can be used with differential interference contrast (DIC) and all brighter techniques. In such a case we would use VEC microscopy image improvement, but moderate analogue enhancement may suffice. If uneven shading occurs in the image then an analogue shading correction is usually sufficient at this level of magnification. Background correction by digital processing, or digital enhancement is usually unnecessary. Typical applications include the observation of whole cells, such as changes in cell form, cell division, or movement of large organelles.

1.2.3 *Visualization of the smallest objects possible*

If it is desired to visualize the smallest objects possible, we would also use VEC microscopy, preferentially employing DIC or anaxial illumination techniques. Visualization of microtubules (25 nm in diameter) or vesicles with diameters of 50 nm or less can be achieved (8,9). For this purpose we would need the following functions of electronic image improvement: high analogue enhancement, high performance polarized light microscopy [DIC or POL microscopy according to Allen (3,4) or Inoué (1)], and digital image processing including real time background subtraction, and digital enhancement. If the resulting image is noisy, real time averaging over two or four frames or real time digital filtering might be employed.

For most users VEC and VIM microscopy are complementary since the former reveals the intracellular details while the latter, with the use of fluorescent tags such as dyes or antibodies, is needed to determine the identity of the objects depicted.

For the microscopist the most important requirement of the equipment is that it operates rapidly, that is the processor must operate sufficiently fast to give real time changes in the image. The keying in of commands has to be restricted to the striking of only a few keys.

1.3 The different video-microscopic techniques

1.3.1 *Video-intensified microscopy*

Video-intensified microscopy has greatly extended the capabilities of the light microscope (*Figure 1*) and has provided the technical vehicle for the development of several important new techniques. VIM has made possible the observation and recording of images too weak to be seen by direct viewing or film recording. Further, it has provided a mechanism to study living cells for extended periods without disrupting normal metabolic activity or bleaching photosensitive molecules.

Examples of naturally occurring low light phenomena are widespread. Moreover, the use of exogenous luminescent and fluorescent molecules as probes of cellular structure and function has become an important tool in many areas of research. The application of VIM to these areas has been reviewed (7,10,11).

Video-intensified microscopy is particularly useful in situations where:

(i)   the total number of photons available for imaging is limited by the nature of the event, as in bioluminescence or in fluorescence where the number of labelling sites is small;

(ii)   low intensity illumination is required to avoid interfering with the biological process(es) under investigation or to avoid phototoxic effects;

(iii)   there are rapid changes and the amount of light available in the time required is small;

(iv)   the long exposure times necessary with film prevent the recording of dynamic processes;

(v)   fluorescence excitation is minimized to reduce photobleaching;

(vi)   labelling is intentionally limited to avoid biological interference.

The procedures of analogue contrast enhancement (Section 1.3.2) or digital image processing will not be required routinely but are often very helpful.

1.3.2 *Analogue contrast enhancement*

The introduction of analogue enhancement resulted in the remarkable breakthrough which led to a new level of performance of light microscopy. Digital enhancement,

Table 1. The various sources of straylight.

Bright-field microscopy

Excessive condenser aperture
Uncoated lens surfaces
Reflected light from tube inner surfaces

Polarized light and interference microscopy

Optical rotation at lens surfaces
Strain birefringence in lenses
Light scatter due to dust, lens cement etc.
Surface imperfections in lenses
Defects (holes) in polarizing filters
Sub-maximal compensation

Fluorescence microscopy

Autofluorescence of any material in the light path
Non-specific localization of fluorochromes

as discussed later, is often useful as an addition but it must be emphasized that it cannot replace analogue enhancement.

Understanding the image manipulations required for analogue contrast enhancement is complicated in that a basic understanding of contrast, straylight and resolution is required. A short review of each of these aspects is therefore included here.

(i) *Straylight.* Light distributed evenly over the image and not contributing to image detail is called straylight. Some of its origins are seen in *Table 1*. In many cases straylight prevents the use of otherwise optimal settings of the microscope. For example, the resolution achievable is often sacrificed by closing the condenser diaphragm, thus reducing the numerical aperture (NA), in order to avoid too bright an illumination. When polarized light is used there is usually an annoying contribution of unpolarized straylight, even at the highest extinction settings of the polarizers or prisms. In the video image such straylight can be removed electronically by applying a negative DC voltage to the camera signal, the applied voltage being called offset or pedestal. By applying suitable gain to the camera the contrast is enhanced, by using offset the camera signal is shifted to the appropriate region of grey levels for best visual contrast on the video screen (*Figure 2b* and *3*). In *Figure 2* the improvements achievable by image processing are shown for each step of the technique. Obviously, *Figure 2f*, where digital enhancement is used, shows the detail of this muscle preparation most clearly.

(ii) *Contrast.* The brightness at each point of the optical microscope image is converted into a voltage signal by the television camera. Contrast can be controlled, within a factor of 100 or more by the gain applied to the camera signal, provided the proper offset setting is used. Contrast (C) for the eye is perceived approximately as the difference between the intensity (or brightness) of the background (I_B) and that of the specimen (I_S), divided by the intensity of the background:

$$C = \frac{I_B - I_S}{I_B} \qquad \text{Equation 1}$$

(iii) *Contrast manipulation.* The manipulation of contrast by gain and offset can be applied to the images in any mode of optical microscopy. With techniques involving

225

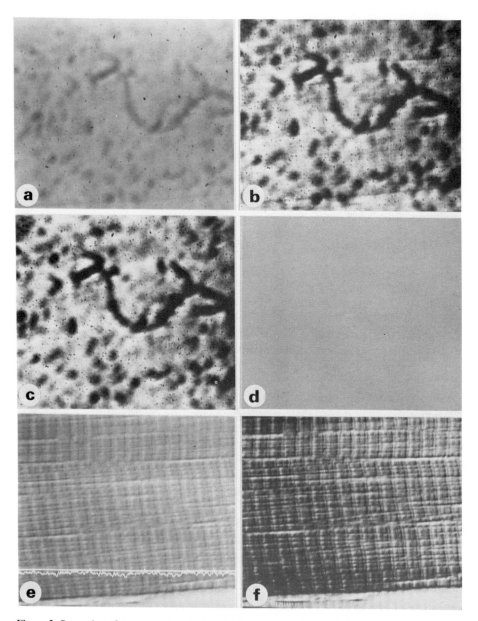

Figure 2. Processing of a poor contrast image. Specimen is an unstained electron microscope (EM) thin section of striated muscle viewed in DIC. The setup was put together and used prior to cleaning of the optics to demonstrate the procedure with unusually heavy mottle. (**a**) in-focus, not enhanced; (**b**) in-focus, analogue-enhanced; (**c**) out-of-focus, with mottle; (**d**) out-of-focus, mottle subtracted; (**e**) in-focus, mottle subtracted; (**f**) digitally enhanced. Microscope, Zeiss IM 405, Plan Neofluar ×63 NA 1.4, ×16 eyepiece, 63 mm camera lens (see *Figure 9a*); processor, Hamamatsu ARGUS 100; frame width, 42 μm.

polarized light, considerable additional contrast can be achieved by adjusting the compensator to a higher bias retardation (AVEC microscopy, see Section 2.3). The resulting images are usually of inadequate visual contrast because the denominator of

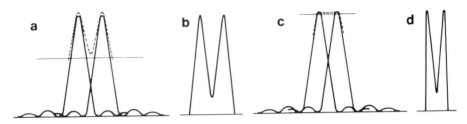

Figure 3. The diffraction pattern (Airy pattern) of a very small object is characterized by a central zero order maximum and smaller maxima of 1st, 2nd and higher orders. The overlapping images of two closely adjacent objects (pinholes) with their summed intensity distribution (dashed) are shown in (**a**). The two objects are resolved according to Rayleigh's criterion since the central depression is sufficiently deep to be perceivable. If the contrast is manipulated by redefining the low intensity (black) end at the position indicated by the horizontal line by either analogue (applying offset) or digital enhancement, and subsequently amplifying the signal (applying gain) a much better image results (**b**). In (**c**) the same objects are somewhat closer so that they are not resolved according to Rayleigh's criterion. However, if contrast is enhanced as for (**b**), even in this situation an image can be obtained which shows the two objects separated (**d**). Sparrow's limit of resolution is reached when there is no trough between the two peaks.

Equation 1 is too high due to excessive straylight (see Section 2.1 and *Figure 10*). However, in the electronic image, the offset voltage applied to the video signal acts in a manner analogous to a 'negative brightness or intensity' (I_V), which is subtracted from the denominator. Video contrast (C_V) is then expressed as Equation 2.

$$C_V = \frac{I_B - I_S}{I_B - I_V} \qquad\qquad \text{Equation 2}$$

Once the straylight has been compensated for by the offset (pedestal) voltage, the analogue gain of the camera can be adjusted once more to utilize the full range of grey scales in the unprocessed image.

(iv) *Contrast enhancement and resolution of objects.* One reason why resolution can be somewhat increased by contrast enhancement is that the Rayleigh criterion of resolution that is defined as a 15% drop between the two peaks that the eye can perceive (*Figure 3a*) is replaced by the Sparrow criterion (12) (*Figure 3c and d*) which the video camera can detect. This is applicable to electronic images because they can be enhanced, so that even a slight trough in the intensity distribution of the two unprocessed images (*Figure 3c*) can be enhanced to give good separation (*Figure 3d*). The gain in resolution is about 2-fold. Using the best lenses, where aberrations are negligible, image quality is only limited by the size of the spreading of each image point due to diffraction, that is by the size of the Airy pattern.

(v) *Contrast enhancement and visualization of objects.* Objects smaller than the limit of resolution create Airy patterns whose amplitude (intensity) is small. Their size cannot be reduced. By applying video enhancement such invisibly weak Airy patterns can, however, be visualized, although if the distance between the two objects is less than the limit of resolution their diffraction images will still fuse. Hence, by using contrast enhancement such objects can be visualized although they are not resolved. Using Nomarski-DIC and VEC microscopy, biological material of between 15 and 20 nm in size can be visualized, while other materials, such as colloidal gold, can be visualized down to sizes of 5 nm and less. Diffraction patterns from many very small objects existing at distances less than the limit of resolution cancel out and nothing is visible. This situation is met in very small cells and in nerve endings crowded with organelles.

(vi) *Advantages of analogue contrast enhancement.*

(a) Straylight is removable in the video situation by offset (I_V).

(b) The practical resolution is increased by a factor of about two. This is partly because it is possible to use the highest working NA since the resulting excessive image brightness due to straylight can be suppressed electronically with offset, and partly because Rayleigh's criterion of the limit of resolution is replaced by Sparrow's criterion.

(c) The gain in contrast is sufficient to visualize structures in living cells that are around one order of magnitude smaller than could be resolved or detected previously under the same conditions (*Figure 1*).

(d) The AVEC conditions reduce the diffraction anomalies that produce spurious detail and contrast in polarized light-based techniques, whether they are caused by depolarization at lens surfaces or by residual strain birefringence in the lenses (see Section 2.1).

(vii) *Limitations of analogue contrast enhancement.*

(a) Electronic noise is amplified along with the video signal in the enhancement process and may have to be subsequently reduced (see Section 2.3 and 3.2).

(b) If the optical system (including the slide and coverglass) contains dust, dirt, or manufacturing imperfections, these will create a fixed pattern of mottle that is enhanced along with the image. This can only be removed by digital processing (Section 2.3).

(c) If the illuminating system is poorly designed or incorrectly aligned and focused (for Köhler illumination), the field may be unevenly illuminated. For optimal results in VEC, the requirements for even illumination are much more stringent than with photomicroscopy. Within certain limits, however, uneven illumination can be treated as fixed pattern noise of mottle and removed by digital subtraction (*Figure 4*) (see Section 2.3).

1.3.3 *Digital image processing*

Many of the digital image processing routines were available long before their value was recognized by microscopists (e.g. 5,13). With the rapid development of faster computers many of these routines became available at video frequency. The principles of digital image processing and some of the procedures to be employed have already been described in Chapter 7. In VIM and VEC microscopy they are used for rapid preprocessing, that is improvement of the images prior to storage on tape or disc. It should be noted, however, that individual frames, once stored, can also be subjected to further digital processing employing essentially the techniques and image processors discussed in Chapter 7. Most of the processors mentioned there are basically different from the ones discussed here as only the latter ones perform their operations at video frequency thus allowing manipulation of the live video image.

Following analogue contrast enhancement, the analogue TV signal (a pattern of voltage changes), is digitized so that it can be manipulated by an arithmetic logic unit (ALU) in an image processor. In the most suitable processors, the instructions necessary to carry out a number of different arithmetic manipulations are controlled by firmware or by packages of macro software routines so that an operator can easily learn to process images using the most suitable procedure for his/her particular operation. The operations themselves are specified manually in seconds, and are carried out in 'real time', that is repetitively during the intervals between two consecutive frames (i.e. 25 or 30 times per second depending on the TV standard used).

During digitization the image is subdivided into 512 or more lines of 512 or more picture elements (pixels) each. Usually, each pixel can assume one out of 256 (8 bit) grey levels with zero being defined as black and 255 as bright white (see Section 4 of Chapter 7). Depending on the complexity of the operations to be performed, one to three frame memories (usually 512×512 pixels) are required. If averaging or other multiframe procedures are intended these frame memories need to be more than 8 bit deep (preferably 16 bit).

(i) *Rolling average or jumping average.* The rolling average function computes the average of the last incoming image and the previously stored average. This procedure results in a weighted average with the most recent frames dominating (recursive filtering). In jumping average mode a pre-defined number of frames is averaged and displayed for the duration of the accumulation of the next set of frames. Both modes diminish electronic noise in the video signal by the square root of the number of frames averaged. The former smears and de-emphasizes any motion present, while the latter accentuates slow motion. Both are generally used for VIM and are advisable for VEC microscopy when high enhancement is used, that is when electronic noise becomes annoying.

(ii) *Mottle subtraction.* Patterns of image imperfections (mottle) remain in the image when the specimen is defocused or moved out of the field of view (*Figure 2c*). Consequently, mottle can be frozen in a video frame memory and then subtracted from each frame of the incoming video signal (*Figure 2d*). This operation (mottle or background subtraction) results in a 'clean' image lacking mottle (*Figure 2e*). The same operation also eliminates inhomogeneities in background brightness (*Figure 4*), if their contrast does not exceed the 'window' of 256 grey levels which can usually be displayed.

(iii) *Digital enhancement.* The analogue-enhanced, mottle-free image may not have sufficient contrast. In this case the image can be digitally enhanced, for example by stretching the histogram of grey levels (*Figure 2f*). The procedure is analogous to analogue enhancement but the selection is made digitally by choosing the restricted region of the grey levels containing the image information and expanding it to stretch the entire distance from black to white, that is 256 grey levels. This is done by assigning new grey levels to the original ones (defining an output look-up table). Please note that analogue enhancement (Section 1.3.2) cannot be replaced by digital enhancement.

(iv) *Enhancement of motion by sequential subtraction or interval subtraction.* The analogue-enhanced image can be subjected to sequential subtraction in order to observe and detect only moving elements. This is done by freezing a reference image without taking the specimen out of the field or out of focus and subtracting it from all incoming frames. Subtraction of a (stored) image from very similar, following (incoming live) images results in empty images unless moving elements cause image differences and

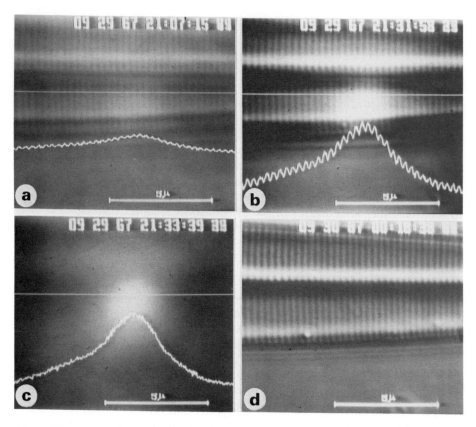

Figure 4. Correction of uneven illumination (shading) in analogue-enhanced images by background subtraction. If after fixing potential flaws in the optics (step ii in Section 2.3) and possibly correcting the shading in analogue mode (Section 1.4.1) an image as in (**a**) appears, this will show a very annoying hot spot after analogue enhancement (**b**). Subtraction of a specimen-free mottle image (**c**) results in an evenly illuminated image (**d**). This sequence also demonstrates the usefulness of a calibratable scale bar, the timer (months to 1/100 sec) and the intensity measurement along a line (Section 4.2). Specimen is the diatom *Amphipleura pellucida* with a known line spacing of 250 nm. Microscope, Zeiss Axiophot, Planachromat ×100 NA 1.25; processor, Hamamatsu C 1966 Photonic Microscope System.

so make their presence known (*Figure 5*). This is an extremely sensitive means of motion detection, but only works satisfactorily with a very stable microscope stand under good temperature control. Any drift in focus or pressure applied to the stage or microscope body may restore a distorted image of the whole object. This mode gives both the position of the moving object at time zero (frozen and in negative contrast, i.e. a 'missing object') and the live position of the moving object (*Figure 5*). Distance measurements for velocity calculations can be very conveniently obtained by this technique. Interval subtraction is a mode of sequential subtraction that is programmed to refresh itself after pre-determined intervals, that is a new 'background' image to be subtracted from incoming video images is automatically picked after a certain, pre-selectable number of frames.

(v) *Pseudocolour display.* This process allows the user to systematically or arbitrarily

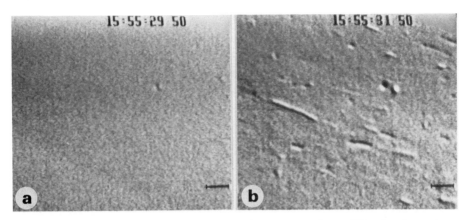

Figure 5. Selective visualization of moving objects by sequential subtraction. (**a**) When the reference image taken at time 29:40 sec is subtracted from the incoming video image at time 29:50 sec, i.e. 0.1 sec or five frames later, almost no contrast is visible because all objects remained close to their original location. (**b**) 2 sec later (31:50 sec) the moving organelles became visible. Each organelle is depicted twice, once appearing as a depression and once in positive contrast. The depression marks the organelle's location at commencement, i.e. the locus where the organelle is now missing, while on the video screen the actually moving objects become visible in positive contrast. In this sequence in a bundle of pike olfactory nerve axons of 0.25 μm diameter the movement of mitochondria (elongate) and lysosome-like organelles (round) is observed. Most organelles except one mitochondrion near the centre can be seen moving to the lower right. Microscope, Polyvar, Reichert/Cambridge Instruments, Planachromat ×100 NA 1.32; processor, Hamamatsu C 1966 Photonic Microscopy System; scale bar, 2.7 μm.

assign colours to various grey levels. This can be very helpful in discerning patterns or small differences in intensity.

Additional functions which prove particularly useful for VIM include the following.

(vi) *Frame (or beam) blanking.* This is another method for improving signal-to-noise ratio (S/N) during image acquisition. Here, the charge pattern on the faceplate of the camera is allowed to accumulate for an extended period of time rather than being 'read out' every 1/25 or 1/30 of a second (video rate). This greatly reduces the amount of readout noise. For example, a typical 64 frame integration sums 64 images, each containing a certain percentage of readout noise. If, however, the image is allowed to integrate on the camera target for a period equivalent to 16 frames before being read out the amount of readout noise for a similar 64 frame integration is reduced to 1/16th.

(vii) *Spatial filtering.* By applying various convolution operations to the image the image can be 'filtered' in the spatial domain. This process allows for suppression of noise or the accentuation of high frequency information as in edge detection or image sharpening (see Chapter 7 for details).

(viii) *Arithmetic operations.* This function permits the application of the basic arithmetic operations to a single image or between multiple images. Ratio imaging, where one image is divided by another, is a typical example (see Section 3.3.2).

(ix) *Image superimposition.* The ability to superimpose one image upon another is very useful in many applications. Examples include combining a pseudocoloured fluorescence image with the corresponding black-and-white transmitted light image or combining two images, each of which represents a different fluorescent label, to evaluate their relationship to one another.

It is clear that a multitude of additional procedures to accentuate specific features can be imagined and verified with programmable digital image processors.

1.4 **Electronic equipment for video microscopy**

Once you have determined the functions which will be essential for your research (see Section 1.2) you need to select the appropriate type of video microscopical equipment. A generalized scheme is set out in *Figure 13*. Cameras and image processors are discussed in this section, while comments on suitable ancillary equipment such as recorders and monitors will be found in Section 5.

1.4.1 *Cameras and camera control units*

For VEC microscopy high resolution vidicon cameras are used, mainly of the Chalnicon, Newvicon or Pasecon type. For low light applications special cameras are required which will be discussed in Section 1.4.2. If special spectral requirements such as high sensitivity in the red or UV range, or extremely low lag are desired, other suitable cameras may be selected from various manufacturers, such as COHU Inc., Hamamatsu Photonics K.K., DAGE-MTI Inc. and Zeiss West Germany. Until recently the use of CCD cameras (charge-coupled devices) was not to be recommended due to their lower resolution and sensitivity, but now they are available at a quality matching that of camera tubes, having very low geometrical distortion and even higher dynamic range. Only at very low light levels is cooling required (Section 1.4.2).

Black-and-white (B/W) cameras (3/4- or 1-inch tubes) are used exclusively if the images are to be processed digitally. Colour cameras have three separate outputs for red, green and blue (RGB signal) and would, therefore, require three parallel image processors. When selecting a camera make sure that it can be used *without* automatic gain control because gain and offset (pedestal) have to be set by the user manually [see Section 2.3(v)]. Only if this is possible can analogue contrast enhancement be performed.

The non-uniformity of response across the camera target is called shading. In some low light level cameras this effect can be as high as 30% from one region to another. A number of commercially available cameras offer 'shading correctors' which allow the user to introduce various combinations of linear and parabolic waveforms to compensate for this non-uniformity. Two examples of such cameras and controls are shown in *Figure 6*. In general, the control units which can be used with different camera heads [such as vidicons, silicon intensifier target (SIT) cameras etc.] are to be preferred.

1.4.2 *Camera systems for video-intensified microscopy*

Historically, a diverse number of camera technologies have been applied to low light level imaging. For a variety of reasons most of these technologies have given way to two general classes of low light detectors: intensified video cameras and cooled solid state cameras. From an application standpoint these detectors are differentiated from one another by the fact that the former, being intensified, is capable of viewing dynamic specimens; while the solid state device, being an integration type detector, is, at very low light level situations, suitable only for very slowly moving or static samples.

(i) *Intensified video cameras.* These devices, as implied by the name, consist of two separate functional components—an image intensifier and a video camera. The image intensifier serves to detect the image, amplify its intensity and present the resulting image to the video camera so that it may be 'readout' in a systematic format.

The most common low light level camera currently utilized in microscopy is the silicon intensifier target (SIT) camera. This design combines an electrostatically focused image intensifier with a silicon target camera tube within a common glass envelope. SIT cameras can provide sensitivities up to 100 times greater than the silicon target camera alone, which itself is considered a sensitive camera. Based on light intensity, the SIT typically produces 300–600 TV lines of resolution.

The SIT camera is also available in a double intensified configuration known as the ISIT (intensified silicon intensifier target). This camera utilizes an additional intensifier which is fibre-optically coupled to the photocathode of the SIT. This combination provides sensitivity approximately 20–30 times greater than the SIT camera and allows for operation very close to the limits of human vision. Both the SIT and ISIT employ a multialkali photocathode which provides spectral sensitivity from 300–850 nm.

An alternative approach to the SIT/ISIT design is to optically couple an image intensifier to a video camera. In contrast to the SIT, most image intensifiers employ a phosphor window as the output element, thereby reconverting the electron image back to an optical image for viewing by the video camera. By way of lenses or a fibre-optic coupling the image at the phosphor is focused onto the faceplate of the video camera. A major practical advantage of this design is that it provides flexibility in selecting image intensifiers and video cameras with performance characteristics for specific applications.

The major differences among image intensifiers revolve around the focusing mechanisms employed and the method of amplification. The simplest configuration is the wafer, or biplanar, type. Its greatest attributes are small size and lack of distortion. Higher performance in terms of gain and image quality can be obtained by incorporating a focusing mechanism and both electrostatic and electromagnetic focusing systems have been employed. The electrostatically focused type is more compact, lightweight and less expensive. For extremely high intensification requirements these tubes can be configured so that the output phosphor of one is optically coupled to the input photocathode of another. Such cascaded intensifiers can realize luminous gains in excess of 10^6 when constructed of three or four stages.

High gain can also be achieved by providing for electronic amplification within the image tube itself (14). This is made possible by placing a microchannel plate (MCP) between the photocathode and output phosphor. A single MCP can provide electron gains of 10^3 and multiple MCPs may be used for higher gain requirements. MCP image intensifiers offer similar performance to multiple stage electrostatic focused systems with smaller size, lower distortion and decreased power requirements (*Figure 6a*).

(ii) *Cooled solid state detectors.* Recent advances in the development of CCDs have been very promising with regard to their application in low light microscopy. While technically not an intensified camera, it has the ability to obtain images at low light levels. Being inherently high quantum efficiency devices (the ability to convert photons into electrons), high sensitivity is achieved primarily by cooling and slow readout. Cooling the CCD (-25 to $-125°C$) dramatically reduces the dark current as a noise component. The slow readout further reduces the noise associated with high band width

233

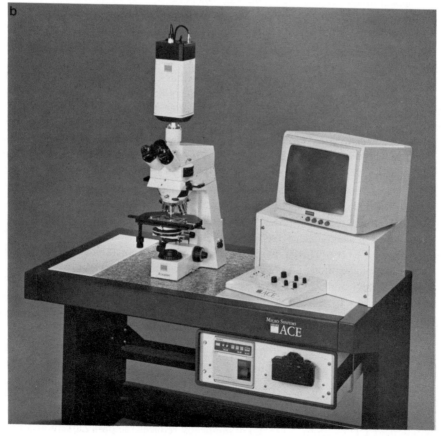

electronics. Low light images are integrated directly on the chip in much the same way as an extended photographic exposure. High quality CCDs offer excellent geometry, photometric accuracy and large dynamic range and they will undoubtedly find increasing applications in microscopy for the quantitative imaging of static samples.

(iii) *Practical considerations in choosing a low light level camera system.*

(a) Sensitivity. Clearly the most important consideration in selecting a camera for low light imaging is sensitivity. A wide variety of cameras is available with sensitivities ranging from applications in DIC and phase-contrast microscopy down to the single photon level.

Sensitivity, or responsivity, is typically measured by illuminating the camera target with a known quantity of tungsten light and measuring the resulting output current of the camera. The data is expressed in amperes/lumens and is plotted in a log−log fashion as the 'light transfer characteristic'. This provides a useful way to compare various systems but this definition can be misleading when applied to low light cameras. It does not take into account the noise component of the output signal and, being based on tungsten light, is heavily biased toward red-sensitive photocathodes. *Figure 7a* illustrates the typical sensitivity range of a number of low light level cameras.

(b) Spectral response. The spectral response of intensified cameras is determined by the window material and the type of photocathodes. The most common photocathode is the multialkali type (S20) which has peak sensitivity at 420 nm and provides usable sensitivity to approximately 800 nm. The spectral responses of three common photocathodes, multialkali, bialkali and S1, are shown in *Figure 7b*. It should be noted that the extended red-sensitivity of the S20 and S1 types is accompanied by higher noise levels (thermal noise). While not a problem in most situations, at very low light levels one is advised to select the bialkali type unless this extended sensitivity is required.

(c) Resolution. As noted, high resolution and high sensitivity tend to be mutually exclusive characteristics. This relationship is due to noise. Noise in low light level systems can generally be classified as signal-independent or signal-dependent. Signal-independent noise is essentially present in a fixed amount, in both the absence or presence of light. It arises from thermal noise of the intensifier, the camera target, the videoamplifier, etc. This is the predominating type of noise at relatively high light levels. As light levels decrease signal-dependent noise becomes the dominant component. At low photocathode illumination levels, resolution is primarily limited by the finite number of photoelectrons released at the photocathode—the so-called photoelectron, or quantum, noise. It is this latter 'noise' which accounts for the decrease in resolution as light levels decline. Resolution, therefore, should be evaluated as a function of light intensity and

Figure 6. Camera control units which can be used for analogue contrast enhancement. (**a**) The Hamamatsu C 2400 vidicon camera (in the back) and its control unit (lower box) with gain, offset and analogue shading correction capability. The picture shows also a highest sensitivity photon counting camera consisting of a two-stage microchannel plate device with the low lag vidicon attached to its rear (courtesy of Hamamatsu Photonics K.K.). (**b**) The Zeiss ACE System for analogue contrast enhancement also allows for analogue shading correction. A video printer and a photocamera aiming at a separate high resolution monitor are situated underneath the isolation table (courtesy of Zeiss West Germany).

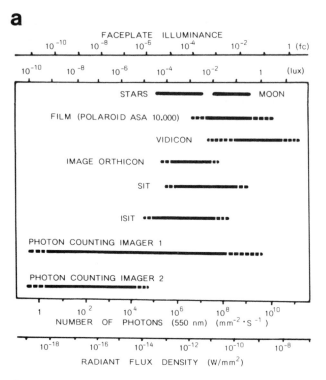

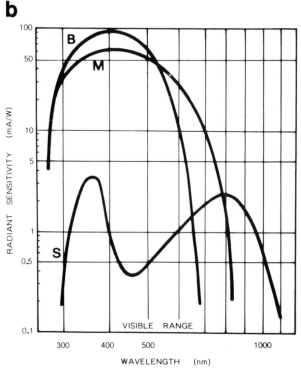

preferably at a specific wavelength when comparing for possible photocathode differences.

(d)　Lag. Lag, or dynamic response, describes the camera's ability to respond temporally to changes in light intensity. If the system is to be used for the imaging of rapidly changing specimens, a camera with low lag characteristics should be selected. Lag in an intensified camera is due primarily to the readout video camera, therefore utilizing a low lag camera such as a saticon or solid state will greatly improve this characteristic.

1.4.3 *Compact digital image analysers*

For those who require digital image processing in real time in order to obtain the desired image quality, but do not intend further digital image analysis of single frames (or plan to perform this using other equipment, see Chapter 7), a number of compact digital image processors is available at very reasonable cost. Be aware, however, that some of these do not include the analogue enhancement feature, which is indispensable. The processors most suitable for our purpose should include digital enhancement, real time subtraction with simultaneous frame averaging functions, plus additional potentially useful features. Within this group the SIGMA II (Hughes Aircraft Co.), the DVS 3000 (Hamamatsu Photonics K.K.), or the Multicon (Leitz) are examples (*Figure 8*).

1.4.4 *Video microscopy image processing systems*

A few dedicated systems are available which allow all essential real time functions and provide the user with a fair number of measurement functions and some flexibility to manipulate and analyse a set of stored images. These systems also allow the user to add software for specific applications. As in the preceding category these systems usually include a camera control and analogue enhancement unit. They can also be used with cameras and control units from other manufacturers. Similarly, previously recorded sequences played back from a recorder can be input into these systems. Examples in this category are the QX-7 (Quantex Corporation), the ARGUS 100 of Hamamatsu (being the successor of the first microscopy dedicated processor C-1966 which was built in 1984 by Hamamatsu in collaboration with R.D.Allen), the Image I/AT (Universal Imaging Corporation), the Sapphire (Quantel), and the BioVision (Perceptics Corporation).

1.4.5 *Video processor boards*

A number of manufacturers offer computer boards for image storage and processing which fit either personal computers of the IBM-AT type or VME-bus systems. Since these are usually not ready-to-use-systems, a customer-tailored processor may need to be put together. To develop a working unit a specialist in image analysis soft- and

Figure 7. (a) Sensitivity range of vidicons and low light level cameras relative to incident light intensity. Very sensitive film is included for comparison as well as illuminance levels of moonlit and starlit scenes. Photon counting imager type 1 is with phosphor screen output while type 2 is with semiconductor-based position sensitive detector output. fc, foot candles or lumens per square foot; lux, lumens per square metre. (b) Spectral responses of the three most common types of photocathodes. B, bialkali type; M, multialkali type (S20); S, S1 type.

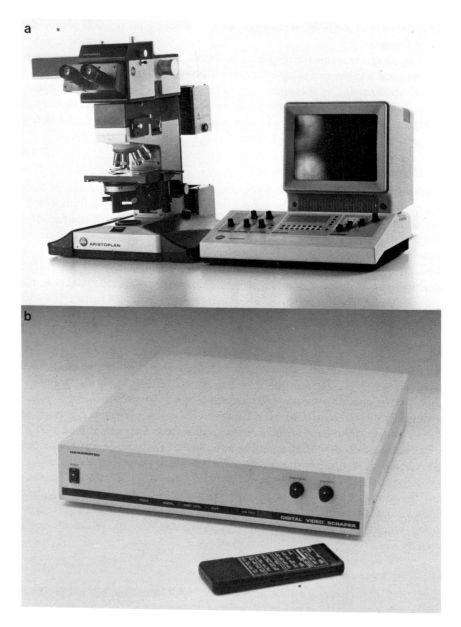

Figure 8. Compact image processors for analogue contrast enhancement and digital processing. The Leitz Multicon, here working with a lightweight solid state camera (**a**) and the Hamamatsu DVS 3000 (**b**) are examples for systems permitting real time image quality improvement with a great variety of both analogue and digital functions (courtesy of E.Leitz and Hamamatsu Photonics K.K., respectively).

hardware may sometimes be required. Examples of companies producing such boards are Datacube Inc., Matrox Electronic Systems Ltd, Data Translation Inc., and Imaging Technology Inc.

1.4.6 *Single function processors*

While the above-mentioned devices are more or less multifunctional, sometimes only one or two functions may be required. In this case it may be cheaper to obtain unifunctional, hard-wired instruments for analogue enhancement, distance measurement, perimeter measurement, time/date generation, or intensity measurement along a scan line etc. Such devices can be obtained from, for example For-A Company Ltd, Colorado Video Inc., or Hamamatsu Photonics K.K.

1.4.7 *Outlook*

In the near future the general-purpose computers (workstations) will be fast enough to perform more and more of the real time functions required for video microscopy by utilizing less hardware and applying more software. In parallel, the image analysis software offered by those manufacturers, or by specializing companies, will soon reach the point where it replaces some of the hard- or firmware solutions covered in Sections 1.4.3 and 1.4.4. The flexibility of workstations is generally much greater so that all-purpose and multi-user systems for image preprocessing, processing and analysis can be realized. However, since specialists will be required to operate these systems, and the prices are well in excess of those for most of the above-mentioned instruments, such systems cannot be discussed here in detail.

1.5 **Considerations on the microscope**

For all types of video microscopy, a research grade microscope should form the basis of the optical imaging system. A wide range of techniques may be applied in VIM and VEC microscopy.

All VIM and most of the high magnification VEC microscopy applications are characterized by the limited amount of light available. Therefore, by far the two most important considerations involve optimizing the microscope for light gathering ability and the efficient transmission of light through the optical train.

The light gathering ability of the microscope is a direct function of the NA of the objective lens. While factors such as an objective's working distance often necessitate using a lens with lower than maximal aperture for a specific magnification, whenever possible the highest available NA should be utilized. The importance of this point is clearly appreciated when one considers the relationship of magnification, NA and light intensity. In a system where the illuminating aperture is equivalent to the objective's acceptance aperture (such as in epi-fluorescence), intensity (I) is proportional to the fourth power of the NA.

$$I \propto (NA)^4 \qquad\qquad \text{Equation 3}$$

Further, light intensity is inversely proportional to the square of the magnification:

$$I \propto \frac{1}{\text{mag}^2} \qquad\qquad \text{Equation 4}$$

Therefore,

$$I \propto \frac{(NA)^4}{\text{mag}^2} \qquad\qquad \text{Equation 5}$$

239

In comparing two ×40 objectives with apertures of 0.9 and 0.5, one sees that the higher NA objective captures over 10 times more light than its lower NA counterpart.

As a general rule, in lenses of similar design, the NA tends to increase with magnification. The above relationship clearly dictates that when using additional magnifying optics, such as optovars or projection eyepieces, one should maximize the NA and magnification of the objectives and minimize the magnification of the intermediate optics since these optics contribute nothing to the light gathering ability of the system. Therefore, if faced with the choice of using a ×40 objective in combination with a ×10 camera relay lens versus a ×25 objective and ×16 relay lens, the former combination will prove far more efficient despite the fact that both systems deliver a final magnification of ×400.

Regarding the second major consideration, we have to remember that the efficiency of light transmission within the optical system, is primarily a function of the transmission properties of the objective lens and the number of optical elements in the system. Interestingly, there are often significant differences in the transmission characteristics of objectives of different designs (e.g. fluorite versus apochromatic) and from different manufacturers, even when the NA and magnification are identical. If possible, it is recommended that a number of lenses be evaluated with regard to the specific wavelength(s) that will be utilized.

Of equal, if not more, importance, is the issue of intermediate optical elements. Uncoated lens surfaces typically reflect 4−5% of the light incident upon them. And while it is true that research quality microscopes generally employ antireflection coatings, these cannot completely eliminate losses due to reflection and the cumulative loss associated with the complex light paths can be substantial. Therefore, all unnecessary optical elements should be removed from the light path. If this is not possible, the use of a microscope with a simple light path should be considered.

When planning fluorescence work, make sure the optics transmit the short wavelength light which is required for the excitation of dyes, for example 340 nm for FURA-2 used for video measurements Ca^{2+} concentrations (see Section 3.3 and *Figure 12*). The requirements of the optics are sometimes even less stringent, since in most cases monochromatic light is recommended anyway, and since only the central area of the field (~1/3 to 1/2) is picked up by the video camera. Wide-field optics are usually of no particular advantage.

When planning high magnification VEC microscopy xenon or mercury arc lamps are required, in most cases, to provide enough light, so that the video camera works near the saturation end of its dynamic range. In some microscopes halogen or tungsten lamps may be sufficient even for VEC-DIC if all diffusers are removed from the light path and the lamp is optimally adjusted. We have seen the best results with a combination of the HBO 50W DC mercury lamp, which has a short intense arc, in combination with the Axiomat (Zeiss), the inverted microscopes IM and ICM (Zeiss), and the Polyvar (Reichert/Cambridge Instruments), but the 100 or 200 W mercury arc lamps also worked in most cases (*Figures 2, 4* and *5*). Due to the fact that after contrast enhancement minute changes in lamp intensity may result in transitions from well-modulated to bright white or black images, stabilized DC power supplies are strongly recommended. With arc lamps it is advisable to use a narrow-band green interference filter (e.g. the Hg-line 546 ± 10 nm for mercury lamps) for optimal results and to protect cells from blue light. In this case it will also be essential to protect the interference filter, the polarizers and the cells from heat and UV light with at least one piece of each of the following filters: a UV filter, a heat-reflecting filter and a heat-absorbing filter.

A heavy stand for the microscope is recommended in order to reduce vibrations and internal movements resulting from temperature changes. This is especially so when heavy cameras are used like those designed for low light level work and when using the highest magnifications. It should be noted that an internal displacement equivalent to a change of 1 μm is registered on the video screen at typical magnifications for VEC microscopy as 1 cm. In some cases vibration isolation tables may therefore also be required.

The contrast modes to be used for video microscopy of unstained, living biological specimen are discussed in Section 2.5.

1.6 How to interface the two

One should ensure that 100% of the light can be directed to the camera. If the microscope has a fixed 'split' between the binocular and camera port, it may be possible to have a 100% reflecting mirror installed in place of the existing one.

For video microscopy we need considerable additional optical magnification. This is due to the fact that because of the video lines the video image has a spatial resolution which is much lower compared to that of a photomicrograph or that of the microscope optics. In order to fully utilize the resolution of the microscope we have to make sure that it be the limiting component rather than the video system. This means that we have to magnify the specimen onto the target of the video camera far beyond the magnification which is normally considered useful in conventional light microscopy. Whenever possible this should be reached by the use of higher power objectives, but when subresolution objects are to be visualized with the ×63 and ×100 oil-immersion objectives, additional magnification of ×4 to ×6.3 is required. This can be achieved either by a zoom system as in the Zeiss Axiomat and the Zeiss Axio series, or by a ×2 or more additional magnification changer (optovar type) and/or a high foto eyepiece (×16 or ×25) plus projective lens (50−63 mm) as in most other microscopes [e.g. IM, ICM, and Axio series (Zeiss), Orthoplan and Aristoplan (Leitz), and Polyvar (Reichert/Cambridge Instruments)]. Alternatively a combination of two ×2 or one ×4 extension tubes may be used. As a rule of thumb, when objects at or below the limit of resolution are to be observed on a medium sized monitor, the final magnification should be such that the field of view portrayed on the monitor is 15−30 μm wide, or the final magnification is around ×8000 to ×15 000.

The system projecting the image into the camera target classically consists of the ocular and the camera objective lens in close contact (*Figure 9a*). Both consist of several lens elements and often some surfaces are close enough to the image plane to introduce very disturbing mottle from dust and non-perfect coatings. Oculars are, however, indispensable in all cases where part of the correction of optical aberrations is taken care of in the objective lens and part in the ocular.

Several manufacturers have switched to totally internally corrected objective lenses (e.g. Zeiss and Reichert/Cambridge Instruments), so that video microscopists have much more freedom to use alternative arrangements. In such optical systems the tube lens directly projects an intermediate image onto the TV-target through an empty connecting tube (e.g. Zeiss Axio series) (*Figure 6b*). This yields very good images with little mottle. For highest magnifications, this arrangement requires that the additional magnification of ×4 to ×6.3 is reached by a Galilean-type telescope in the parallel, infinity corrected, light path or by an additional magnification changer ('optovar'), or by a combination of the two.

Another connection which creates little mottle but requires considerable bench space is to put the camera onto a lateral exit with a high power eyepiece and to project the intermediate image directly onto the camera target. The desired magnification is adjusted by sliding the camera to and from the microscope at a 10–30 cm distance.

2. HIGH RESOLUTION: VIDEO-ENHANCED CONTRAST MICROSCOPY

2.1. Different types of VEC microscopy

Video contrast enhancement of microscopic images obtained by bright-field, dark-field, anaxial illumination or fluorescence techniques is very straightforward and generally described by the term VEC microscopy. Allen (3,4) and Inoué (1) simultaneously described procedures of video contrast enhancement for polarized light techniques which differed considerably in their approach but yielded very similar results. There is a need to distinguish clearly the two strategies in order to avoid confusion.

Allen named his techniques Allen video-enhanced contrast differential interference contrast and polarization (AVEC-DIC and AVEC-POL, respectively) microscopy. This

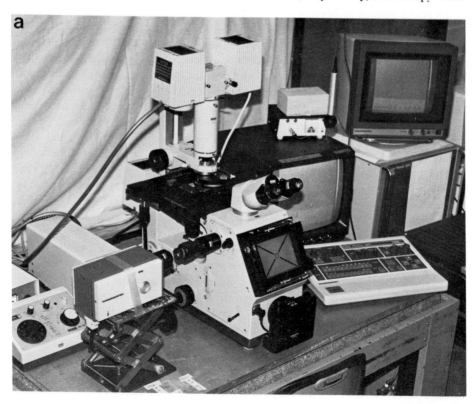

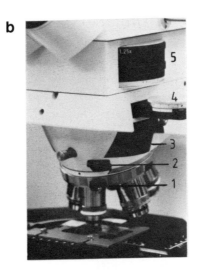

Figure 9. (a) Video microscopy setup used by the authors for temporary work at a marine biological laboratory. From left, Chalnicon camera at the lateral exit of the Zeiss inverted microscope IM 405. A ×25 eyepiece and a 63 mm camera lens are used. (For permanent setups a more stable camera support is recommended.) The microscope is equipped with tungsten and a 50 W stabilized DC mercury arc lamp for transmitted light, and a 100 W mercury arc lamp for epi-fluorescence. The 1/4 wave plate for de Sénarmont compensation is mounted in the fluorescence filter holder (not visible). The monitor for the processed image is seen behind the keyboard of the C 1966 image processor of Hamamatsu. A second monitor displaying the raw image (for focusing) is located on top of the image processor. Far right: colour monitor for pseudocolour display and photographing and U-matic recorder. (b) Parts for AVEC-DIC microscopy with de Sénarmont compensation as built into a Zeiss Axiophot microscope. (**1**) Objective Wollaston prism; (**2**) 1/4 wave plate; (**3**) ×1.6 magnification optovar; (**4**) rotatable calibrated polarizer used as analyser; (**5**) additional magnification changer ×2.5. With this setting an empty camera attachment tube can be used (*Figure 6b*). The polarizer and Wollaston prism in the condenser are not visible.

technique involves setting the analyser and polarizer far away from extinction in order to gain high signal-to-noise ratio [see Section 1.3.2 (iii)]. Allen suggested the use of a de Sénarmont compensator setup (3,4,15) which includes a 1/4 wave plate (for the wavelength used) in front of a rotatable analyser (*Figure 9b*). He recommended a bias retardation of 1/4 to 1/9 of a wavelength away from extinction with 1/9 as the best compromise between high S/N and minimal diffraction anomaly of the Airy pattern. The enormous amount of straylight (I_B) introduced at this setting (*Figure 10*) is removed by the appropriate setting of analogue and/or digital offset (Section 1.3.2).

The technique recommended by Inoué, which is called IVEC microscopy for distinction in this chapter, aims to reduce straylight and diffraction anomaly arising from curved lens surfaces (*Table 1*) by employing extremely strain-free objectives and the rectifying lenses developed by Inoué (16). The latter are commercially available only for a few microscopes, such as some lines of Nikon and are expensive. Inoué's special optimized microscope (2,16) is used at a polarizer setting very close to extinction, which cannot be used with many other instruments because of insufficient light for saturation of the camera.

The AVEC technique improves images of low contrast due to non-optimal optical arrangement, while in IVEC microscopy no compromise is made regarding the optics and consequently less demanding electronic steps are required to rescue the image. The

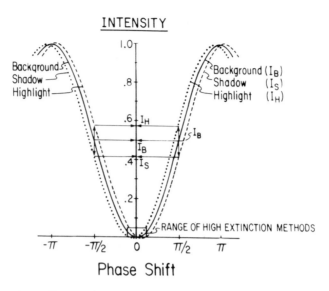

INTENSITY

Phase Shift

Figure 10. Dependence of intensity (image brightness) and contrast (see Section 1.3.2) of DIC images of phase objects on phase retardation. Phase retardation is introduced and varied by laterally displacing a Wollaston prism or by setting the de Sénarmont compensator. This converts positive and negative phase gradients of specimens to contrast, thus producing highlights (I_H) and shadows (I_S) relative to a neutral grey background (I_B). A phase shift of $\pi/2$ radians corresponds to $\lambda/4$, i.e. 1/4 of a wavelength. It can be seen that theoretically the image contrast is highest where $\pi/2$ (vertical arrow). Due to straylight the image background (I_B) is of considerable intensity at this setting and has to be compensated for by offset. (Reproduced with permission from ref. 4.)

The AVEC technique is, however, the one which can be used with any good research microscope equipped with commercial polarizers. The proper compensator setting can be experimentally evaluated between 1/100 and 1/4 of a wavelength within the limits of your illuminating system. Best resolution has been found at 1/9 of a wavelength (17) and best visualization (highest contrast) has been reported for 1/15 of a wavelength in some microscopes (18). According to Allen the dimensions of an object imaged at the latter setting may be different for different orientations due to the diffraction anomaly of the Airy pattern (3,4).

2.2 Sample preparation

Samples used in conventional light microscopy can also be used for video microscopy. Live cells from tissue cultures should preferentially be grown on the coverslip. The specimen's region of interest should be close to the coverglass surface, where the best image is obtained. If high magnifications are intended it may be found that the optics can only be set for Köhler illumination at this surface and a few tens of micrometers below (using an upright microscope), since high magnification objectives are usually designed for optimum imaging of objects at a distance of 170 μm from the front element. Unless you work with dry objectives (oil-immersion is, however, to be preferred) it is therefore recommended to use No. 0 coverglasses (80−120 μm thick) instead of regular ones (∼170 μm thick) (e.g. Gold Seal from Clay Adams Co., or O.Kindler GmbH). Similarly thinner slides may be useful (0.8−0.9 instead of 1 mm).

Aqueous samples have to be prevented from drying out by completely sealing the coverglass with nail polish or, if 'live' specimens, such as microtubules, extruded cytoplasm or cultured cells are observed, with VALAP. This consists of equal parts by weight of vaseline, lanolin and paraffin ($51-53\,^{\circ}$C) and liquefies at around $65\,^{\circ}$C. It is applied around the coverglass with a cotton tip applicator. If the specimen is in suspension, no more than $5-10\ \mu$l aliquots should be used with regular size coverglasses in order to produce very thin specimens ($\sim 10\ \mu$m) for best image quality.

If working with an inverted microscope, the slide has to be applied with the coverglass pointing downward. With most microscope stages this will interfere with the VALAP sealant and flat positioning of the slide will not be possible. It is recommended instead to use a metal frame of the size of a regular slide and $0.8-1$ mm thick. The centre half is cut out so that a U-shaped holder is obtained. A larger coverglass is then attached with adhesive tape on the upper side so that it covers the opening. If necessary, this has to be the No. 0 coverslip since its top surface is the one best suited for observation. After the sample has been applied, the top coverglass of regular size and thickness is added and sealed.

If thicker specimens such as tissue slices, vibratome sections or nerve bundles are to be observed, only DIC or anaxial illumination (19) techniques are recommended. It will be only the first 10 or 20 μm next to the objective front lens which will yield good serially sectioned images, the quality degrading quickly if you focus deeper into the tissue.

2.3 Procedure for image generation

Because VEC requires some steps which are different from conventional microscopy, image generation is discussed here in some detail. Steps i−v of the procedure yield the image if only analogue enhancement is required. The continuation leads to highest resolution and visualization of submicroscopic objects. The procedure chosen is basically that used for AVEC-DIC, but if DIC is not required step iv can be disregarded; if the user does not want to follow the procedure given by Allen (AVEC), he may find the comments in Section 2.1 useful.

(i)   Find the specimen preferably by looking through the oculars or, alternatively, by looking at reduced magnification at the monitor. If the entire specimen consists of subresolution size material (density gradient fractions, microtubule suspensions, unstained EM sections, see *Figure 2*) it will be difficult to find the specimen plane. Use a relatively dark setting of the condenser diaphragm and/or polarizers or prisms and look for contaminating larger particles. If there are none, routinely apply a fingerprint to one corner of the specimen side of the coverglass and use this for focusing.

(ii)   Adjust Köhler illumination (see Chapter 1). After finding a coarse setting for the illumination, the desired plane for the specimen is selected exactly. Then the condenser is finely adjusted, but now in relation to the image on the monitor (make sure the light is reduced to avoid damage of the camera!). The field diaphragm must be centred on the monitor and opened until it becomes just invisible. If the field diaphragm is opened too much, most microscope−camera adapter tubes or high power projectives and oculars will create a very annoying central hot spot (*Figure 4*). If this persists at the adjustment for proper illumination, close the projective diaphragm, or insert a self-made diaphragm to cut the peripheral light at the microscope exit. Note that at the high magnifications and NAs used here, Köhler illumination has to be re-adjusted once you change the focus more than a few micrometers.

Since we will apply extreme contrast enhancement later, we have to start out with as even an illumination setting as possible. Proper centering of the lamp and setting of the collector lens is therefore important. At high magnifications as much light as possible needs to be collected. Some workers have for this purpose used critical illumination instead of Köhler illumination, that is focusing the light source onto the specimen plane (18). This is counter to good microscopical practice and can lead to very uneven illumination since the filament or arc will be superimposed onto the image of the specimen and has subsequently to be subtracted digitally by mottle subtraction. Critical illumination might be useful, however, in those cases where the illuminating light is made extremely homogeneous by light scrambling with a light fibre device (2,18,20).

(iii) Open the condenser diaphragm fully in order to utilize the highest possible NA to obtain highest resolution. Also the iris diaphragm (if present) of the objective should be fully opened. Be careful to protect the camera from high light intensity prior to this step. The result of opening the condenser diaphragm will usually be that the optical image worsens because it becomes too bright and flat for the eye. This setting will result in a small depth of focus, especially with DIC (optical sections of 0.3 μm or less with ×100 oil objectives). If a large depth of focus is required (e.g. when viewing dilute suspensions), the condenser diaphragm can be closed down as desired but preferably to not smaller than 1/4 of the aperture of the objective.

(iv) (Polarized light techniques only.) Set the polarizer (AVEC-POL) or the main prism or compensator (AVEC-DIC) to about 1/9 λ (read Section 2.1 first in order to know whether you want this setting). The optical image, that is that seen in the oculars, will disappear due to excessive straylight. Reduce the illumination to protect the camera (but not by closing diaphragms).

If you have the accessories for de Sénarmont compensation as recommended by Allen *et al.* (3,4) (*Figure 9b*) set them at 20° off extinction. The basic setting up of de Sénarmont compensation is done as follows.

(a) Remove both Wollaston prisms and 1/4 wave plate from the light path.
(b) Set the analyser and polarizer to the best extinction.
(c) Insert a 1/4 wave plate at 0° (best extinction).
(d) Insert the Wollaston prisms and set the adjustible one to the best symmetrical extinction (if possible, check with a phase telescope for symmetry of the pattern).
(e) Use the rotatable analyser as compensator and set it as desired (1/9 of a wave is 20°, 1/4 is 45°).

If you do not have such a calibrated system, first determine the distance between extinction (0°) and maximum brightness (90° or 1/2 λ) by moving the adjustable Wollaston prism, then estimate and select the 1/9 of a wave or 20° position (*Figure 10*). Many microscopes equipped with DIC for biological applications do not allow a phase shift of 90° and some may not even allow 20° since for observation by eye phase shifts of a few degrees yield good contrast. Microscope manufacturers will, however, have the proper parts in their mineralogy programmes.

At this point you have to make sure that the camera receives the proper amount of light to work near its saturation end. Some manufacturers have red and green LEDs built in to indicate this. Otherwise you should see a moderately modulated image on the monitor while a very flat or no image indicates insufficient light. In this case re-adjust the illumination, if possible, while observing the image on the screen. If necessary, remove any diffusers or use brighter lamp types. Opening the crossed polarizers beyond 20° will not improve the image. If you have excessive light, reduce it (if using adjustable lamp types) or, (with arc lamps) use neutral density grey or other filters, or go a few degrees closer to extinction.

In IVEC microscopy straylight is not admitted, that is the polarizers stay close to extinction and the special rectifying optics further reduce the straylight. Filters to reduce brightness will not be required, but much brighter lamps will most probably be necessary to saturate the camera.

(v)   Analogue enhancement. Increase the gain on the camera to obtain good contrast. Then apply offset (pedestal). Always stop before you lose parts of the image that become too dark or too bright. Repeat this procedure several times, if necessary and of help. Make sure that the monitor for watching the changes is not set to extreme contrast or brightness, and is terminated properly (see Section 5.1). Analogue enhancement improves the contrast of the specimen but unfortunately also emphasizes dust particles, uneven illumination and optical imperfections. These artefacts, termed 'mottle', are superimposed on the image of the specimen and may in some cases totally obscure it (*Figures 2* and *4*). Disturbing contributions from fixed pattern noise (mottle) or excessive amounts in unevenness of illumination can be tolerated if digital enhancement is performed later (*Figure 2*).

If digital processing is not possible stop enhancement just before the mottle or uneven illumination becomes annoying. Apply analogue shading correction and other types of analogue image improvement if your camera control unit offers these features (Section 1.4.1 and *Figure 6*). Thorough cleaning of the inner optical surfaces of the microscope, especially the surfaces in the projecting system to the camera (ocular, camera lens) usually results in images which allow the application of considerably higher analogue contrast enhancement.

Finding dust: When the imaged dust particles or the mottle pattern rotate when the camera head is rotated, they are located in the optical path before the camera. Immobile dust is to be found on the camera faceplate. Rotate the ocular or camera objective lens to find dust located there. Dust should be removed with a low-pressure air gun or an optical cleaning brush. If this does not help use lens paper or a fat-free cotton tip applicator (wooden stick, not plastic) with ethanol or ether. Work from the centre to the periphery in a circular fashion while carefully avoiding applying pressure. Dust or mottle which is defocused if the specimen is defocused is part of the specimen.

(vi)   Try removing the specimen laterally out of the field of view or (when using DIC) defocus to render it just invisible (preferably towards the coverglass). The result is an image containing only the imperfections of your microscope system (mottle pattern) (*Figures 2c* and *4c*). This step will not be satisfactory, however, with such techniques as phase-contrast.

(vii) Background (mottle) subtraction. Store (freeze) the mottle image, preferably averaged over several frames, and subtract it from all incoming video frames. You should see an absolutely even and clean image, which may, however, be weak in contrast. If there are regions 'missing' that are grey and flat there is too much contrast in the raw image to perform proper background subtraction. Reduce gain, adjust offset and repeat the procedure (*Figures 2d; e* and *4d*).

(viii) Perform digital enhancement in a similar manner to step v, that is alternate between stretching a selected range of grey levels (setting 'width') and shifting the image obtained up and down the scale of grey levels (setting 'level') until a pleasing result is found (*Figure 2f*). If it is available, display the grey level histogram and select the upper and lower limits which are to be defined as bright white and saturated black, respectively. If the image is noisy (pixel noise) go to step ix.

(ix) Use an averaging function in a rolling (recursive filtering) or jumping mode over two or four frames. This will allow the observation of movements in your specimen, but very fast motions and noise due to pixel fluctuations will be averaged out. Averaging over longer periods of time will filter out all undesired motion such as for example distracting Brownian motion of small particles in suspension. The image will then be of immobile objects, exclusively.

Please note that not all image processors are capable of performing background subtraction and rolling average simultaneously. If this is the case, averaging generally yields the better image improvement for VIM, while background subtraction is more advantageous in VEC microscopy, although this should be determined experimentally. Alternatively, background subtracted or averaged scenes (plus empty scene for later background) can be stored on video tape or disc and then subsequently be played back into the processor for further processing.

(x) There are a number of procedures for spatial filtering available, which can be used to reduce noise, to enhance edges of objects or to reduce shading. These have been described for the analysis of single images in Chapter 7 but some image processors offer such filters at video rate so that live sequences can be accentuated by filtering prior to recording.

2.4 Interpretation

Unlike EM images, which truly resolve the submicroscopic objects depicted (*Figure 11a*), the sizes of objects seen by AVEC-DIC microscopy may not necessarily reflect their real size. Objects smaller than the limit of resolution, that is 100−250 nm, depending on the optics and the wavelength of light used, are inflated by diffraction to the size of the resolution limit. The orientation of birefringent objects may also somewhat affect their apparent thickness if they are oriented at angles very close to 45° or 135° (*Figure 11b*). Whereas the size of the image does not enable a decision

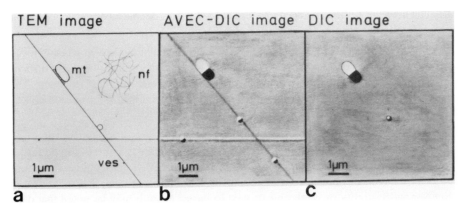

Figure 11. Schematic representation of the visualization capabilities of different kinds of microscopy. (**a**) In transmission electron microscopy (TEM) all membraneous and cytoskeletal elements of the cytoplasm can be resolved and imaged at their true size. TEM is, however, in most cases impossible on living or hydrated specimens. (**b**) AVEC-DIC microscopy permits visualization of objects smaller than the limit of resolution if they are larger than 10−20 nm. These objects, however, would not appear at their real size, but inflated by diffraction to about the size of the resolution limit. (**c**) Differential interference contrast (DIC) microscopy images are characterized by their typical shadowcast appearance. The smallest objects visible are of an apparent size in the order of the theoretical limit of resolution. ves, axoplasmic or synaptic vesicles of about 60 nm diameter; mt, mitochondrion; nf, neurofilaments; the straight lines represent microtubules (25 nm in diameter); the larger vesicular organelle is about 250 nm in diameter. (Reproduced with permission from ref. 9.)

on whether one or several objects of a size smaller than the limit of resolution are present, the contrast sometimes permits this judgement to be made. A pair of microtubules would, for example have the same thickness as a single one, but the contrast would be about twice as high. If large numbers of subresolution objects are separated by distances of less than 200 nm from each other (e.g. vesicles in a synapse), they will remain invisible, but they will be clearly depicted if they are separated by more than the resolution limit. Also remember that, if in-focus subtraction (*Figure 5*) or averaging is used, the immobile or the moving parts of the specimen, respectively, may have been removed from the image.

2.5 Typical applications and limitations

2.5.1 Bright-field microscopy

(i) *Advantages.* Toxicity is often less of a problem at low concentrations of dyes. Low contrast due to low concentrations of natural chromophores or especially vital stains can be greatly enhanced. Observations at the wavelength of maximum absorption increase contrast. Using appropriate cameras with quartz windows, UV microscopy is feasible and yields high resolution and high contrast images (21).

Collodial gold, especially when coupled to antibodies, is an important tool in immunocytochemistry at the EM level (22). Collodial gold in sizes down to 10 nm can also be detected using bright-field VEC-microscopy and localized to around 200 nm. Different sizes of colloidal gold will appear the same size (0.1−0.2 μm depending on focus), but occupy different grey levels. By pseudocolour conversion of these grey levels, they can however be differentiated into size classes. Thus, double labelling with different sizes of collodial gold is feasible (See also Section 2.5.7).

(ii) *Problems with enhanced bright-field*. Phase objects exhibit minimal contrast in focus, and show opposite contrast above and below focus. It is, therefore, often very difficult to store the out-of-focus mottle image without contributions from the specimen. This limits the use of high analogue contrast enhancement in bright-field microscopy.

2.5.2 *Dark-field microscopy*

(i) *Advantages*. Dark-field images provide very high contrast, and this can be further enhanced by applying gain to the video signal. Genuine straylight from nearby or out-of-focus scatterers can be subtracted by using offset to a limited extent, thereby increasing contrast and visibility of weak objects next to larger ones. It may be noted that dark-field images can also be subjected to video-intensified microscopy.

(ii) *Disadvantages*. The specimen must be thin and contain relatively few scatterers. Most annoyingly out-of-focus scatterers do not produce true straylight because the scattered light is not evenly distributed over the image. Because for transmitted light dark-field microscopy the objective NA must be limited to less than about 1.1, dark-field is not a very high resolution method. Nevertheless, it can *detect* the presence of structures in the nanometer range in thin preparations. In epi-illumination dark-field mode it is possible, however, to work at maximum NA and resolution.

2.5.3 *Anaxial illumination*

Anaxial (or oblique) illumination is a collection of methods that illuminate the condenser aperture unevenly to produce a differential (shadowcast) image. This is usually done by excluding the undiffracted light to a varying extent.

In Abbe's original method, the iris is moved to one side of the front focal plane of the condenser. In Hoffman's modulation contrast method (23) (Modulation Optics Inc.) undiffracted light is partially excluded by a trizonal plate, which also serves to accentuate high spatial frequencies. The optional rotatable polarizer makes it possible to vary the excentricity of the undiffracted light. The single sideband technique as developed by Ellis (24) yields very good results with video microscopy. It is also possible, using a tungsten filament lamp with a hemispherical mirror, to displace the mirror to illuminate only half of the aperture (19). For video-enhanced contrast microscopy, these methods work very well, because their main limitation of providing low intensity images can be overcome by enhancement.

(i) *Advantages*. Phase details are observed in directionally shadowed differential images (similar to Nomarski-DIC). High specimen birefringence does not interfere as in DIC microscopy. For example, vertebrate axons can be better seen with enhanced anaxial illumination than with DIC because their highly birefringent myelin sheaths create excessive contrast in DIC. The equipment is cheap: only a bright-field microscope is required. It has been reported that observation in deeper layers of tissue is still satisfactory where the quality of DIC images would be much poorer (19). In principle, anaxial illumination does not reduce working aperture or resolution as long as not less than half the entrance pupil is illuminated.

(ii) *Disadvantages.* Anaxial illumination is sometimes less sensitive than Nomarski-DIC, and the resulting image may have some anisotropy (for test images see ref. 25). The optical sectioning is thicker and less precise.

2.5.4 *Phase-constrast*

Phase-contrast is an adequate method for viewing very thin specimens and very small phase objects, especially living cells. Video contrast enhancement is often of great value, especially in thin specimens and if analogue enhancement is sufficient.

Disadvantage. Thick specimens imaged in phase-contrast cannot be optically sectioned, because each image plane contains spurious contrast contributed by details in other levels of the specimen, and because halos are seen around phase details. Enhancement rapidly raises a very disturbing mottle pattern. It can sometimes be subtracted from a plane approximately 0.25 μm above or below the plane of interest if an area free of specimen is used. Out-of-focus mottle images of the specimen cannot be obtained in most cases without disturbing contributions from the out-of-focus specimen. Even extreme cleaning of the projecting optics is usually not sufficient to remove all the annoying mottle pattern.

2.5.5 *Polarizing microscopy*

For visual observations of most biological objects that require small bias retardations, the microscope equipped with simple polarizers is rather inadequate. The image is flooded with straylight owing to a combination of depolarization at lens surfaces and strain birefringence in one or more lens elements (*Table 1*).

For sensitive visualization or photomicrography of birefringement structures, it is better as a rule to use a polarizing microscope with a rectifier (Nikon) that eliminates not only straylight due to depolarization at lens surfaces, but also a disturbing diffraction anomaly that can lead to spurious contrast and resolution (26). The rectifier consists of a zero power meniscus lens and properly oriented half-wave plate (2,16).

Using video-enhanced polarization microscopy, however, relatively sensitive observations can be made with a simple polarizing microscope even if the lenses are slightly strained. One can, in fact, in most instances convert a bright-field microscope into a video equivalent of a relatively high extinction microscope by adding inexpensive film polarizers (e.g. Polaroid HN 22 polarizing material which comes in sheets and can be cut down to any desired size; Polaroid Inc.). In an analogy to what has been said concerning AVEC-DIC (Section 2.1 and *Figure 10*), at 1/9 λ bias retardation (AVEC-POL), the anomalous clover leaf Airy disc diffraction pattern seen at extinction (zero retardation and crossed polars) has been converted into a normal Airy disc, with the result that there is no longer any spurious resolution or contrast. In addition, the diffraction anomaly due to weak strain birefringence in any of the lens components also disappears (3,4).

The use of highest quality optical components for polarized light microscopy together with video enhancement (AVEC-POL and IVEC-POL, Section 2.1) visualizes extremely weak birefringent objects, such as individual microtubules, bacterial flagella etc. (2,25).

2.5.6 *Differential interference contrast microscopy*

(i) *Advantage.* This is the technique which is, in most applications, best suited for video contrast enhancement. The Nomarski-DIC method gives in-focus, high contrast, shadowcast images of phase details in which the detection of shadowing is opposite for phase advancing and retarding details (26). The generation of contrast at high working aperture is limited to a very shallow depth of field, with the result that this technique is unique in its ability to render high contrast optical sections of only 200−300 nm in thickness. Amplitude contrast can also be obtained (26). With the AVEC- or IVEC-DIC method the sensitivity of detection is increased to the level that phase objects 15 nm and larger can be detected under best conditions.

Increasing the bias retardation (*Figure 10*) causes a disproportionate increase in the contrast due to small phase retardations in comparison to large retardations. For example, if low contrast details of cytoplasmic transport are obscured by bright organelles, such as, for example starch grains, the contrast due to the latter can be diminished by using a dimmer light source and increasing the bias retardation more towards $1/4 \lambda$ (*Figure 10*).

(ii) *Disadvantage.* Perhaps the only disadvantage of this method is that contrast due to refraction in some highly birefringent objects (e.g. striated muscle or myelinated axons) can be masked by the contrast due to birefringence. In this case anaxial illumination is recommended as a substitute. Because contrast is directional, the specimen should be mounted on a rotatable stage and examined at more than one orientation to avoid missing linear features aligned parallel to the direction of shear (see *Figure 11*).

2.5.7 *Reflection contrast microscopy*

Surface reflection interference from epi-illuminated specimens is used to image zones of cellular attachment to glass (27,28). To reduce the inevitable straylight caused by reflections from lens elements, Ploem suggested the use of antiflex lenses designed for metallurgical microscopes containing a 1/4 wave plate on the front surface of the objective. Using video enhancement, simple epi-reflection microscopy (i.e. no polarizing elements or antiflex lens), can be useful for several purposes.

(i)   To enhance the observation of zones of cell attachment to glass. (Surface reflection interference at high illumination aperture.)

(ii) To enhance the observation of contour-mapping interference fringes to document the shapes of attached cells or the topology of surfaces of cells above a flat glass surface. (Surface reflection interference at low illuminating aperture.)

(iii) To render visible colloidal gold or other metals down to a diameter of 5 nm as bright objects. Epi-polarization even further improves the visibility (29,30).

In all these applications the use of Antiflex objectives with $1/4 \lambda$ plate (Zeiss) as in the original technique (27,28) will further improve image quality.

2.5.8 *Fluorescence microscopy*

Fluorescence images usually need intensification, especially if the level of illumination has to be kept low to delay bleaching. Often a brighter low power fluorescence image obtained with high NA objectives can be seen with an enhancement camera when gain is applied to the video signal. Any straylight arising from residual autofluorescence in lenses or the mounting medium can be removed by offset. However, light from out-of-focus fluorescent structures usually cannot be treated as straylight because of its non-random distribution. Often fluorescent details can be seen better when their contrast has been reversed by inverting the television signal (negative image).

Fluorescent specimens have usually to be viewed with video intensification (see Section 3) but this will inevitably lead to reduced resolution when compared to images obtained by VEC microscopy at the same magnification. If higher optical magnification can be achieved, the obtainable resolution is theoretically the same. With some applications, especially in the case of strong fluorescence it may pay to try the use of a high resolution, high sensitivity camera (Newvicon or Ultricon) in the rolling average mode (over one or more seconds). Similarly accumulation of images for a few seconds may bring about a useful image.

Since in fluorescence microscopy self-luminous objects are depicted, images contain a strong contribution from out-of-focus objects. Images can however be limited to a depth of focus of about 1 μm by digital blur-deconvolution (31). This relatively complicated method is presently being replaced by the advent of confocal microscopes (32) (Bio-Rad Lasersharp Ltd, Heidelberg Instruments GmbH, Tracor Northern, and Zeiss West Germany) which were also mentioned in Chapter 6.

2.5.9 *Examples of biological and biochemical applications*

Specimens best suited for AVEC-DIC microscopy do not include stained material or other objects, which already have high contrast. Specimens which are extremely weak in contrast or even invisible by conventional microscopy are best. Examples in this class are micelles, liposomes and single or double-layer membraneous material, colloids (33), live, actively transcribing rDNA genes (34), synaptic and other small cytoplasmic vesicles (8,9), artificial latex particles of 70 nm and smaller (unpublished observations), and cytoskeletal elements such as microtubules (9,25) and actin bundles (35,36). The process of microtubule gliding was discovered by AVEC-DIC microscopy (8), its ATP-dependence and the enzyme kinetics of the translocating ATPase are assayed microscopically (37), and even molecular events such as microtubule subunit assembly and disassembly can be measured (38). An overview of what can be seen by this method in the living cell has been published by Allen (25).

The AVEC-POL technique will also visualize very weakly birefringent objects such as individual microtubules (25). When applied to bright-field or epi-polarization microscopy, the VEC technique visualizes 5−20 nm diameter colloidal gold particles as used in immuno-electronmicroscopy (29,30). Besides observing the behaviour and redistribution of the antigens tagged with gold particles in living cells, VEC microscopy can be used to screen gold-labelled EM specimens quickly in the light microscope. The same is true for semi-thin and sometimes thin, unstained, plastic-embedded EM sections (*Figure 2*).

Techniques to improve images of *moving objects* can be generated with the aid of digital processors. The 'trace' operation adds frames at predetermined intervals to the frame memory thereby generating images showing multiple positions of moving objects. Jumping averaging over a few seconds visualizes processes too slow to be detected otherwise, such as cell growth, chromosome movements and cell locomotion. Rolling average can be used as a filter to remove velocities greater than pre-selected velocities. Subtraction of in-focus images and scenes can be used to view moving objects only, while stationary ones are absent from the image (*Figure 5*). The interval subtraction mode [see Section 1.3.3(iv)] can be used to filter out slowly moving objects. The length of the interval before the next background image is grabbed determines whether objects of a given velocity will become visible. Very long intervals will be required to detect slower moving objects also.

3. LOW LIGHT: VIDEO-INTENSIFIED MICROSCOPY

3.1 Introduction

This technique requires the specially sensitive cameras or the image intensifier units mentioned in Section 1.4.2. Once the image has been analogue enhanced and is picked up then the digital techniques are used as in VEC microscopy (Section 2.3). First, and most importantly, they improve the S/N during image acquisition by, for example integration, averaging or digital filtering. Secondly, they remove any 'fixed pattern noise' such as mottle, uneven illumination, background fluorescence or any pattern emanating from the camera or the digitizing process (background/mottle subtraction). Thirdly, they can be used for digital contrast manipulations, while enhancement or suppression of motion is the fourth and least important step for VIM. The major difference relative to VEC microscopy in the application of these functions to VIM is that generally only static or slowly moving specimens can be examined. These require the processing described in Section 3.2.2, while a dynamic specimen (Section 3.2.3) can be treated similarly to those viewed by VEC microscopy.

3.2 Procedure for image generation

3.2.1 Microscope considerations

There are no special or unconventional adjustments of the microscope necessary to perform VIM. One should, of course, adhere to good microscopic practices such as cleanliness of the optics, careful alignment of the illumination system and of the camera to the light path. There are, however, a number of practical issues which will help ensure the best results when performing VIM.

(i)   If using a camera with extended red-sensitivity (e.g. multialkali photocathode) for viewing in visible wavelengths it is suggested to use at least one high quality IR cutoff filter in the camera's light path. While the human eye and photographic film are not sensitive to these longer wavelengths, they can seriously degrade image quality. This is true even if good fluorescence filters are in place since these are notorious for passing through light over 700 nm.

(ii)   The camera image should be as parfocal with the eyepieces as possible. When working with very low intensity images it can be very difficult to accurately assess exact focus from the live image on the monitor.

(iii)   Provisions should be made for working in a dark room or establishing some method to shield the objective lenses from *all* extraneous light. This light is easily picked up by highly intensified cameras and image quality will consequently suffer. It is often necessary to cover the eyepiece of the microscope with, for example a black cloth.

(iv) The illuminator should be equipped with a regulated power supply if low intensity illumination will be utilized (e.g. to reduce photobleaching in fluorescence) or if quantitative measurements are planned. Small fluctuations in line current are greatly magnified by intensified cameras so that stabilized power supplies are recommended.

(v) It is advisable to equip the microscope in such a way that a good optical image (bright-field or DIC) can be obtained for positioning and focusing the sample prior to directing the low intensity light to the camera. For applications in fluorescence it is a good idea to work with non-excitatory light before moving into the fluorescence mode to minimize photobleaching effects.

(vi) Since relatively long integrations or averages may be required, the microscope should be as stable as possible to prevent any image blurring. Additionally, because many low light cameras are relatively large, it is often good practice to stabilize the camera by some means other than just the C-mount.

(vii) Care should be taken to obtain slides, coverslips, immersion liquids, etc. that are free of, or extremely low in, autofluorescence.

(viii) The microscope should be equipped with a series of neutral density filters. These can be used to reduce illumination of the specimen as well as to protect the camera from excessive light.

3.2.2 *Acquisition of static images*

(i) Focus the specimen through the eyepiece ensuring that no light is directed to the camera.

(ii) With the camera sensitivity reduced to its minimum setting, direct the light to the camera. If light to the camera is excessive *immediately* direct the light away from the camera or block the illumination path. It will be necessary to insert neutral density filters into the light path before trying again.

(iii) With light to the camera and an image on the monitor, focus the specimen, if necessary, and increase camera sensitivity until some portion of the image begins to saturate. Now reduce sensitivity to just eliminate any saturation. This procedure ensures that the video signal is of full amplitude (1 V peak-to-peak) and will provide the best S/N. An alternative approach to this is to increase the light level rather than the sensitivity. Obviously, this is not possible in those cases where light is limiting or where light levels are intentionally kept low in order to reduce photobleaching, phototoxicity, phototropic effects, etc.

(iv) Adjust the monitor for best picture quality.

(v) If the camera is equipped with gain and offset, adjust these parameters for the best overall image quality or until that part of the specimen which is of interest displays the most information (see step v in Section 2.3).

(vi) Integrate the image for a suitable number of frames. Depending on the quality of the 'raw' or analogue image this can vary anywhere from 2 to 512 frames. The value chosen should be based on the user's qualitative judgement or some established criteria such as S/N ratio. As S/N is proportional to the square root of the number of frames, a point of diminishing returns is reached between 128–256 frames. It should be pointed out that the maximum number of frames integrated may be restricted by the 'bit depth' of the digital image memory. Systems for low light imaging should be equipped with memories at least 12 bits and preferably 16 bits deep to allow for extended integration.

(vii) Establish and maintain a background image. Two possibilities exist for this background image. The best is to find a suitable background area adjacent to the specimen. This area will contain all the components of the image which should ideally be removed. This includes any shading contributions or fixed pattern noise of the camera, illumination irregularities, optical defects, autofluorescence in the system and any digital noise. If such an area is not available, the second alternative is to block light to the camera and use this 'dark image' as the background. This will correct for any camera-related phenomena (*Figure 12*).

In almost no case should an out-of-focus image of the specimen be used as the background image. With few exceptions (e.g. DIC) defocusing cannot eliminate all specimen-based image information and one runs the risk of erroneously subtracting non-background information.

(viii) Now, with all settings identical (camera sensitivity, illumination, etc.) integrate a background image into a different digital memory than that of the specimen and subtract it from the stored specimen image analogously to step vii in Section 2.3. Continue with steps viii and x of Section 2.3 to manipulate image contrast for the most pleasing or informational image.

3.2.3 *Acquisition of dynamic images*

(i–v) As in Section 3.2.2.

(vi) Acquire a background image as in Section 3.2.2, step vii. Ideally, this background image should be integrated or averaged for the same period as the rolling average utilized in the next step. This image should be placed in a memory which can be subtracted from each incoming video frame as in VEC microscopy.

(vii) Return to the specimen image and begin video rate background subtraction as in Section 2.3, step vii. If background is high, this subtraction process may result in the image losing intensity. If so, add digital offset to the image to restore intensity or brightness.

(viii) Apply a 'rolling' or 'exponentially weighted moving' average to the background subtracted image as in Section 2.3, step ix. This type of average offers similar S/N improvement (S/N = $\sqrt{(2N-1)}$, N = number of frames) as integration, which is used for static scenes [Section 3.2.2 (vi)], but offers the advantage of being applicable to dynamic specimens. Additionally, unlike image integration, in averaging the image is normalized, therefore the digital memory does not overflow or saturate.

(ix) Digitally manipulate image contrast as in Section 2.3, steps viii and x.

3.3 **Typical applications**

3.3.1 *Fluorescent analogue cytochemistry*

In fluorescent analogue cytochemistry (FAC), the molecule or organelle of interest is isolated, fluorescently labelled and reintroduced into the living cell (40). Ideally, images

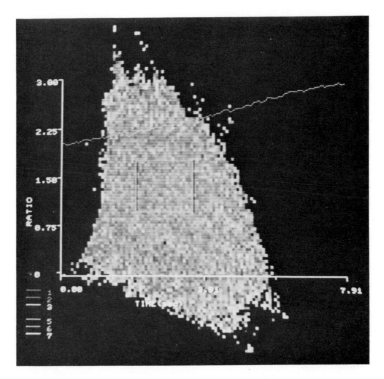

Figure 12. Temporal analysis of intensity changes in a predefined region of the image. Specimen, human fibroblast loaded with the Ca^{2+}-dependent fluorochrome FURA-2. FURA-2 is used to measure changes in Ca^{2+}-concentration over time by the ratio imaging technique (see Section 3.3. and ref. 41). The specimen is excited alternately at 340 and 360 nm and the resulting fluorescence images emitted at 500 ± 10 nm are divided. The resulting ratio image has the information on Ca^{2+} concentration coded as intensity. The ratio values of 0−3.00 which cover the range of biological Ca^{2+} concentrations (~0.05−5 μM) are assigned to the grey levels 0−256 (ordinate). An increase in image brightness thus means a rise in free Ca^{2+} concentration. The average intensity in the frame placed over the cell centre is measured and plotted versus time (8 sec, abscissa). In this example 100 frames at a sampling frequency of 12.5 Hz were processed at reduced spatial resolution (128 × 128 pixels). Microscope, Polyvar, Reichert/Cambridge Instruments equipped with quartz collector, 200 W mercury arc and ×100 planfluorite oil objective; processor, ARGUS 100 of Hamamatsu; frame width, 62 μm.

of these 'analogues' can reveal the distribution and organization of the analogous native molecule or organelle and how they change over time. The rate and polarity of the incorporation of actin and tubulin analogues into *in vivo* structures has been examined using FAC as well as the changes in the cytoplasmic distribution of a variety of molecules.

3.3.2 *Ratio imaging*

Ratio imaging is a powerful technique in analysing the spatial and temporal dynamics of ions, molecules and organelles in living cells (see Section 4 and *Figure 12*). By generating the ratio or quotient of two images, this technique normalizes the cell for pathlength and accessible volume, thereby allowing for subcellular quantitative evaluations. Fluorescence ratio imaging microscopy has been successfully applied to studies involving the dynamics of intracellular Ca^{2+} and pH (41−43). These studies were made possible by the use of fluorescent probes which exhibit differential excitation wavelength sensitivity to Ca^{2+} (FURA-2) or pH (BCECF). The ratio image calculated from the two images obtained at the different wavelengths is converted to an image representing the concentration of free intracellular Ca^{2+} or H^+ ions (pH) using calibration curves constructed from dyes in buffer.

3.3.3 *Fluorescence recovery after photobleaching* (FRAP)

Studies of translational diffusion and lateral motion in membranes have benefited greatly from the application of video-intensified imaging. In this method, an area of cytoplasm or membrane containing a fluorescent probe is bleached using a laser or other strong illumination source. The rate and pattern of the reappearance of fluorescence in the bleached region can be used to analyse the contribution of isotropic diffusion, anisotropic diffusion and bulk flow to lateral transport phenomena in living cells (44,45). This technique has been coined 'TRSP' (time-resolved spatial photometry).

3.3.4 *Molecular imaging*

A number of macromolecules, including DNA and actin, have been covalently labelled with fluorescent groups and studied using VIM (46, 47). The visualization of single DNA molecules in solution has allowed the study of changes in molecular conformation. Chromatin structure in isolated nuclei and intact cells has also been studied. Similarly, the conformation of single actin filaments and their movement on immobilized myosin has been studied (47).

3.3.5 *Low light video microspectrofluorometry*

The use of intensified cameras as an alternative to photomultiplier tubes offers the advantages of whole image, spatially resolved spectral analysis (48). Since many of the fluorochromes utilized to monitor physiological changes in living cells exhibit changes in quantum yield, spectral shape and spectral moments, this technique shows great promise for *in vivo* analyses.

3.3.6 *Luminescence*

The recent commercial availability of cameras with single photon sensitivity has made it possible to image the extremely low intensity emission associated with several luminescent systems. Calcium transients during fertilization have been visualized using aequorin luminescence (49). Another, particularly exciting, application of VIM in this area has involved the direct visualization of gene expression in living cells using the lux operon as a 'reporter' gene. This gene, which codes for the enzyme luciferase, can be incorporated into cells in such a way that its transcription is controlled by the promoter of the gene under investigation; hence, activation of the promoter results in the simultaneous emission of light (50,51).

3.3.7 *Neurobiology*

In neurobiology, a number of VIM applications have opened unexpected experimental approaches. By selecting especially non-toxic and non-metabolizable dyes mammalian nerve cells and the stability of their connections can be monitored over time in living animals. Appropriate dyes stain, for example neuromuscular end plates for several months, so that the innervation of superficial muscle can be reinspected during small surgery (52). Voltage-sensitive fluorescent dyes can be used at low magnification to monitor the electrical activity of the vertebrate brain with much better spatial resolution as was possible with EEG analyses (53). At high magnification one yields information on membrane potentials in tissue sections and single cells (54).

The detailed information on spectral properties and the potential application of all dyes for the above and related studies are compiled in ref. 55. The dyes can be obtained from companies such as Eastman-Kodak, Sigma Chemicals, and especially from Molecular Probes Inc. (Eugene, OR) which specializes in fluorescent compounds.

4. VIDEO-BASED TECHNIQUES FOR QUANTITATIVE ANALYSES IN LIVING CELLS

Extracting quantitative data out of microscope images was a relatively tedious process before the advent of video technology. As discussed in Chapter 7, Section 4, digitization of an image and processing of the numerical representation of the image in a digital image processor has made a great number of quantitative parameters accessible with relative ease. The video microscopist has the further advantage that his images are already in video format either on tape or optical disc and can be analysed as such by analogue devices (see Section 1.4.6) for determining for example intensity distribution along a given line (*Figure 4*) or distance information. Such analogue devices are however less versatile by far than digital image analysers.

Most digital processors for video microscopy are considered real-time image preprocessors, that is their output is an analogue video signal destined to a monitor or recorder. If one wants to get access to the enormous variety of techniques for digital image analysis, usually the special equipment as described in Chapter 7 will be required and the analogue video signal is directly used as input signal. Only a few of the more expensive processors capable of real time processing (see Sections 1.4.3, 1.4.4 and 1.4.7) are also able to perform most of the functions required for image analysis within their own processor. The third alternative would be to use a direct computer (DMA, direct memory access) interface between the digital memory of the real time processor and some additional digital image analyser.

As a first level of the analysis, a single frame can be taken up and analysed (Chapter 7). However, since video microscopy has newly opened the field of vital microscopy we want to be able to analyse changes in *sequences* of video images. Depending on the complexity of the algorithms required to extract the desired features it may or may not be possible to achieve this at video rate, that is at 25 or 30 times a second. In the latter case the life images from the video microscope or from tape are sampled such that the image analyser grabs only every 2nd, 4th, 5th or so image, extracts the desired detail and stores the information for later display as a function or plot (e.g. *Figure 12*).

4.1 **Spatial measurements**

It is possible to extract such parameters as size, length, width, area, perimeter, the co-ordinates of the centre, for one or more objects which can be differentiated according to their specific grey shade. This is achieved by setting an upper and lower threshold (binarization of the image) which can be done automatically and in real time by several processors. In addition to conventional measurements, this makes accessible the analysis of motion, such as growth of objects (nuclei, cells, microtubules) or movement of individual organelles in intact cells or along free microtubules (8,9,29,30,36−39).

The tracking of microtubule ends for measurement of subunit assembly/disassembly and of moving organelles is performed in our laboratory as follows.

(i)   Derive the scenes of AVEC-DIC microscopy from the real time image processor and store on video tape [storage of sequences of images in a computer memory would be possible only for a limited number of frames (a $512 \times 512 \times 8$ bit image takes up ¼ Mbyte of memory)].

(ii)   Play back the sequences through a X,Y-tracker (C-1055 from Hamamatsu Photonics K.K.). The tracker detects by thresholding, a bright object in a small user-defined and -positioned frame which can be moved as a cursor in the image on the monitor.

(iii)   The frame automatically takes a position over the centre of the particle. Read its co-ordinates out through a serial or parallel interface to a personal computer. This can be done at video rate or more slowly, as desired.

(iv)   Further analyse the time series of positional data in the personal computer using a user-specified software for motion analysis (38,39).

(v)   Plot the x co-ordinate versus the y co-ordinate to obtain the trace of the moving object.

(vi)   Plot position versus time to get the velocity function. Similarly acceleration behaviour and directional changes can be displayed as functions of location or time.

(vii)   Analyse the movement for regular features by the standard techniques of time series analysis, such as autocorrelation and fast Fourier transform for oscillations, or cross-correlation for similarity of the motion of different organelles.

(viii)   If the object is not easily distinguishable by contrast, but visible by eye, perform an interactive single frame analysis. To this end, use a tape recorder with a 'single frame advance' feature and from each consecutive, or 5th or 10th frame extract the co-ordinates by manually moving the cursor to the particle or, for example the microtubule end (38).

Simultaneous analysis of a multitude of moving objects such as organelles, swimming micro-organisms, or sperm cells is achieved by sophisticated equipment for motion analysis (e.g. from Motion Analysis Inc. or Mitec GmbH).

4.2 Intensity-derived measurements

The intensity of each picture element is represented by a number, in images digitized to 8-bit accuracy, one out of 256. Intensity in a microscope image means absorption, phase retardation, fluorescence intensity or birefringence, depending on the technique used. If we ensure that only one of these contributes to the image, we can quantify these properties in the different objects comprising the image. This is easily done by using specific fluorochromes in fluorescence and by applying monochromatic light in bright-field (absorption image), while the other parameters are more difficult to isolate.

It should be made clear that photometers for work with cuvettes or built into microscopes use photomultiplers to measure light intensities in a given slit or diaphragm at high accuracy, typically at a resolution of several thousands. Measuring intensities in the TV-image, on the other hand, although giving a best resolution of 256 grey levels, has the advantage of providing this information spatially resolved over 512×512 pixels. Intensity in a video image is the result of many steps and cannot necessarily be assumed to still be linear or to follow Beer−Lambert's law. It is, therefore, essential, if absolute measurements are intended, to calibrate with known samples to be put under the microscope. This and the relatively low resolution in the intensity domain are the two grave drawbacks of these techniques.

The distribution of absorbing and fluorescent compounds in biological samples are accessible both in the spatial and temporal domain. Devices to read out the intensity of a given pixel or along a line (*Figure 4*) or in a frame (*Figure 12*) are available. In the spatial domain such measurements provide information on diffusion and transport of solutes which have either been taken up by cells, or have been micro-injected, or of endogenous molecules. If light is shone on the specimen, or an area of it, this can be used to measure photobleaching or FRAP of fluorescent compounds (44,45).

While most measurements give only estimates of amounts of the compounds in the light path, it is possible in some cases to measure concentrations. This is if images are divided by each other, such as in ratio imaging (41−43) which is used to determine the Ca^{2+} (*Figure 12*) or the H^+ concentration (pH) (41,43) in living cells.

These techniques, which aim toward a 'biochemistry with the microscope', that is assaying amounts and concentrations of specific molecules and ultimately also enzymes, are presently in the process of rapid development. Hundreds of fluorescent markers are available for specific organelles, cells, enzymes and membranes, which can be used in living cells to detect and analyse specific features (55) (available from, for example Molecular Probes Inc.). The procedures for the different applications are beyond the scope of this chapter so that the reader is referred to the publications mentioned in Sections 3.3 and 8.3.

5. THE DOCUMENTATION AND PRESENTATION OF VIDEO MICROSCOPY DATA

Only the expert use of video technology allows profitable working of the video microscopy laboratory. It is the aim of this section to provide the beginner with the necessary basic knowledge of this technology which may be new to many biological laboratories. As the images improved or generated by video microscopy cannot be seen in or photographed from the microscope directly, we have to learn how to properly record and then how to obtain hard copies such as photographs, drawings or films, from the video monitor.

5.1 Video recording and editing

Video recorders are indispensable for the storage of video microscopic images. They are a comfortable way of storing information because of their easy handling and they offer the possibility of recording up to 6 h on one tape. The recordings can be played back and examined immediately without processing (unlike movie film).

Before purchasing video equipment one should know about the standards and formats available and weigh up the advantages and disadvantages of particular systems.

5.1.1 *Video standards*

There are several colour television standards in the world. Many of the European countries, Australia and many African countries use the West German PAL standard (PAL = phase alternation line). France, some African, and most East European countries use the French SECAM system (SECAM = Séquentielle à Mémoire). The NTSC standard (NTSC = National Television System Committee), developed in the USA, is the video standard in North America, many countries of Middle and South America and Japan. Complete lists of the TV standards of all nations are available in video shops.

The American standard differs from the European ones mainly in the scan rate. The NTSC standard displays 30 pictures per second; each is scanned by 525 horizontal lines. Because each frame is dissected into two interlaced fields ('half pictures') the frequency is 30 frames/sec (f.p.s.) or 60 fields/sec. Therefore, the NTSC scan rate is defined as 525/60. The two European standards have a scan rate of 625/50, that is 25 f.p.s. scanned by 625 lines.

These different scanning rates imply that it is impossible to copy a PAL recording onto NTSC equipment or vice versa. To transcribe PAL/SECAM tapes onto NTSC or vice versa requires special equipment, available only in larger commercial video studios or at TV stations. This transformation can be done with very little loss of quality but the cost is relatively high. A special chip for NTSC/PAL conversion is presently under development and may become available soon.

The terms PAL, SECAM and NTSC characterize especially the colour modes of the standards. However, they are fully compatible with the corresponding black-and-white (B/W) standards, namely PAL and SECAM with CCIR (50 Hz) and NTSC with EIA (60 Hz). The acquisition of colour equipment (monitors and recorders) is recommended, because this allows one to record B/W sequences and false-colour (pseudocolour) images (see Section 5.1.7) which are becoming more and more popular in video microscopy.

It is important that every part of the video equipment is compatible. This means that all sets must have the same TV standard. You must not connect an American NTSC video camera to European video tape recorders (VTRs) of the PAL system. Most recorders are compatible to only one standard while most monitors accept 50 Hz as well as 60 Hz signals. Some professional VTRs, however, can be used with all three standards, for example the ¾ inch recorders SONY VO-5630 or JVC 158 MS. In order to exchange tapes with colleagues overseas such a multistandard video set is necessary.

5.1.2 *Video tape formats*

For use in scientific laboratories two different video tape formats are practical. These are ½ inch (½") and ¾ inch (¾") tapes. Both tapes are available in cassettes, while ½" tapes are sometimes also used in reel-to-reel form.

(i) *The ½ inch format.* The ½" format is the well-known home video standard. Cassettes of the VHS format are not compatible with those of the Beta format, which is likewise obtainable. The VHS format is expected to become the ½" standard in the world. The advantage of the ½" format are the low prices for recorders and tapes. The tape, however, is thinner and narrower than ¾" tapes, resulting in more dropouts (small white flashing spots) and noisy jams after repeated recordings than occur with ¾ " tapes.

Most of the high quality ½" VTRs offer the possibility of assembly editing. This means that you can add scenes consecutively in a very easy way without interferences between the scenes [see Section 5.1.8(iii)]. Some high grade recorders (e.g. by JVC) even offer insert editing.

Recently in the USA and Japan an improved VHS system called Super-VHS (S-VHS) became available. Super-VHS uses the same cassette format as standard VHS but offers much better resolution (400−430 lines vertically). Another advantage is that the quality when copying is comparable to that with ¾ " VTRs. The new recorders are compatible with standard VHS tapes but not vice versa. Nevertheless, when purchasing new VHS equipment it is highly recommended to test this new system. It is thought that Super-VHS will be available in Europe in 1989.

(ii) *The ¾ inch format.* Video tape recorders of the ¾ " format (U-matic cassettes) are wide-spread in scientific laboratories and among other semiprofessional users. Both the VTRs and the cassettes are about five times as expensive as the ½" ones. Their value is in the higher quality of the recordings on the thick or, more stable tape material. This is important if the user often needs still frames or search mode, because these modes put a severe strain on the tape. Likewise, copies made from ¾ " onto ¾ " show nearly the same quality as the original. The larger format offers a vertical resolution of 340 lines instead of only the 250−300 lines of ½" format (B/W modus).

The ¾ " VTRs are available either as 'low-band' or as 'high-band' recorders. The more expensive high-band type permits (due to its higher carrier frequency) higher resolution and crisper images (especially in colour mode). B/W tapes are interchangeable between the two types of recorders while high-band colour tapes can only be played back on high-band VTRs.

The ¾ " format is considered a good compromise between the ½" home video format and the full-professional 1 and 2" formats used at TV stations. These large tapes require very expensive and massive VTRs, and usually a trained technician for operation.

There is no difficulty in copying from one format to another provided that the two VTRs have the same video standard (e.g. NTSC). However, bear in mind that you will lose quality when copying from ¾ onto ½" while transferring from ½ to ¾ " tape will not gain any further resolution.

5.1.3 *Video tape quality*

There are very different qualities of tape material obtainable. For video microscopy where high density and low noise recording is required only high quality tapes should be used. In addition, there are special tapes designed for still-frame operation, a technique very often used in video microscopy. These tapes have fewer dropouts and are made from very sturdy material. ¾ " cassettes with such tapes are marked 'KCA' or 'KCS', highest quality tapes with the addition 'Broadcast Quality'. Especially for colour recordings these professional tapes are recommended. Furthermore, high-grade tapes are more wear-resistant and protect the video heads.

The maximum recording time per VHS cassette (½") is 6 h but the stronger 1 or 2 h tapes are recommended. U-matic cassettes (¾") are available with running times of 20, 30 and 60 min.

5.1.4 *Video tape recorders*

VTRs used in video microscopy should have several special features. For taking photographs off the monitor, for example sometimes a good still-frame capability is required. For this most VTRs have such a PAUSE button. The displayed field should be contrasty and stable without jumping or flickering. A frame-by-frame advance will provide smooth transition to the next field. Several VTRs offer a (speed) search forward and reverse, moving the tape back or forward to other scenes without the long procedure of stop, rewind, and play back.

To add comments later to a video film without erasing the recordings you need a VTR with AUDIO DUB capability. Most of these recorders have two audio channels which can be played back separately or synchronously. It is also convenient to have a remote control, especially if the VTR is out of reach. The control unit may be connected to the recorder by line or be operated by IR light.

5.1.5 *Time lapse recorders*

Various microscopic specimens such as, for example slow particle motions or the progress of cell division, require time-lapse recording. Some motions indeed only become apparent after speeded up play-back. A time-lapse recorder may be used in parallel with a normal speed VTR so you have the same sequences both in real time and time-lapse for comparison. Previously recorded real time scenes may also be speeded up with a time-lapse recorder. However, you have to ensure the SYNC pulse signal is correct [see Section 5.1.8(ii)].

There are several time-lapse VTRs available but all of them are strictly speaking 'long-time recorders'. They are designed to record for up to 400 h or more onto a regular 3 h tape for observation and surveillance applications. Played back at normal speed the recordings of most of these VTRs show considerable interferences and noise bars. Only the modes of eight times or less speeded up replay are satisfactory in most cases. From all time-lapse recorders clear still frames can be obtained.

Before purchasing a time-lapse recorder one should test all available ones because each has strong and weak points. Most are designed for the ½″ format, for example the ones from SONY, GYYR, Panasonic and Hitachi. At present, the only one using ¾″ tapes is made by JVC. Really high grade time-lapse VTRs are only available for the 1 and 2″ formats.

Time-lapse recorders use regular ½ or ¾″ cassettes but recorded tapes cannot be played back on a standard speed VTR of the same format. During time-lapse recording the tape moves very slowly past the incessantly rotating video head, thus straining the tape. Therefore, only new and best quality tapes (see Section 5.1.3) should be used. If possible, important time-lapse sequences should be copied onto a ¾ or ½″ standard speed tape to protect the original tapes from further stress.

Sometimes animation control units (e.g. EOS AC 580 from EOS Electronics AV Ltd, or IT-2 from AVT GmbH), are alternatives to time-lapse recorders. These units are connected to an edit VTR (e.g. the SONY VO-5850) and permit recording of one or several frames at intervals of $10-999$ sec. Played back in real time such recordings show at least a $\times 100$ time-lapse effect, but they are free of image distortions. These units are well-suited for the documentation of *very* slow movements such as cell growth, locomotion, or division (56).

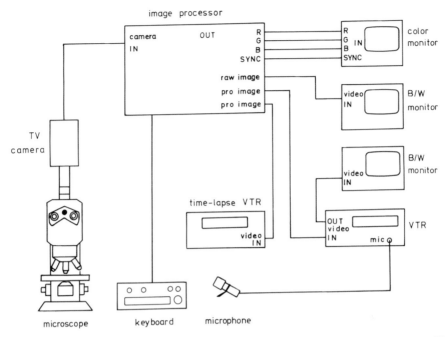

Figure 13. A possible circuit diagram for video microscopy. At least one B/W monitor is required. The colour monitor and time-lapse recorder are optional. Pro image = processed image.

5.1.6 *Recording video microscopic sequences*

Prior to recording it is important to check the equipment, ensuring all plugs and switches are in their proper positions. A suggestion as to how to connect the system is shown in *Figure 13*. The VTR should receive the signal directly from the video processor and not 'at second hand' from a monitor. For this the signal has to pass the recorder prior to display on the monitor. This is the E-to-E (electronics to electronics) mode picture. With some recorders this E-to-E mode picture can only be obtained when the RECORD button of the VTR is pressed.

If the monitor is the last set to receive the video signal, it has to be 'terminated'. This means that the 75 Ω switch at the back of the monitor must be set to '75 Ω' position to avoid disturbances. With monitors operating in the E-to-E mode the switch is set to 'HI-Z' position. Likewise, this must be done for all the other pieces of equipment. If there is no switch 75 Ω/HI-Z the VIDEO OUT connector of this set has to be terminated with a plug containing a built-in 75 Ω resistance.

During image processing brightness and contrast of the image are altered electronically. It is strongly recommended to set the brightness and contrast controls of the monitor to their standard positions where they should always be kept. This measure will provide scenes of the same brightness level which later will fit together properly when they are edited or processed further.

The timer and the scale bar are added to the video image by most image processors. To estimate the correct magnification, especially when photographing off the monitor (see Section 5.2) record the scale of an object (stage) micrometer both in horizontal

265

and vertical position. This is used to calibrate any scaling function or scale bar (see *Figure 4*) of the processor and to adjust the horizontal and vertical axes of the monitor (see Section 5.2.2).

Sometimes you will notice blemishes in the recorded sequences which were not recognized while recording. It is therefore good practice to also record from time to time the background (out-of-focus) image of the corresponding sequences. It will then be possible to remove the disturbances later by additional image processing.

Too high a level of the video signal gives rise to snowy and noisy images when scenes are replayed from the VTR. It is, therefore prudent to check this signal (at the VIDEO OUT connector of the image processor or VTR) with the aid of an oscilloscope. The peak-to-peak voltage of the composite video signal is standardized to a maximum of 1.0 V (the picture level amounts to approximately 0.7 V and the synchronization signal to 0.3 V). A video level higher than 1.0 V will cause distortion in the bright regions of the image. The recording should be checked by playing back the tape. If the video signal is too high, the service personnel should be asked to reduce the output video signal to a maximum possible value of 1.0 V.

Although the video recording level is controlled automatically we have to adjust the audio recording level manually on some recorders. Most VTRs, however, are equipped with an audio level limiter which minimizes audio distortion at the peaks.

It is very helpful to record comments during working at the microscope, for example such events as changes of optics, the position of the specimen or the focal plane (note presence of microphone in *Figure 13*). If the VTR offers two separate audio channels and the audio dubbing feature one should use channel 2 for these comments, as one can add a second commentary afterwards (dubbing) to the previously recorded scenes only onto channel 1.

Recorded tapes can be protected against accidental erasure. For that purpose the red button on the bottom of U-matic cassettes is removed, or respectively, the plastic tab at the back of ½″ cassettes is broken off. If you later decide to record again onto these tapes, replace the button or affix adhesive tape over the safety tab slot. The unintentional use of erasure protected tapes is the likely cause if the VTR stops working when the RECORD mode is activated.

Some frequent mistakes which should be avoided are as follows.

(i)   It is very annoying to the photographer of a video scene or to the person who evaluates the images if the sequences are of short duration.

(ii)   Some scientists tend to try to improve the focus of the microscope continually while recording, or to shift the specimen around, incessantly expecting better positions. These faults become conspicuous especially when using a time-lapse recorder. During recording one must remember that such mistakes cannot be improved afterwards and one ought to keep in mind the designed application of the scenes. For example, scenes which are to be used for motion analysis or other evaluation should last at least 5 min without changing focus or stage position.

5.1.7 *Recording pseudocolour or false-colour sequences*

The pictures obtained by the camera and fed into image processors are usually monochrome (B/W). For special purposes, such as, for example, to enhance the contrast by adding colours or simply to get more aesthetical pictures, many image processors permit assignment of different colours to the different grey values. Thus all areas of the same grey value are coloured equally. All generated colours are coded as a mixture of the three basic colours red, green, and blue (RGB) and split up into the corresponding colour channels. This RGB signal can be displayed at high image quality but only on monitors with RGB input capability.

To record coloured images a composite video signal of one of the international standards PAL, SECAM or NTSC is needed. Several image processors provide only analog RGB output. These four or five signals (R, G, B, composite SYNC, or horizontal and vertical SYNC) have to be encoded into the composite standard colour video signal. For this purpose a colour-encoder for either PAL, SECAM or NTSC is required. This encoder is connected between the RGB outputs of the image processor and the video input of the VTR. The encoder, however, requires a very regular and exact SYNC signal, as can only be supplied by a TV camera. It is not difficult to add colour to original microscopic scenes coming directly from the TV camera, through the image processor, and to record these scenes after encoding the RGB signal.

To colour previously recorded sequencers from a tape, the VTR as input source for the image processor may not provide as regular a SYNC signal as would be required. In this case, a time base corrector (TBC) may help and is connected either to the TBC socket (if available) of the first VTR or to the encoder. Using a TBC is often advisable as it always provides the best possible playback picture, whether for copying, editing, mixing two signals or additional processing.

5.1.8 *Copying and editing video tapes*

Usually tapes recorded in the laboratory are not directly suitable for presentation to an audience. A demonstration tape with scenes of suitable sequence and length will need to be compiled.

(i) *Preparatory considerations for editing.* To copy video sequences two recorders of the same standard (PAL, SECAM, NTSC) are required. There are no problems with different formats (¾ and ½"), although there may be loss of quality when copying from ¾ onto ½" (cf. Section 5.1.2). It is even possible to copy from time-lapse VTRs with variable recording speeds onto standard speed recorders. Another special effect is to record a still frame (or field) from one VTR onto another.

Before beginning to edit it is profitable to compose a script where all titles and scenes are put together consecutively. For titles schedule about 5 sec, for long titles the time allowance should be sufficient to enable them to be read slowly. Scenes should run for periods of time not shorter than 20 sec because the audience needs a certain time to grasp the new situations demonstrated. Only singular scenes may be longer than 1 min, important ones can be repeated.

If the image processor does not allow writing into the image, titles can be added either with a commercial video typewriter or with a suitable personal computer. You may also record title slides with a video camera as described in Section 5.3.

(ii) *Obtaining a correct video signal.* For the copying process endeavour to improve the video signals as much as possible. For this purpose the same VTR as used for the original recording should be used, or another set which will play back this tape well.

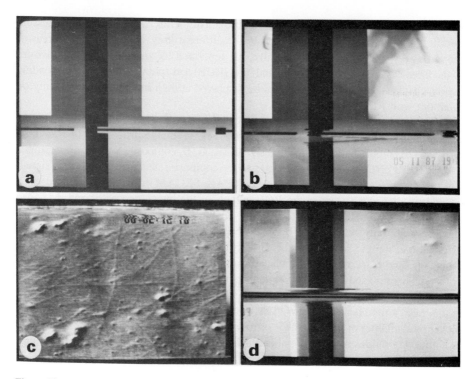

Figure 14. (a) A correct pulse-cross display derived from a TV camera. Due to their low voltage the SYNC pulses appear as dark stripes. Vertical on the monitor, H SYNC pulses; horizontal, V SYNC pulses. (b) Pulse-cross display of a video signal from a video tape recorder. The vertical SYNC pulse shows a distortion. (c and d) Photograph of a time-lapse scene and the corresponding pulse-cross display. The 'flagging' in the upper part of the picture is caused by a jammed SYNC signal.

Even recorders of the same standard and format are not always completely compatible. This is caused above all by different mechanical adjustment (e.g. the stretch of tape) or jamming suppression of the different sets. If several VTRs are available establish which one supplies the best recording quality and which the best play-back quality and use them accordingly.

In the following description the source VTR with the original tape is called VTR I, the receiving recorder with the editing tape is VTR II. The VIDEO OUT and AUDIO OUT terminals of VTR I are connected by line to the VIDEO IN and AUDIO IN sockets of VTR II. A monitor is connected to the VIDEO OUT and AUDIO OUT terminals of VTR II, where the image is displayed on the screen. With some recorders, this E-to-E mode picture (cf. Section 5.1.6) can be seen only if the RECORD button of VTR II is pressed. With a monitor providing pulse-cross display we may control and improve the SYNC signal of VTR I. In pulse-cross mode such a monitor displays the four corners of the picture and the blanked region of the frame which normally is hidden (see *Figure 14*).

Figure 14a and *b* show a correct video signal from a TV camera and one from a video tape recorder with a discordant time base for the vertical SYNC signal. Often one can improve the signal slightly by turning the TRACKING control of the VTR

268

Table 2. Assembly editing of video tapes for smooth transitions between scenes (Procedure I).

VTR I (original tape)	VTR II (editing tape)
	Rewind the tape to the start. Run the tape for about 5 sec in PLAY mode as a leader. Press the PAUSE button, then the PLAY and RECORD button simultaneously without releasing PAUSE.
Locate the prospected scene or title and start the tape (PLAY mode) about 5 sec before beginning of this scene.	
	At the beginning of the scene to be recorded release the PAUSE button, press it again to stop the recording.
Change the cassette or wind to the next sequence; start the tape about 5 sec before the scene starts.	
	At the end: Press the STOP button and rewind the tape to check it.

or sliding the SKEW lever. The SKEW control adjusts tension of the tape while the TRACKING control minimizes the tracking variances between different recorders. Even the strong distortion (flagging) of time-lapse scenes seen in the upper part of the screen (*Figure 14c* and *d*) may possibly be reduced by altering the TRACKING control. This may also improve the image if the video level of the original tape is inadvertently more than 1.0 V (cf. Section 5.1.6). Normally (especially during recording) the TRACKING control has to be set to the 'fixed' position.

To correct the signal from VTR I use a TBC (cf. Section 5.1.6). A TBC video signal is not only required for copying but is also useful for adding a timer (from a separate time date generator or an image processor), a scale bar or other overlay to the video signal from a VTR.

(iii) *Procedure for editing video tapes.* Before copying advance the source tapes to the beginnings of the required scenes, so that one scene after another can be added quickly onto the second tape. The original cassettes should be protected from accidental erasure (see Section 5.1.6).

To obtain the more pleasing black frames instead of noise and noisy bars on parts of tape II which will not be recorded on, we should first prepare a 'black' tape by running the tape (new or used) completely through in recording mode.

Many VTRs (½" and ¾") assure a smooth transition between recorded scenes. The procedure for such a transition is outlined in *Table 2*.

Sometimes it may take a few minutes or longer to prepare the next sequence to record. VTR II, however, should not stay in the PAUSE mode for such a long period of time. Some recorders switch over to STOP automatically if a PAUSE lasts longer than several minutes. We should do likewise with non-automatic recorders. Stopping the recorder protects the tape and the video head from abrasion. The next scenes can be smoothly added to those already recorded, see *Table 3*. Thus the last seconds of the preceding scene will be erased, but the scenes are smoothly added without frame breakdown at the splicing point.

Using this sequence of assembly editing, nearly perfect changes between the video film sequences can be obtained.

Table 3. Assembly editing of video tapes for smooth transitions between scenes (Procedure II).

VTR I (original tape)	VTR II (editing tape)
The next sequence is ready to start.	
	Rewind the tape briefly to inspect the last scene. A few sec before its end press the PAUSE button, then the RECORD and PLAY button simultaneously without releasing the PAUSE button.
Start the tape (PLAY) about 5 sec prior to the beginning of the desired scene.	
	At the beginning of the scene to record release the PAUSE button.

It is not possible to insert or exchange one scene for another between sequences already recorded (i.e. insert editing) in the same way without noisy frames resulting at the end of the scene. This interference is due to the distance between the erasing and the recording heads of the recorder. With some high grade VHS recorders, however, insert editing can be done as easily as normal recording.

Fully professional assembly editing and insert editing with perfect invisible changes between scenes can be done with an automatic editing control unit and two edit VTRs. It may be done in audiovisual centres in many universities or in commercial video studios. These studios can be rented for hours or days at reasonable prices. There one usually finds several VTRs, a TBC, title generator (video typewriter), video cameras and other professonal equipment.

5.2 **Photographs off the video screen**

Some effort is necessary to obtain good photographs off the monitor to illustrate scientific papers or book articles since poor photographs can negate all previous efforts at image improvement. Photographs are likewise necessary to archive all recordings. For the latter purpose the quality of a video printer may be sufficient, but for publication there is no alternative to the still photograph.

Video printers (e.g. SONY, Mitsubishi, Javelin) are connected to the VIDEO OUT socket of the VTR. The video signal passes the printer in the E-to-E mode to a monitor. When the PRINT button is pressed the present image is stored and then printed on B/W thermographic paper within several seconds. The print quality depends on the resolution of the printer, the number of its grey levels (16−64) and on the quality of the paper used. The advantages of such video printers are the instant availabilty of the pictures and the moderate cost of the paper. The latest video printers attain almost photograph quality and have even become available in colour mode.

Microscope manufacturers are starting to offer documentation units specially designed for the need of video microscopists. The Analog Contrast Enhancement microscope system (Zeiss) includes besides a built-in video printer also a photographing unit which is an automatic camera attached to a separate high-resolution monitor (*Figure 6*).

5.2.1 *Equipment*

If no automatic photographing device is available, then use a 35 mm single-lens reflex camera. There is no value in using a larger format camera as the limiting factor of image resolution is the monitor and not the film material or the negative format.

The camera should be equipped with a 50 mm lens (and possibly a close-up lens) or a 50 mm macro lens. For 6 × 6 cameras the lens range from 80 to 100 mm is the best. Automatic exposure is not always an advantage so a camera capable of manual exposure is appropriate.

To shoot B/W pictures we need moderately fast negative film such as Agfapan 100, Kodak Plus-X or Ilford FP4. Such a film is easy to process in the dark-room where its speed and contrast can be manipulated to obtain special effects. The negatives may not only be used for prints but also for copying onto high-contrast material (e.g. Agfaortho 25) to get slides. Due to their contrast, low sensitivity films (e.g. Agfapan 25, Ilford Pan F) are suited for reversal processing with a commercial kit (produced e.g. by Tetenal GmbH) to obtain slides directly without negatives.

For colour slides and colour prints load the camera only with reversal (slide) film. This is because commercial labs will be overtaxed to print correctly the mostly very unnatural colours of a false-colour display from negative film. Using reversal film you can point out that the colours of the print should correspond to those on the slide. Colour reversal films for best results are the Fujichrome 50, Ektachrome 64 or Agfachrome 50, all of them in the professional type.

For shooting photographs it is advisable to use a high resolution monitor with a small screen with a diagonal length of 21 cm or less. To remove the video scan lines, which are very annoying especially on large scale prints, a Ronchi grating may be fixed at the front lens of the camera (1). A Ronchi grating (obtainable from optics supply companies such as Rolyn Optics Company) is a glass plate with fine parallel lines of a defined interval. The best line interval depends on the visual angle from the camera to the monitor. Good results will be obtained with a 50 lines/inch Ronchi grating combined with a 21 cm screen. Using this simple technique of optical filtering, the image will be slightly blurred vertically, thus eliminating the scan lines but without removing any image details (see *Figure 15b* and *c*). Instead of the Ronchi grating being mounted at the front camera lens it may be fixed to the objective of the enlarger in the darkroom. The easiest, but not fully satisfactory way to eliminate the scan lines without a Ronchi grating is to defocus the enlarger objective slightly. Another method to be used in the darkroom is to shift the photographic paper vertically after half the exposure time by half the distance of the scan lines (57).

Good results photographed off a high resolution colour monitor designed for computers, without using a Ronchi grating are given in *Figure 15*. Because these monitors show less contrast than B/W monitors the images may need additional enhancement.

5.2.2 *Special problems when photographing from the monitor*

Before beginning to photograph first darken the room to increase the contrast of the image and to eliminate reflections on the screen or use a black hood between screen and camera. Secondly, since the screen is permanently covered with a layer of fine dust due to the static electricity, it should be cleansed with a slightly moistened cloth.

It is important to adjust the monitor in the optimal way, that is to terminate the set (see Section 5.1.6). Contrast and brightness should be set so that it is comfortable to the eyes, preferably a little less contrasty than too hard an image. Remember that the

Video microscopy

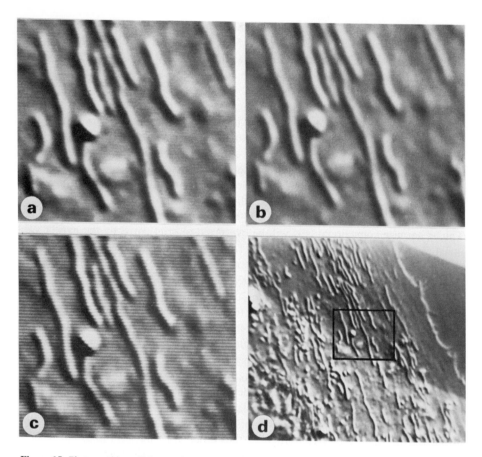

Figure 15. Photographing off the monitor. Specimen, mitochondria of a MDCK cell; microscope, Polyvar (Reichert/Cambridge Instruments), ×100 planapo/1.32 NA. (a) Highest resolution colour monitor, (b) normal B/W monitor, using a Ronchi grating, (c) normal B/W monitor without Ronchi grating, (d) photograph with a shutter speed of 1/30 sec. The section shown in a, b and c is indicated in d.

film material does not cover a range of brightness levels as large as that of your eyes. With a test pattern (including circles and other geometrical figures) adjust the monitor to correct for horizontal and vertical distortion. This adjustment may be done easily by the two potentiometers in the front or at the back of the monitor. CAUTION—*if the back cover has to be removed to gain access to the screws, however, this should be done only by trained personnel.*

Mount the camera on a firm tripod and adjust it at right angles to the screen. The level of the camera lens should correspond to the centre of the screen. The picture in the viewfinder may show either the full monitor with some dark background around it or only a section of the screen. Prior to attaching the Ronchi grating focus the lens; if necessary add a close-up lens. The Ronchi grating should be mounted as close as possible to the front lens. Do not forget to orientate the Ronchi grating lines parallel to the scan lines (possibly with the aid of a spirit-level), otherwise a moiré pattern may arise. If this arrangement is needed often, fix the monitor and the camera reproducibly on a wooden board. To reduce camera vibrations use a cable release. A winder attached to the camera will facilitate the film transport without disturbing the setup.

Check the exposure with the built-in TTL (through the lens) exposure meter of the camera or a separate exposure meter. Try photographing with a shutter speed of 1/8 sec or longer. Shorter times (even 1/15 sec) will result in heterogeneous pictures because of the video scan rate (*Figure 15d*). The aperture control may be set manually or automatically (time-preferred automatic exposure) to the corresponding f-stop. If the TV screen is curved make sure that the f-stop number used guarantees sufficient depth of field. Using aperture-preferred automatic exposure remember the shutter time should not be shorter than 1/8 sec. Programmed automatic exposure is thus unsuitable for photographing off the monitor.

Most of the latest cameras are supplied with an ultra-fast photocell which is even affected by the running scan ray which is invisible to the human eye. Therefore the information from the exposure meter fluctuates widely and pictures taken with automatic exposure would be either too dark or too bright. Exceptions are the Olympus OM-2 and OM-4 cameras which do not fix the exposure at the moment the shutter is released but measure continuously during the entire exposure (off-the-film). The RICOH XR-X camera offers a special TV mode to take pictures with full-automatic exposure off the monitor. Likewise worth mention are cameras with very fast vertical shutters like the Nikon F3 which are said to enable one to shoot TV pictures with 1/15 sec without problems.

We obtain good results with an exposure compensation of $-1/2$ to $-2/3$ f-stop when using B/W negative film. Reversal films (B/W and colour) should be exposed without compensation (provided that there is no dark margin around the screen and that the picture shows normal brightness levels).

When shooting the running image at 1/8 sec exposure time the camera effects a kind of averaging by integrating about four frames. This reduces the noise but may be totally inappropriate for scenes with rapidly moving objects. In these cases it is better to shoot still frame pictures. Pressing the PAUSE button of the recorder provides usually only one field, that is half the number of video scan lines as the running picture. The consequence is an image with clearly visible scan lines. It is better to store (freeze) a whole frame (two fields) in the image processor. Additional opportunities of improving the image, for example with the gain and offset controls are then provided. In any event a photograph of a still video picture always shows more pixel noise and is rougher than one of a running scene. It should therefore only be taken if fast movements would otherwise be blurred.

If possible, photographs should be taken always from the primary video image coming directly from the TV camera or the image processor. The pictures played back from the recorder usually are somewhat poorer and show less resolution. However, the original tape should be handled with care since long periods of PAUSE or excessive searching back and forth decrease the quality of the recording.

Note down the exposures including all important data required in order to optimize the settings and remember to identify the pictures taken.

5.3 Transfer of video recordings to movie film

The optimal way to present video film to a large audience is by means of a video projector. Usually this equipment provides a large and bright image on a regular or,

Figure 16. The 16 mm film equipment used by the authors to film video scenes off the monitor, the Arriflex 16 SR camera, the control unit of the phase shifter (centre) and the induction sensor (in front).

even better, on an incandescent screen. Due to their high price video projectors are, however, not available in every institute. Where B/W tapes are to be shown a RGB colour video projector yields much less contrast and cannot be used for projecting the image to large screens.

Movie 16 mm film projectors, however, are to be found in almost every conference room. Thus in many cases a 16 mm film (same standard all over the world!) is much easier to present and more efficient. A video film requires a VTR and at least one large TV monitor, which in most cases can only be seen by viewers near the monitor well enough to follow the video sequences. The projected movie film on the other hand yields an image large and brilliant enough to be presented to the whole audience.

When done properly the transfer from video to movie film does not reduce the image quality. Filming off the TV screen generally presents similar problems as still photography. Equipment such as a Ronchi grating or a higher resolution monitor are therefore similarly required (see Section 5.2.2).

B/W or colour, negative reversal film may be used. Reversal films however do not permit copies to be made if the original should be damaged or lost. In addition, reversal emulsions must be exposed exactly as they are very sensitive to over- or under-exposure.

The use of negative film (B/W or colour) is, however, more expensive than reversal film. We obtain best results with Kodak Double-X (Eastman) and Gevapan 36 (Agfa-Gevaert) B/W films, exposed 2/3 American Standards Association (ASA) steps more than effective film speed (corresponds to a −2/3 f-stop compensation). After processing the lab returns the negative and a preliminary positive copy. This copy can

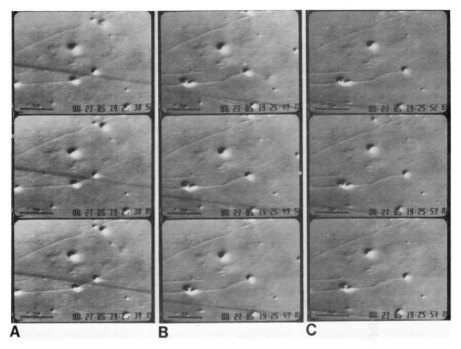

A　　　　　　　**B**　　　　　　　**C**

Figure 17. A sequence of pictures of a 16 mm film taken off a monitor. (**a**) No synchronization at the beginning of the scene. (**b** and **c**) Using the inductive synchronizing device, the black stripe is shifted out of the image within a short period of time.

be cut and edited with a simple film splicer. At the beginning and the end of the film we attach 1.5−2 m black or coloured film (green at the beginning, red at the end). The quality of this preliminary copy is usually already suitable for presenting, but if we want a better copy we have to cut the negative according to the edited film. This work should only be done with professional equipment to prevent damage to the original negative. From the cut negative as many copies as desired can be made in top quality, whereby even incorrect exposures can be compensated to a great extent.

The main problem when transferring from video to movie film is to synchronize the shutter of the movie camera with the video scan (i.e. 25 or 30 f.p.s.). Incorrect synchronization causes a dark horizontal or diagonal stripe to pass up or down the screen. To yield a controlled and constant shutter speed we need an electronically controlled camera, preferably with a quartz crystal oscillator such as the Arriflex 16SR (Arnold and Richter Cine Technik GmbH) or the CP 16-GSMO (Cinema Products Corporation). These cameras may be triggered externally by the video frequency taken from the VIDEO OUT of, for example, a VTR or wireless by an inductive synchronizing device. In the latter case the induction sensor of a special 'phase shifter' is placed near to the monitor or, if the monitor has a metal case, near its opened back. The control unit of the phase shifter is connected between the sensor and the external SYNC input of the movie camera (*Figure 16*). A switch at the control unit enables one to increase or decrease the shutter speed briefly, causing the black stripe (which may possibly appear at the beginning of a sequence) to be shifted up or down out of the image (*Figure 17*). Subsequently the stripe remains in this outside position unless the next noisy video frames

275

(e.g. caused by incorrect editing) require a fresh 'shifting' with the switch. The control unit of the phase shifter may be adjusted to any video standard (50 Hz for PAL/SECAM; 60 Hz for NTSC). The equipment may be found at audiovisual centres of larger institutions or can be rented on a daily basis at rental companies for cinematographic equipment (~ $300/day).

Just like the still photo camera, the movie camera is mounted on a stable tripod and oriented exactly to the screen. We have had good results with an Arriflex 16SR equipped with a Zeiss 1.3/25 mm lens and the Ronchi grating. The required aperture has to be determined with a separate or a built-in exposure meter. The latter enables one to control the aperture while filming, thus compensating brightness changes of the video scenes.

As the projection speed of movie films is 24 f.p.s., the transfer from video to movie film causes a slow-down of all recorded motions (from PAL/SECAM: 4% slower, from NTSC 20% slower than the original). Transfer without this effect is carried out by some commercial video studios. Films (16 and 35 mm) of high quality from video recordings are produced by Image Transform Inc. utilizing a wobbling scanning electron beam. As the negatives are free of scan lines and show good contrast they can also be used to make photographic prints.

A time-lapse effect can be achieved by the use of a manual control unit for variable frame speeds connected to the cine-camera. For example, to yield a ×5 speeded-up film an exposure speed of 10 f.p.s. (PAL/SECAM) or 15 f.p.s. (NTSC) is needed. The reduced speed must be compensated by closing the aperture to prevent over-exposure. Of course, a stored picture (or still frame) can also be filmed off the monitor. Possible dark stripes can be removed as described above.

Film titles may be simply produced. Using a typewriter with clear letters write the text on a white sheet of paper. Photograph onto a high contrast B/W negative film (e.g. Agfaortho 25) to produce negatives showing white letters on dark background. These mounted negatives can be projected onto a screen and filmed with normal speed (24 f.p.s.). The length of the title scenes should correspond to their message [cf. Section 5.2.8(i)].

5.4 Making drawings from the monitor

Sometimes clear drawings show more than poor photographs. To get a drawing from the monitor it is possible to tape transparent sheets on the TV screen for transcription. However due to the thickness of the glass used for monitor tubes, this may introduce considerable geometrical distortion. A better method is achieved by the following procedure.

A thin (1 − 2 mm) glass plate (such as for thin-layer chromatography) is mounted in a 45° angle on a clamp stand in front of the centre of a vertically oriented screen. The lower side of the plate points at the monitor. The centre of the plate has to be equi-distant from the screen and to a sheet of paper placed at the table below the plate. Looking from above at the plate both the reflected image of the monitor and the paper are simultaneously visible. Similarly one can look horizontally through the plate onto the monitor. It is essential to finely adjust the glass plate (especially if the TV screen is not vertically oriented) in the following way.

(i)   Vary the height and angle of the plate until the screen is projected *exactly* into the plane of the paper. With curved screens the four corners will appear to be located slightly underneath and the centre slightly above the paper.

(ii) Only after this adjustment of the plate may the head be moved without relatively displacing the images of the screen and the one on the paper. It is useful to fix the drawing-paper with adhesive tape.

(iii) For comfortable working dim the roomlight and/or illuminate the paper with an adjustable lamp. The result is an upside-down image of the screen in the scale 1:1.

The same system can be used by replacing the paper with a digitizing tablet in the absence of comfortable digitizing equipment for the transfer of positional data from the TV screen to a computer.

6. ACKNOWLEDGEMENTS

This article could only be written because Robert D.Allen (1927−1986) so generously shared his enormous experience and knowledge in this field during collaborations with one of the authors (DGW) in the summers of 1984 and 1985. Similarly, part of this article greatly benefited from notes of the many lectures Bob gave on this topic. DGW is also greatly indebted to Shinya Inoué and the other colleagues of the Woods Hole video microscopy community for numerous discussions and advice.

7. REFERENCES

1. Inoué,S. (1981) *J. Cell Biol.*, **89**, 346.
2. Inoué,S. (1986) *Video Microscopy.* Plenum Press, New York.
3. Allen,R.D., Travis,J.L., Allen,N.S. and Yilmaz,H. (1981) *Cell Motil.*, **1**, 275.
4. Allen,R.D., Allen,N.S. and Travis,J.L. (1981) *Cell Motil.*, **1**, 291.
5. Allen,R.D. and Allen,N.S. (1983) *J. Microsc.*, **129**, 3.
6. Willingham,M.C. and Pastan,I. (1978) *Cell*, **13**, 501.
7. Reynolds,G.T. and Taylor,D.L. (1980) *BioScience*, **30**, 586.
8. Allen,R.D., Weiss,D.G., Hayden,J.H., Brown,D.T., Fujiwake,H. and Simpson,M. (1985) *J. Cell Biol.*, **100**, 1736.
9. Weiss,D.G. (1986) *J. Cell Sci. Suppl.*, **5**, 1.
10. Arndt-Jovin,D.J., Robert-Nicoud,M., Kaufman,S.J. and Jovin,T.M. (1985) *Science*, **230**, 247.
11. DiGuiseppi,J., Inman,R., Ishihara,A., Jacobson,K. and Herman,B. (1985) *BioTechniques*, **3**, 394.
12. Hecht,E. and Zajac,A. (1974) *Optics.* Addison-Wesley Publishing Co., Reading, MA.
13. Walter,R.J. and Berns,M.W. (1981) *Proc. Natl. Acad. Sci. USA*, **78**, 6927.
14. Wick,R.A. (1987) *Applied Optics*, **26**, 3210.
15. Bennett,H.S. (1950) In *Handbook of Microscopical Techniques.* McClung,C.E. (ed.), Harper & Row (Hoeber), New York, p. 591.
16. Inoué,S. (1961) In *The Encyclopedia of Microscopy.* Clark,G.L. (ed.), Reinhold, New York, p. 480.
17. Conchello,J.A., Hansen,E.W. and Allen,R.D. (1987) *Proc. 13th Northeast Bioeng. Conf. Philadelphia*, p. 4.
18. Schnapp,B.J. (1986) *Methods Enzymol.*, **134**, 561.
19. Kachar,B. (1985) *Science*, **227**, 766.
20. Ellis,G.W. (1985) *J. Cell Biol.*, **101**, 83a.
21. Lichtscheidl,I. and Url,G.W. (1987) *Eur. J. Cell Biol.*, **43**, 93.
22. De May,J. (1983) In *Immunocytochemistry.* Polak,J.M. and Van Noorden,S.W. (eds), Wright-PGS, London, p. 92.
23. Hoffman,R. (1977) *J. Microsc.*, **110**, 205.
24. Ellis,G.W. (1978) In *Cell Reproduction. In Honor of Daniel Mazia.* Dirksen,E., Prescott,D. and Fox,C.F. (eds), Academic Press, New York, p. 465.
25. Allen,R.D. (1985) *Annu. Rev. Biophys. Biophys. Chem.*, **14**, 265.
26. Allen,R.D., David,G.B. and Nomarski,G. (1969) *Z. Wiss. Mikr. Mikrotech.*, **69**, 193.
27. Izzard,C.S. and Lochner,L.R. (1976) *J. Cell Sci.*, **21**, 129.
28. Beck,K. and Bereiter-Hahn,J. (1981) *Microsc. Act.*, **84**, 153.
29. De Brabander,M., Nuydens,R., Geuens,G., Moeremans,M. and De Mey,J. (1986) *Cell Motil. Cytoskel.*, **6**, 105.

30. Geerts,H., De Brabander,M., Nuydens,R., Geuens,S., Moeremans,M., De Mey,J. and Hollenbeck,P. (1987) *Biophys. J.*, **52**, 775.
31. Mathog,D., Hochstrasser,M. and Sedat,J.W. (1985) *J. Microsc.*, **137**, 241 and 253.
32. White,J.G., Amos,W.B. and Fordham,M. (1987) *J. Cell Biol.*, **105**, 41.
33. Kachar,B., Evans,D.F. and Ninham,B.W. (1984) *J. Coll. Interface Sci.*, **100**, 287.
34. Trendelenburg,M., Allen,R.D., Gundlach,H., Meissner,B., Tröster,H. and Spring,H. (1986) In *Nucleocytoplasmic Transport*. Peters,R. and Trendelenburg,M. (eds), Springer-Verlag, Berlin, p. 96.
35. Allen,R.D. and Allen,N.S. (1981) *Protoplasma*, **109**, 209.
36. Kachar,B. (1985) *Science*, **227**, 1355.
37. Cohn,S.A., Ingold,A.L. and Scholey,J.M. (1987) *Nature*, **328**, 160.
38. Weiss,D.G., Langford,G.M., Seitz-Tutter,D. and Keller,F. (1988) *Cell Motil. Cytoskel.*, **10**, 285.
39. Weiss,D.G., Keller,F., Gulden,J. and Maile,W. (1986) *Cell Motil. Cytoskel.*, **6** , 128.
40. Taylor,D.L., Amato,P.A., Luby-Phelps,K. and McNeil,P. (1984) *Trends Biochem. Sci.*, **9**, 88.
41. Tsien,R.Y. and Poenie,M. (1986) *Trends Biochem. Sci.*, **11**, 450.
42. Bright,G.R., Rogowska,J., Fisher,G.W. and Taylor,D.L. (1987) *BioTechniques*, **5**, 556.
43. Bright,G.R., Fisher,G.W., Rogowska,J. and Taylor,D.L. (1988) *Methods Cell Biol.*, **30**, in press.
44. Peters,R. (1985) *Trends Biochem. Sci.*, **10**, 223.
45. Kapitza,H. and Jacobson,K. (1986) In *Lateral Motion of Membrane Proteins. Techniques for the Analysis of Membrane Proteins*. Ragon and Cherry (eds), Chapman and Hall, London, p. 345.
46. Yanagida,M., Morikawa,K., Hiraoka,Y., Matsumoto,S., Uemura,T. and Okada,S. (1986) In *Application of Fluorescence in the Biomedical Sciences*. Taylor et al. (eds), Alan R.Liss, New York, p. 321.
47. Kron,S.J. and Spudich,J.A. (1986) *Proc. Natl. Acad. Sci. USA*, **83**, 6272.
48. Wampler,J.E. (1986) In *Application of Fluorescence in the Biomedical Sciences*. Taylor et al. (eds), Alan R.Liss, New York, p. 301.
49. Yoshimoto,Y., Iwamatsu,T., Hirano,K., Hiramoto,Y. (1986) *Dev. Growth Differ.*, **28**, 583.
50. Schauer,A., Ranes,M., Santamaria,R., Guijarro,J., Lawlor,E., Mendez,C., Chater,K. and Losick,R. (1988) *Science*, **240**, 768.
51. O'Kane,D.J., Lingle,W.L., Wampler,J.E., Legocki,M., Legocki,R.P. and Szalay,A.A. (1988) *Plant Mol. Biol.*, **10**, 387.
52. Purves,D. and Voyvodic,T. (1987) *Trends Neurosci.*, **10**, 398.
53. Blasdel,G.G. and Salama,G. (1986) *Nature*, **321**, 579.
54. DeBiasio,R., Bright,G.R., Ernst,L.A., Waggoner,A.S. and Taylor,D.L. (1987) *J. Cell Biol.*, **105**, 1613.
55. Haugland,R.P. (1988) *Handbook of Fluorescent Probes and Research Chemicals*. Molecular Probes Inc., Eugene, OR, 2nd edn.
56. Allen,T.D. (1987) *J. Microsc.*, **147**, 129.
57. Wolf,R. and Fuldner,D. (1986) *Mikrokosmos*, **75**, 375.

8. FURTHER READING

Video microscopy

Reference 2.
de Weer,P. and Salzberg,B.M. (eds) (1986) *Optical Methods in Cell Physiology, Society of General Physiologists Series*, Vol. 40. Wiley-Interscience, New York.
Shotton,D.M. (1988) *J. Cell Sci.*, **89**, 129.
Optical Approaches to the Dynamics of Cellular Motility (1988) Special Issue of *Cell Motil. Cytoskel.*, **10**, no. 1/2.

VEC microscopy

References 18 and 25.

VIM

References 2, 43 and 55.
Taylor,D.L., Waggoner,A.S., Murphy,R.F., Lanni,F. and Birge,R.R. (eds) (1986) *Applications of Fluorescence in the Biomedical Sciences*. Alan R. Liss, New York.

Digital image processing

Baxes,G.A. (1984) *Digital Image Processing*. Prentice-Hall, Englewood Cliffs, New York.
Image Analysis. Principles and Practice. (1985) Published by Joyce-Loebl; distributed by IRL Press.

Instrumentation

Reference 2.
Guide to Biotechnology Products and Instruments (1988) *Science*, **239**, G73 and G164–G180.

CHAPTER 9

Chromosome banding

A.T.SUMNER

1. INTRODUCTION

Chromosome banding can be defined as the use of special staining procedures to induce patterns of longitudinal differentiation along chromosomes. According to this strict definition, visible structural differentiaton, such as the bands of polytene chromosomes and the chromomeres of pachytene chromosomes, is excluded, although in the latter case there is good evidence for a similarity between the pattern of chromomeres and the pattern of certain bands on mitotic chromosomes (1). Nevertheless, the essential feature of chromosome banding is the demonstration of patterns on chromosomes in the absence of any obvious structural differentiation. Chromosome banding is evidently one aspect of chromosomal organization, but its importance lies in the uses which can be made of different types of banding, particularly in clinical cytogenetics and in evolutionary studies.

A very large number of chromosome banding methods have been described and new ones are still being invented. In practice, however, these methods fall into a limited number of categories, a classification of which will be given in Section 2. The strategy adopted in this chapter is to describe a small number of chromosome banding methods in detail, rather than to give superficial information on a wide variety of techniques. At the same time, more specialized techniques, which might be valuable in certain situations, will be referred to without giving full practical details.

2. CLASSIFICATION OF CHROMOSOME BANDS

Chromosome bands can be classified into one of four groups: heterochromatic bands, euchromatic bands, nucleolar organizers (NORs), and kinetochores (1). These are listed, together with relevant banding techniques, in *Table 1*. It should be noted that a particular banding technique may stain distinctively more than one class of bands; for example, both heterochromatic and euchromatic bands stain distinctively with quinacrine (see Section 4.4); thus Q-bands may be heterochromatic or euchromatic. As an aid to understanding and interpretation, the characteristics of each class of bands will be described briefly.

2.1 Heterochromatic bands

Heterochromatin is those parts of the chromosomes that do not decondense at telophase, but remain condensed throughout interphase (2). This material is normally strongly stained by C-banding methods, although there are a few reports of regions of chromosomes that behave as heterochromatin but do not C-band (3,4). Heterochromatin

Table 1. Classification of chromosome bands.

Class of bands	Principal methods	Other banding techniques
Heterochromatic	C-banding	G11; Q-banding; N-banding; distamycin/DAPI; 5-methylcytosine immunofluorescence
Euchromatic	G-banding Q-banding R-banding	T-banding; various fluorochromes
NORs	Ag−NOR staining	N-banding
Kinetochores	Immunofluorescence with CREST serum	Cd-staining; silver staining

generally contains highly repeated fractions of DNA, usually with a distinctive base composition (1). Other banding techniques usually stain only a subset of heterochromatic bands, often apparently on the basis of affinity for DNAs of different base composition. C-banding itself appears to show no DNA base preference, and thus stains the majority of heterochromatic bands.

Heterochromatic bands represent material additional to the euchromatin and the amount of heterochromatin (and therefore the size of heterochromatic bands) may vary. Such polymorphisms of heterochromatic bands appear to be universal, providing useful markers for determining parental origin of chromosomes, and in the laboratory mouse are characteristic of different inbred strains. No clear phenotypic effects of these polymorphisms have been found, in accordance with the belief that heterochromatin is genetically inert.

2.2 Euchromatic bands

With G-banding, Q-banding, or R-banding, a series of bands can be demonstrated throughout the length of the chromosomes of higher vertebrates (reptiles, birds and mammals). These euchromatic bands form patterns characteristic of each chromosome pair and are indispensable aids for identifying chromosomes. There have been very few reports of euchromatic bands in organisms other than higher vertebrates but it is not yet clear whether this is due to genuine differences in chromosome organization, or is merely a consequence of technical difficulties.

The precise nature of euchromatic bands is not yet established but a number of correlations between these bands and other chromosomal properties are listed in *Table 2*. The correlations with DNA replication patterns, the distribution of genes and the arrangement of pachytene chromosomes appear to be particularly significant.

Euchromatic bands are particularly important in identifying both normal and rearranged chromosomes and are valuable in evolutionary studies, where the same (homologous) chromosomes can be recognized in related species.

2.3 Nucleolar organizers

Nucleolar organizers (NORs) are the sites of the repeated genes for 18S and 28S rRNA. These sites are routinely stained in chromosome preparations using silver, although the material stained is actually protein associated with the NORs (5). The amount of

Table 2. Properties of euchromatic bands.

Positive G-bands	Negative G-bands
Positive Q-bands	Negative Q-bands
Negative R-bands	Positive R-bands
Pachytene chromomeres	Interchromomeric regions
Early condensation	Late condensation
Late replicating	Early replicating
A + T-rich DNA	G + C-rich DNA
Tissue-specific genes	'Housekeeping' genes
Long intermediate repetitive DNA sequences	Short intermediate repetitive DNA sequences

silver staining is a reflection, not only of the number of rRNA genes at a particular site, but also of the activity of those genes. In humans, only 5−9 of the 10 possible NORs are stained; each person has a characteristic pattern of NOR staining. As with heterochromatic bands, NOR staining forms a set of polymorphisms useful for tracing the parental origin of particular chromosomes (6).

2.4 Kinetochores

The kinetochores are the sites of attachment of the spindle fibres to the chromosome and are located at the primary constriction, or centromere, of the chromosome. Techniques were developed for staining them in standard chromosome preparations using Giemsa or silver staining. Since the discovery that auto-immune serum from patients with the CREST syndrome contained antibodies that bind to kinetochores (7), the preferred method for labelling kinetochores has been the use of such sera in conjunction with immunfluorescence techniques.

3. CHROMOSOME CULTURE AND FIXATION

For reasons of space it is not possible to give a full account of procedures for making chromosome preparations. Chromosome preparations are required from so many different kinds of material that it would be impossible to describe procedures appropriate for all. Nevertheless, the methods of making chromosome preparations are a vital part of the complete banding procedure. Without good quality chromosome preparations, good quality banding cannot, in general, be obtained, and in some cases pre-fixation treatments, while the cells are still in culture, are an integral part of the banding procedure.

A standard source of human and other mammalian chromosomes is blood lymphocyte culture. Essentially, a blood sample is incubated in a suitable culture medium in the presence of a mitogen, usually phytohaemagglutinin (PHA), to stimulate DNA replication and cell division. Shortly before it is desired to harvest the cells, a spindle inhibitor is added to the culture to arrest the cells at metaphase, thereby increasing the proportion of dividing cells. Finally, the cells are treated with a hypotonic solution (usually 0.075 M KCl) for a few minutes to swell the cells and disperse the chromosomes, and fixed with a mixture of methanol−acetic acid (3:1 v/v). Chromosome spreads are made by dropping the fixed cell suspension onto slides. Detailed protocols are described by several authors (8−11) and a typical one is described in Section 3.1.

For insect and plant material it is not usually possible to obtain mitotic cells in suspension and therefore preparative methods radically different from those described above must be used. It is normal to use squashing methods for preparations of insect (10) and plant (14) material. A procedure specifically designed to be compatible with banding of plant chromosomes has been described by Schwarzacher *et al.* (15); this paper points out which stages of the procedure are particularly important for good banding. It has been pointed out above that the retention of cytoplasmic material around the chromosomes is inimical to good banding, yet this is almost impossible to avoid with squash preparations. Squash preparations can be used successfully for staining of heterochromatin with C-banding methods or with fluorochromes but it has not proved possible to demonstrate euchromatic bands in plant and insect chromosomes. It is not certain at present whether this is merely a consequence of preparation by the squashing method, which leaves too much cytoplasm around the chromosomes to permit euchromatic banding, or whether this represents a genuine difference between the organization of the chromosomes of higher vertebrates on the one hand and those of plants and insects on the other.

Meiotic chromosomes of most species are prepared by squashing methods (10). For mammalian meiotic chromosomes air-drying procedures, similar in principle to those used for lymphocyte chromosome cultures, are used but the exact procedure depends on which stage of meiosis it is desired to show. Methods using a shorter hypotonic treatment (16,17) produce mainly first and second meiotic metaphases, with some diakineses and spermatogonial metaphases. Longer hypotonic treatment (18) spreads pachytene chromosomes but destroys most other stages. It is therefore necessary to decide in advance which stages of meiosis to investigate. Meiotic chromosome preparations can be used for demonstrating heterochromatic bands but techniques for euchromatic bands are not successful.

3.1 Protocol for human lymphocyte chromosome culture

(i) Collect adult blood by venipuncture into a sterile container coated with lithium heparin (anticoagulant). Mix gently to prevent clotting. Note that blood should only be taken by a qualified person, and that all culture procedures must be carried out under sterile conditions.

(ii) Incubate 0.8 ml of blood for 72 h at 37°C in a Universal bottle in 10 ml of culture medium, made up as follows:

F 10 medium (Gibco)	8.0 ml
Fetal calf serum	2.0 ml
PHA	0.1 ml
Penicillin, 10 000 U/ml Streptomycin 10 mg/ml	0.01 ml

(iii) 3 h before harvesting, add 0.1 ml of solution containing 10 μg of Colcemid/ml, and shake gently to mix.

(iv) After 72 h, transfer the culture to a conical 10 ml centrifuge tube and centrifuge at about 200 g for 5 min.

(v) Decant off the supernatant and add 10 ml of 0.075 M KCl to the pellet. Mix on a vortex mixer and leave for 8−10 min.

(vi)   Re-centrifuge at 200 *g* for 5 min, and decant off the supernatant. Add 10 ml of freshly prepared fixative (methanol:glacial acetic acid, 3: 1 v/v) while mixing vigorously.

(vii) Re-centrifuge, decant the fixative, and add new fixative twice.

(viii) Re-centrifuge, decant the fixative and add 0.5−1 ml of fixative to produce a slightly milky suspension.

(ix)   Allow a drop of the suspension to fall onto a clean slide from a height of a few cm. Breathe on the slide to spread the cells and allow to dry.

(x)   Examine the preparation by phase contrast microscopy to check the quality and cell density. If the density is too high or too low, adjust the cell concentration by diluting with more fixative, or re-centrifuging and resuspending in a smaller volume of fixative, and make as many slides as required.

Careful attention to detail is required, as good quality chromosome preparations are essential if chromosome banding of adequate quality is to be obtained. For best results it is necessary to have relatively long chromosomes, well spread with few, or, if possible, no overlaps, and a negligible quantity of residual cytoplasm surrounding the chromosomes. A poorly spread preparation such as that shown in *Figure 1* cannot be expected to produce satisfactory banding; compare this preparation with the illustrations

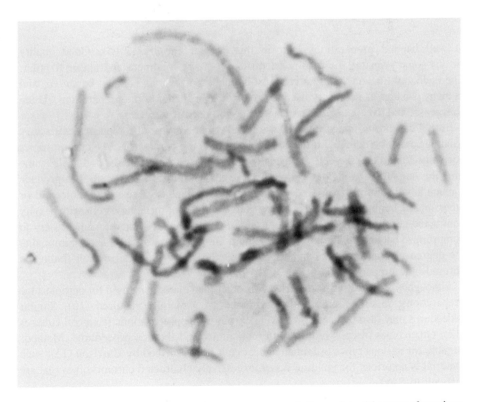

Figure 1. Poorly spread human metaphase chromosomes, surrounded by a substantial amount of cytoplasm. Such preparations will not yield satisfactory banding.

Table 3. Guide to fault-finding in blood lymphocyte chromosome preparations.

Fault	Remedy
Dirty fixed cell suspension with brown lumps of haematin.	Better mixing at first fixation
Cell suspension remains as drop on slide instead of spreading.	Dirty slide; clean slides in acid alcohol (1% conc. HCl in absolute alcohol).
Chromosomes poorly spread and embedded in cytoplasm (*Figure 1*).	Increase time in hypotonic KCl. Improve mixing of fixative. Use more changes of fixative (up to 6).
Chromosomes poorly spread, but no cytoplasm, and slides clean.	Use cold slides, or slides dipped in distilled water (without drying), or increase height from which cell suspension dropped.
Metaphases broken, with scattered chromosomes	Omit or reduce breathing on slide. Reduce hypotonic treatment. Drop cell suspension from lesser height.
Chromosomes over-condensed (short and fat).	Reduce time in Colcemid.

of well banded preparations later in this chapter (*Figures 2–5*). Good quality chromosome preparations require adequate hypotonic treatment and ample fixation. The methanol–acetic acid fixative must be made up fresh and cannot be stored, and as many changes as necessary used to produce clean chromosome preparations. Three changes of fixative is normal but more may be used if necessary.

Even if the chromosomes are well spread and essentially free of cytoplasm, satisfactory banding cannot be obtained if the chromosomes are too contracted. The chromosomes normally contract as the cell proceeds from prophase to metaphase, and at the same time the numerous bands present at early stages of mitosis fuse together to produce a much smaller number of bands at metaphase. Metaphase arrest with Colcemid or other spindle inhibitors tends to produce a large number of metaphase cells with heavily contracted chromosomes, which are useless for banding. It is therefore necessary to compromise between obtaining the maximum number of mitotic cells and minimizing the degree of contraction of the chromosomes (10,11). A guide to fault-finding in chromosome preparation is given in *Table 3*.

Although blood cultures from vertebrates other than man can be used for chromosome preparations, modifications of the technique are generally required (10). Similar procedures may also be used for making chromosome preparations from cell cultures (12). Other types of cells require different culture and fixation procedures. Methods suitable for various types of human cancer cells are described by Harrison (13); such material is notorious for yielding fuzzy, excessively contracted chromosomes that are difficult to band.

It is general experience that chromosome preparations of all types should be aged before any type of banding technique is attempted. Three days to 1 week at room temperature is suitable for most purposes. Prolonged storage results in a decline in the quality of banding, and ultimately even complete loss of bandability. Nevertheless, it may be possible to obtain good G-banding (one of the more capricious techniques) on slides several months old, although this cannot be relied on. If it is necessary to store slides for a prolonged period before banding this should be done in the cold, preferably in a vacuum desiccator. If it is necessary to band chromosome preparations as soon as possible after they are made they can be aged overnight (16 – 18 h) at about 60°C.

4. STANDARD BANDING METHODS

Standard banding methods can be regarded as those methods which would normally be employed in a comprehensive banding study of the chromosomes of an animal. The types of banding covered are C-, G-, R- and Q-banding, and silver staining for NORs (Ag – NOR staining).

4.1 C-banding

C-banding is carried out by immersing standard chromosome preparations successively in acid, alkali, hot salt solution and Giemsa staining solution. The method described here is that published by Sumner (19), which, with variations is the basis for almost all C-banding studies.

Procedure

(i)   Immerse the slides in 0.2 M HCl (17.2 ml of conc. acid/l) for 60 min at room temperature.

(ii) Rinse the slides briefly with distilled or de-ionized water.

(iii) Place the slides in 5% Ba(OH)$_2$ solution at 50°C for 5 min. This solution is best prepared by heating a Coplin jar containing 40 ml of distilled water at 50°C in a water bath; a few minutes before use add 2 g of barium hydroxide octahydrate [Ba(OH)$_2$.8H$_2$O] and stir thoroughly to dissolve. A scum of BaCO$_3$ inevitably forms on the surface, and may be skimmed off if it forms a troublesome deposit on the slides. The solution should be discarded after a short time (not more than 60 min).

(iv) Rinse the slides thoroughly with distilled water to remove as much scum as possible.

(v)   Incubate in 2 × SSC (0.3 M NaCl + 0.03 M trisodium citrate) at 60°C for 60 min. A stock solution of 2 × SSC contains 17.53 g of NaCl and 8.82 g of trisodium citrate/l.

(vi) Rinse with distilled or de-ionized water.

(vii) Stain with Giemsa for 45 min. Add 1 ml of Gurr's Giemsa Improved R66 (BDH) to 50 ml of buffer, pH 6.8, made with Gurr's buffer tablets (BDH).

(viii) Rinse with distilled or de-ionized water and carefully blot dry.

(ix) Allow a few minutes for the slide to dry thoroughly and then mount in DPX mountant (BDH) or a similar polystyrene-based mountant.

A typical result, with heterochromatic regions of chromosomes darkly stained and the chromosome arms paler, is shown in *Figure 2*.

As with all banding techniques, the results of C-banding can be variable and the basic

285

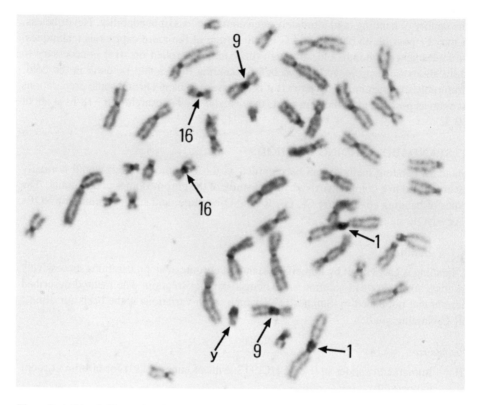

Figure 2. A C-banded human lymphocyte metaphase. The large blocks of heterochromatin on chromosomes 1, 9, 16 and Y are arrowed.

method given above should be modified accordingly. If the staining of the chromosomes is too dark, with indistinct banding, then more extensive treatment is required; conversely, excessively pale staining indicates that the chromosomes have been over-treated. These defects cannot be overcome by altering the length of staining or the concentration of the stain. If satisfactory results are not obtained, it is best to begin by altering the $Ba(OH)_2$ treatment. The length of treatment should be increased if staining is too dark and the C-bands poorly differentiated. In the case of pale staining, the length of $Ba(OH)_2$ treatment can be reduced, but for some sensitive materials, the temperature can also be reduced. For example, the use of a 4% $Ba(OH)_2$ solution at 37°C for 30 sec has been recommended for mouse meiotic chromosomes (20).

If adjustment of the treatment with $Ba(OH)_2$ is insufficient to produce satisfactory results the HCl treatment can also be reduced. In some cases the acid treatment can be eliminated entirely but in general some treatment is advisable. Reduction in the length of treatment from 60 min to 30 or 15 min, can improve the morphology of the chromosomes by reducing swelling, as well as compensating for excessively pale staining.

Gurr's Giemsa Improved R66 has been specified above and has been found to give good results as well as being a stable stock solution. Other makes of Giemsa and related Romanowsky dyes (e.g. Leishman, Wright's, MacNeal's) can probably be used successfully, but each new sample should be tested on material that is known to give good C-banding, and the concentration and staining time adjusted until good results are obtained. The diluted Giemsa solution is not stable, and should be used fresh.

It is important to choose suitable vessels for incubating slides at high temperatures. Ordinary glass Coplin jars frequently break when placed in hot water baths. Coplin jars made from borosilicate glass do not suffer from this problem but seem to be no longer available. Plastic staining jars are available from various manufacturers and, while not heavy enough to stand on the bottom of a water bath, can be suspended in the bath using a metal plate across the top of the bath. Holes are cut in the plate of such a size that the body of the staining jar will pass through and be supported by the wider lip below the screw thread for the lid. Polypropylene staining jars (Azlon) are better than polyethylene ones, as they can be used safely at the highest temperatures required for banding without softening.

4.2 G-banding

G-banding is the standard method for identifying the chromosomes of mammals and other higher vertebrates and thus it occupies a central place in cytogenetics. Because of the importance of G-banding and also because of the capriciousness of the techniques, three different procedures are described in this section. Each has its advantages and disadvantages. The newcomer to these procedures is advised to try them out and discover which one seems to work best on his or her material.

4.2.1 *The ASG method*

This method involves the incubation of chromosome preparations in $2 \times SSC$, followed by staining with Giemsa (21).

(i)   Incubate the chromosome preparations in $2 \times SSC$, preheated to 60°C, for 60 min. See Section 4.1 for the composition of $2 \times SSC$.

(ii)   Rinse briefly with de-ionized or distilled water.

(iii)   Stain with Giemsa (see Section 4.1) for 45 min.

(iv)   Rinse with de-ionized or distilled water and blot dry.

(v)   After allowing the slides to dry thoroughly in air, mount in DPX or a similar mountant.

A human chromosome set, banded by this method, is shown in *Figure 3*. To get satisfactory results with this method, it is essential to use good quality chromosome preparations, in which the surrounding cytoplasm has been removed as far as possible and in which the chromosomes are not too contracted. If good banding is not achieved there are few modifications that can be made to the method to improve results. Some workers increase the time of incubation in $2 \times SSC$ up to 16 h but this tends to produce ragged chromosomes with poor morphology. If banding is not clear because the staining is too dark it may sometimes help to stain for a shorter period, or with a more dilute stain solution.

Once mounted in good quality mountant the staining should be stable for years. If, however, the mountant has deteriorated with age and become acid this will not only cause fading but also alter the banding pattern to one which resembles R-banding. Any suspect mountant should be thrown out and replaced with fresh stock.

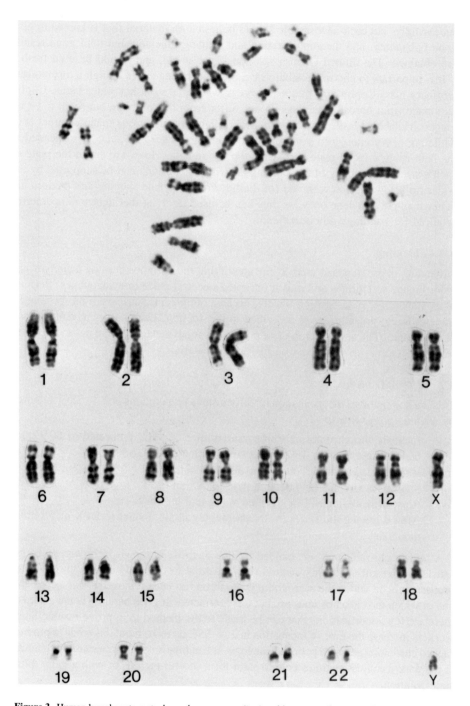

Figure 3. Human lymphocyte metaphase chromosomes (**top**) and karyotype (**bottom**), G-banded by the ASG technique. Reprinted with permission from *Nature New Biology, 232*, pp. 31–32. Copyright© 1971, Macmillan Magazines Ltd.

4.2.2 *Trypsin – Giemsa banding*

This method is essentially similar to that described by Seabright (22); procedures based on this principle are the most popular of all banding techniques. Briefly, the chromosome preparations are digested for a few seconds with trypsin, and then stained with Giemsa.

(i)   Digest chromosome preparations with trypsin for about 15 sec. The trypsin stock solution is prepared by reconstituting a phial of Difco Bacto Trypsin (Catalogue No. 0153 – 59) with 10 ml PBS; this may be prepared from tablets obtainable from Oxoid. The working solution is then prepared by diluting this further.

(ii)   Rinse the slides with distilled water or PBS.

(iii) Stain with Giemsa as previously described (Section 4.1).

(iv) Rinse, blot dry and mount as previously described (Section 4.1).

The critical part of this procedure is the digestion with trypsin and the precise conditions need to be determined by experience. The result depends on the quality of the trypsin, its dilution, the period of digestion and the sensitivity of the chromosomes. For an initial experiment the trypsin stock solution can be diluted ten-fold with PBS but further dilution may well be necessary. In principle, trypsin from any source may be used but the correct dilution and time must be found by experience. It is usual to flood the slide with a few drops of trypsin solution rather than immersing the whole slide in a Coplin jar of trypsin. Trypsin solutions lose their potency with time and reasonably fresh solutions should be used for consistency.

Although a Giemsa solution is recommended for staining, the original method used Leishman's stain (22); this may be worth trying if satisfactory results cannot be obtained with Giemsa.

The actual banding pattern produced by this method is identical with that produced by the ASG method (*Figure 3*), although the overall morphology may be different. Trypsin-treated chromosomes often have a ragged appearance which to some extent results from excessive treatment with trypsin. Another charactersitic appearance often found after trypsin – Giemsa banding is a dark rim delineating the chromosomes. This can be a valuable feature, although it must not be mistaken for the existence of terminal G-bands. With ASG banding, there is no dark rim, and since terminal G-bands are generally pale, it is difficult to define the ends of the chromosomes.

4.2.3 *Gallimore and Richardson's method*

This method is essentially a combination of the ASG method and trypsin – Giemsa banding (23).

(i)   Incubate chromosome preparations in 2 × SSC (see Section 4.1) for 3 h at 60°C.

(ii)   Cool the slides to room temperature by washing in de-ionized water.

(iii) Digest the chromosomes for 90 sec at 10°C in trypsin – saline solution (Difco Bacto trypsin – 0.9% NaCl solution, 1:99, v/v).

(iv) Wash with de-ionized water.

(v)   Stain with Giemsa (Gurr's Improved R66), diluted 1 to 10 with de-ionized water, for 5 min.

(vi) Blot the slides, allow them to dry in air and mount.

Compared with the two previous techniques this one is claimed to give clearer banding

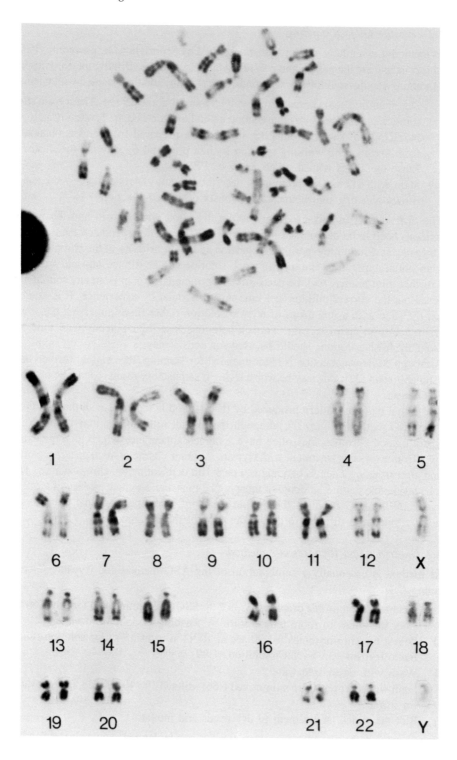

than ASG alone but without the degraded chromosome morphology often produced by trypsin treatment on its own. As with the trypsin—Giemsa method the time and concentration of trypsin should be adjusted to give optimal results. For chromosomes from certain types of cells it may also help to reduce the length of incubation in 2 × SSC somewhat.

4.3 **R-banding**

R-banding produces a pattern complementary to that given by G-banding, that is, positive R-bands are negative G-bands. Except in France, R-banding has not been used much as the primary banding technique for identifying chromosomes, but has particular advantages in two situations. Firstly, it can be used in combination with G-banding on the same chromosomes to define chromosomal breakpoints with greater precision than is possible with either technique alone. Secondly, the terminal bands of chromosomes, which are generally pale with G-banding, are darkly stained with R-banding. Thus the ends of the chromosomes are much more clearly defined and the method is valuable in the study of structural alterations of chromosomes that involve terminal segments. The method described here is that due to Sehested (24).

(i)   Incubate chromosome preparations in 1 M $NaH_2PO_4.2H_2O$ (156.01 g/l, unbuffered, pH 4.0−4.5) at 88°C for 10 min.

(ii)   Rinse the slides briefly in a staining jar of de-ionized water or tap water.

(iii)   Stain for 10 min in Giemsa (Gurr's Improved R66, BDH, 5% in distilled water, diluted immediately before use).

(iv)   Rinse the slides briefly in a jar of tap water.

(v)   Blot dry and mount.

The results of this technique are illustrated in *Figure 4*. Compared with G-banding, R-banding patterns appear to be less detailed in spite of being essentially the reverse of G-banding.

Because of the high temperature used for incubation in this method, staining vessels made of polypropylene, or of some other heat-resistant, inert substance, must be used. The results of the method are affected by the age of the preparations, as with other banding techniques, and the original author recommended using slides 2 days after the chromosome preparations had been made. To avoid the formation of air bubbles on the slide during the incubation at 88°C, with consequent appearance of improperly treated areas on the chromosomes, the slides should be tapped gently against the sides or bottom of the incubation vessel once or twice during incubation, to dislodge the bubbles.

Increasing or decreasing the molarity of the phosphate solution used for incubation results in poor or no banding. Altering the pH however, results in G-banding at pH values between 5.5 and 6.7 (24). If the two types of banding are to be obtained sequentially, the G-banding must be done first; for practical details see Buckton (25).

Figure 4. R-banded human lymphocyte metaphase chromosomes (**top**) and karyotype (**bottom**). Reprinted with permission from Bostock,C.J. and Sumner,A.T. (1978), *The Eukaryotic Chromosome*, copyright Elsevier Science Publishers BV.

4.4 **Q-banding**

Q-banding was not only the first modern banding technique to be devised, but is also one of the most useful. Using this method both euchromatic bands and certain heterochromatic bands can be stained distinctively; it is thus a valuable method for identifying chromosomes in higher vertebrates, and also in organisms such as insects and plants, where chromosome identification depends much more on the characteristics of heterochromatin bands. Heterochromatic bands can show bright or dim fluorescence with quinacrine, and also show size polymorphisms; Q-banding is therefore one of several methods that allow characterization of heterochromatin in more detail than is possible with C-banding alone.

As well as having a wide range of applications, Q-banding is simple to perform and, compared with many other banding techniques, is highly reliable. The disadvantages of Q-banding are the impermanence of the preparations, the fading of the fluorescence under irradiation and, of course, the necessity to use a specialized, fluorescence microscope. Although the original method used quinacrine mustard it is now usual to employ quinacrine, which is cheaper and, perhaps, safer, although both substances must be handled with care because of their potential mutagenic properties. More recently a substance showing greater stability of fluorescence, spermidine bis-acridine, has been proposed for Q-banding. Although this substance has to be prepared in the laboratory, the advantages of a fluorochrome with greater resistance to fading are so great that the synthesis is described here.

4.4.1 *Q-banding with quinacrine*

Quinacrine is available from many suppliers under a variety of names: atabrine, atebrin, mepacrine, quinacrine hydrochloride or dihydrochloride, all of which are essentially the same. The staining procedure is very simple:

(i)   Stain chromosome preparations for 6−10 min in an 0.5% aqueous solution of quinacrine.
(ii) Wash in running tap water for 3 min.
(iii) Rinse in distilled water.
(iv) Mount in distilled water. Press down on the coverslip with filter paper to blot away as much surplus water as possible. Seal the edges with rubber solution; a fairly runny rubber solution, such as 'Pang Supersolution' [Pang (UK) Ltd], is very suitable.

Q-banding of human metaphase chromosomes is illustrated in *Figure 5*. The preparations are observed using a fluorescence microscope and blue−violet light illumination (Section 6.2). The method is very robust, working with a wide range of dye concentrations and staining times. However, as with all banding techniques, the best results are obtained using high quality chromosome spreads.

Various authors have suggested mounting the preparations in a buffer, or buffer plus glycerol, rather than water. There seems to be little advantage in doing so and too much glycerol will result in uniform fluorescence of the chromosomes. The mounted preparations may be stored in a refrigerator overnight or perhaps for a day or two without significant loss of banding or chromosome morphology but they inevitably deteriorate if kept for any length of time.

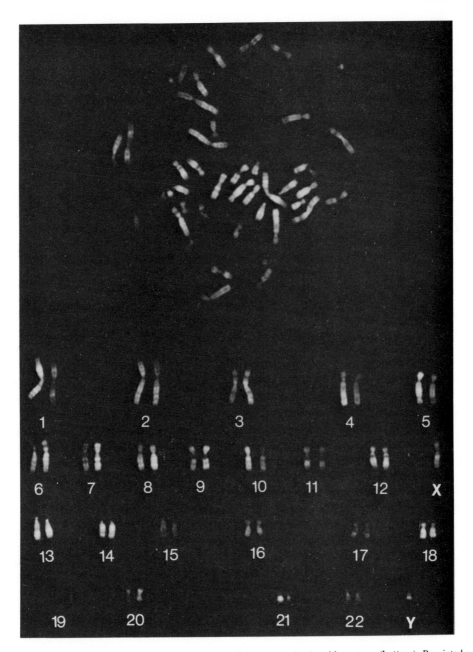

Figure 5. Q-banded human lymphocyte metaphase chromosomes (**top**) and karyotype (**bottom**). Reprinted with permission from Bostock,C.J. and Sumner,A.T. (1978), *The Eukaryotic Chromosome*, copyright Elsevier Science Publishers BV.

4.4.2 *Preparation of spermidine bis-acridine and its use for Q-banding*

Spermidine bis-acridine [(CMA)$_2$S] is essentially a dimeric analogue of quinacrine, which produces brighter, more contrasty, and more stable Q-bands (26). This

293

compound is not available commercially and must be prepared in the laboratory, as follows:

(i)   Dissolve 4 g of NaOH in 60 g of melted phenol at 100°C.

(ii)   Add 14 g of dichloromethoxyacridine (Aldrich) with stirring.

(iii)   Maintain the mixture at 100°C for 1.5 h, then pour it into 500 ml of 2 M NaOH, stir and allow to cool overnight.

(iv)   Filter the precipitate, wash with water, powder and dry. This material is phenoxyacridine.

(v)   Dissolve 8.68 g of phenoxyacridine in 27.6 g of phenol at 80°C in an open vessel.

(vi)   Add 2.0 g of spermidine.

(vii) Increase the temperature to 120°C and continue heating for 1 h.

(viii) Cool and pour into ether.

(ix)   Wash the material with ether, re-dissolve in hot methanol and precipitate it again with ether. Filter through a Buchner funnel, wash with ether and dry. The precipitate is spermidine bis-acridine.

(x)   Store the spermidine bis-acridine in an air-tight container in the refrigerator.

Great care must be taken in performing this synthesis because of the dangerous ingredients. All work should be done in an effective fume cupboard, and suitable protective clothing worn.

The staining procedure using spermidine bis-acridine is essentially similar to that using quinacrine, but a more dilute solution is employed.

(i)   Dissolve 5 mg of spermidine bis-acridine in 1 ml of methanol, add to 99 ml of 0.01 M phosphate buffer (pH 6.5) and stain chromosome preparations in this for 10 min.

(ii)   Wash the slides in phosphate buffer (pH 6.5), two changes, total 1 min, and mount in the same buffer, pressing down the coverslip and absorbing surplus buffer as described above for quinacrine (Section 4.4.1). Observation conditions are also exactly as described for quinacrine.

4.5 Silver staining of nucleolar organizers

The NORs of chromosomes may be stained with silver by a variety of methods. The material stained with silver is proteinaceous material associated with the NORs rather than the genes for 18S and 28S rRNA; the amount of silver staining is related to the degree of activity of the ribosomal genes. Like C-bands, silver stained NORs are often polymorphic (1).

The staining method described here is that due to Howell and Black (27). This method uses two stock solutions, which are mixed immediately before use.

4.5.1 *Stock solutions*

(i) $AgNO_3$: dissolve 4 g of $AgNO_3$ in 8 ml of de-ionized water. This solution is stable if kept in the dark, but should be discarded if significant blackening occurs.

(ii)   Colloidal developer: dissolve 2 g of gelatine in 100 ml of de-ionized water, and add 1 ml of pure formic acid. Stir continuously, with warming if necessary, until dissolved. This solution should not be used for more than 2 weeks after preparation.

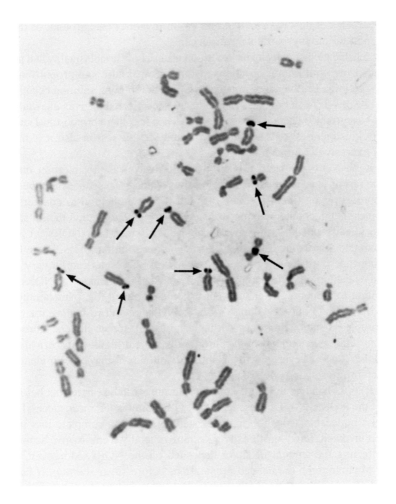

Figure 6. Ag–NOR staining of human lymphocyte metaphase chromosomes. The NORs are arrowed. Note the variability in size of the NORs. Counterstained with Giemsa.

4.5.2 Procedure

(i) Mix 2 drops of colloidal developer with 4 drops of the AgNO₃ solution in a small tube, and pipette on to the slide bearing the chromosome preparations. Cover with a coverslip.

(ii) Place the slide on the surface of a hot plate at approximately 70°C. The staining solution will first turn pale yellow and then golden brown.

(iii) When the staining solution has turned golden brown (1–2 min), rinse off the coverslip and staining solution with a generous flow of de-ionized (or distilled) water.

(iv) Blot the slide and allow it to dry thoroughly. Mount in DPX or a similar mountant.

This method stains the NORs black, while the chromosome arms and any cytoplasmic material are yellow (*Figure 6*). The background should be clean but use of older slides on which particles of dust have settled may result in a dirty background, as the silver may be deposited on any particulate material.

This is a reliable method which works well on material of variable quality and produces good results even with minor variations of time, temperature and proportions of the two stock solutions. Prolonged treatment, in which the staining solution becomes dark brown and begins to form bubbles, should be avoided since this gives uneven staining and a dirty background. There is also a small tendency for other chromosomal structures, particularly kinetochores, to stain with silver and for this reason also it is important not to use excessive treatment.

Although the NORs are stained black it is not always easy to interpret the results because the rest of the chromosomes are so palely stained. To some extent the contrast can be improved using filters (Section 6.1) but it is often helpful to counterstain the chromosomes. If the counterstaining is done with a banding method then it should be possible to identify unequivocally the chromosomes which bear NORs.

For counterstaining without banding, a 5% Giemsa solution, pH 6.8, can be used. The staining time must be determined by experience as results can be very variable after the silver staining. It is important not to overstain as this renders the NORs difficult to see.. A light to moderate magenta colour should be aimed for; overstaining can be corrected by washing out the Giemsa with 70% ethanol and restaining for a shorter period. Simple counterstaining is particularly useful for pachytene chromosome preparations, where the outlines of the chromosomes are difficult to discern after silver staining alone; it is, however, equally useful with mitotic chromosomes if identification of specific chromosomes is not required.

G-, Q-, R- or C-banding to identify the chromosomes can be carried out before silver staining of the NORs (28). Of these methods, Q-banding is the most successful as the others cause noticeable reduction in silver staining, or even complete loss if trypsin treatment is involved. Banding after silver staining generally seems to give better results, without affecting the amount of silver deposited on the NORs. Moreover, it is not necessary to photograph the banded preparations before silver staining; if banding is carried out after NOR staining, both the banding and the silver deposits can be viewed simultaneously. Standard banding methods can be employed, but the times of treatment generally need to be altered (usually increased). Howell and Black (29) obtained good results with both trypsin–Giemsa and ASG methods after silver staining, but obtained poor results with Q-banding. The latter can, however, give good results, although simultaneous observation of the fluorescent Q-bands and the Ag–NORs is less convenient. It is important to try out a variety of banding methods in combination with silver staining of NORs until a procedure is found that gives acceptable and consistent results. The inherent variability of banding techniques, particularly between different laboratories, must not be underestimated and it is therefore not possible to prescribe methods that invariably work without modification.

Note that silver solutions should always be handled with care as they will indelibly blacken everything with which they come into contact. Good laboratory technique should be adequate but the use of disposable gloves and a protective layer on the bench (e.g. Benchkote, Whatman Lab Sales) are useful additional precautions.

5. SPECIALIZED BANDING TECHNIQUES

The methods to be described in this section differ from the 'standard' methods described in Section 4 either in being less widely applicable, or in requiring a special procedure for making the chromosome preparations. The four methods described here are the distamycin/DAPI method, which stains a subset of heterochromatic bands in certain species; replication banding which indicates the order of replication of the DNA in different chromosomal segments; high resolution banding, which provides much more detailed patterns on chromosomes; and an immunocytochemical procedure for labelling kinetochores.

5.1 **Distamycin/DAPI staining**

When chromosomes stained with the fluorochrome DAPI are counterstained with the non-fluorescent antibiotic distamycin, bright fluorescence is restricted to only some of the heterochromatic bands which can be revealed by C-banding (30). Distamycin/DAPI (DA−DAPI) bands have been found in the chromosomes of some but not all mammals, in the chromosomes of certain grasshoppers, but not, so far, in any plant chromosomes. DA−DAPI banding is useful for further characterizing heterochromatin identified by C-banding, and can be used as a marker for those chromosomes which have DA−DAPI bands when such chromosomes are rearranged.

(i) Flood the slides bearing the chromosomes with distamycin A (DA) solution for 15 min. This consists of 0.1 or 0.2 mg/ml of DA (Sigma) dissolved in McIlvaine's citric acid−phosphate buffer, pH 7. The buffer is made up of 82.5 ml of 0.2 M $Na_2HPO_4 \cdot 2H_2O$ (35.6 g/l) plus 17.5 ml of 0.1 M citric acid ($C_6H_8O_7 \cdot H_2O$, 21.01 g/l).

(ii) Rinse the slides briefly with McIlvaine's buffer, pH 7.

(iii) Flood the slides with DAPI (Sigma) solution, and stain for 5−15 min. The DAPI solution contains 0.2 or 4.0 μg/ml of McIlvaine's buffer, pH 7.

(iv) Rinse in McIlvaine's buffer, pH 7.

(v) Mount in the same buffer. Blot excess fluid from under the coverslip, and seal with rubber solution (Section 4.4.1).

The results of applying this procedure to human chromosomes are illustrated in *Figure 7*. The time of staining with DA or with DAPI, or the concentrations of these dyes, can be adjusted to produce optimal results. If the chromosomes show an appearance similar to Q-banding (the pattern that is produced by DAPI alone), stain for longer with distamycin, or use a higher concentration. If the fluorescence is too weak, reduce the distamycin treatment.

Chromosome preparations stained with DA−DAPI fade rather rapidly if exposed to UV light when first made. This results in a pattern similar to Q-banding. If the slides are stored in the refrigerator for a day or so after mounting, the fluorescence is generally found to be more stable.

Distamycin A is not stable in solution, and it is best to prepare solutions freshly when needed. DAPI is stable in solution, and may be stored in the refrigerator at least for several weeks. A stock solution of 0.5 mg of DAPI in 1 ml of distilled water can be

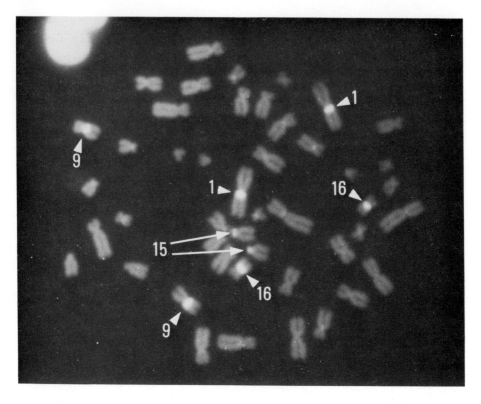

Figure 7. DA–DAPI fluorescence of human female lymphocyte metaphase chromosomes. Only the heterochromatin of chromosomes 1, 9, 15 and 16 fluoresces strongly (arrows); in a male cell the Y heterochromatin would also be bright. Micrograph kindly provided by Mr G.Spowart. Reprinted from Sumner,A.T. (1983), *Science Progress,* **68**, pp. 543–564.

made up and diluted with McIlvaine's buffer as required. Other substances may be used instead of both DA and DAPI to produce comparable results. One of the most useful alternatives is methyl green, which is a much cheaper alternative to distamycin (31). Hoechst 33258 (Sigma) can be used instead of DAPI (32).

5.2 Replication banding

Replication banding is a method for distinguishing chromosomal regions that replicate their DNA early or late in the S phase. The patterns produced are (on mammalian chromosomes) largely similar to G- or R-bands, depending on whether early or late replication patterns are being demonstrated. There are, however, occasional small differences and the late replicating X chromosome in female mammals can be distinguished from the early replicating one and shows quite variable patterns (33).

Replicating DNA is labelled by the incorporation of bromodeoxyuridine (BrdU) into it at the appropriate time during culture. After fixation of the cultured cells in the normal way, DNA that has incorporated BrdU can be differentially stained using appropriate treatments. A commonly used and very satisfactory staining procedure is the FPG method

(35). Using this staining procedure, chromosomal segments that have incorporated BrdU are pale, and unsubstituted segments are darkly stained.

The whole procedure for inducing replication bands consists of two parts: cell culture and staining. While the latter is standard, the culture method must be adapted according to whether early or late replication patterns are required, and according to the cell cycle time of the cells being studied.

5.2.1 Culture method

The methods described here are for human lymphocytes (34). The cultures are set up in the standard way (11) except for the addition of BrdU and certain other constituents (Sigma). These are added to the culture medium at the following concentrations:

BrdU	10^{-4} M
fluorodeoxyuridine (FdU)	10^{-6} M
deoxyuridine (dU)	6×10^{-6} M
deoxycytidine (dC)	10^{-4} M

The addition of FdU and dU increases incorporation of BrdU into the DNA, although satisfactory results can be obtained without the use of these substances.

To obtain a late replication pattern, the BrdU and other substances are added to the culture from the beginning. Then, 5−7 h before harvesting the culture, the medium is changed, and the cells are placed in standard culture medium to which 6×10^{-4} M deoxythymidine is added. This is known as a T pulse, and results in the late replicating DNA being unsubstituted with BrdU.

To obtain an early replication pattern, the cells are grown in standard culture medium until 5−7 h before harvesting, when BrdU, FdU, dU and dC are added at the concentrations given above. This is a B pulse and results in the late replicating DNA incorporating BrdU, while the early replicating DNA is unsubstituted. Because BrdU is light-sensitive, its stock solutions and cultures containing BrdU must be kept in the dark as far as possible. For more detailed protocols for culture in the presence of BrdU, see refs 32 and 34.

5.2.2 Staining procedure

Chromosomes substituted with BrdU are stained differentially using the FPG technique (35).

(i)   Stain the fixed chromosome preparations for 12 − 15 min in a solution of Hoechst 33258 (Sigma; 0.5 µg/ml de-ionized water).

(ii)   Rinse the slides with de-ionized water.

(iii)   Mount chromosome preparations in de-ionized water, seal with rubber solution and leave exposed to daylight for 24 h.

(iv)   Remove coverslips.

(v)   Incubate in 2 × SSC (0.3 M sodium chloride plus 0.03 M tri-sodium citrate) for 2 h at 60°C.

(vi)   Stain in 3% Giemsa (Gurr's Improved R66) at pH 6.8 (Gurr's Buffer Tablets) for 30 min.

(vii)   Blot the slides, allow to dry and mount in DPX.

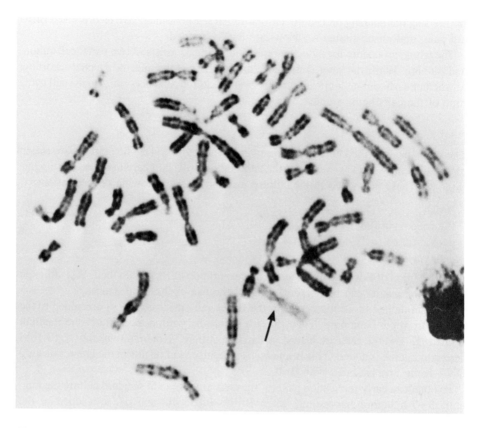

Figure 8. Replication banding of chromosomes from a human female. The DNA in this cell was substituted with BrdU in the latter part of the S phase (B pulse), so that after application of the FPG method, late replicating segments of the chromosomes are weakly stained. Note the late replicating X chromosome (arrowed). Preparation kindly provided by Miss K.E.Buckton. Reprinted from Sumner,A.T. (1983), *Science Progress,* **68**, pp. 543–564.

The results of this procedure are illustrated in *Figure 8*, in which early replicating regions of human chromosomes are darkly stained and late replicating regions, including the whole late replicating X chromosome, are pale. This cell has therefore been given a B pulse in culture.

To produce results more quickly and controllably than in the protocol above, artificial light may be substituted for daylight (32). The length of treatment will depend on the type of lamp used and must be determined by experience. Once this is established the time should be relatively constant, provided of course that the distance from the lamp to the slides is also kept constant. Because Hoechst 33258 absorbs UV light, a lamp having a significant output in the near UV will be most effective.

BrdU substitution of chromosomal DNA, followed by the FPG technique can be used to study not only replication bands but also sister-chromatid exchanges (36) and lateral asymmetry of heterochromatin (37). Although these latter procedures require BrdU incorporation and culture over a longer period of time, so that the BrdU is incorporated throughout one chromatid (in the case of sister-chromatid differential staining), the technology involved is exactly the same as described above except for culture times.

5.3 High-resolution banding

The standard human haploid metaphase karyotype contains about 300 bands. Various procedures can be used to produce more elongated chromosomes which show a greater number of bands and thus permit the detection of smaller deletions and the more precise determination of breakpoints. The production of as many as 2000 bands in a haploid karyotype has been claimed but routinely somewhere in the region of 500–800 bands can be produced by these methods, collectively known as high-resolution banding (HRB).

Methods for producing HRB can be divided into two categories which are not, however, mutually exclusive. In both cases the cultures have to be treated with an appropriate reagent. The purpose of such reagents is either to synchronize the cells in prophase, or to inhibit condensation of chromosomes. It is clear that some of the reagents employed have both effects to some extent. Because there is no reagent which will halt the mitotic process at prophase, as colchicine and other substances will do at metaphase, it is necessary to synchronize the cells in some way, often by inhibiting DNA synthesis. The cells then accumulate at the start of S phase. By changing the medium to one lacking the inhibitor the cells can be induced to start DNA replication together and can be harvested after an appropriate time interval, when the cells will have progressed through to prophase. The timing is quite critical for good results. Procedures involving substances that inhibit chromosome condensation are generally rather simpler, the substance merely being added to the culture in an appropriate concentration.,

None of the methods so far published appears to be universally satisfactory so that instead of describing the detailed protocol the main techniques will be listed with appropriate references. The user should be prepared to experiment to determine the most effective procedure for the material being studied.

5.3.1 *Synchronization with methotrexate (38)*

Cells are cultured for an appropriate time before adding methotrexate (amethopterin; Sigma) to prevent DNA synthesis. The methotrexate block is released by placing the cells in fresh medium supplemented with thymidine, and allowing the cells to grow for a few hours until they have reached prophase. Modifications appropriate for bone marrow cultures are described by Harrison (13). Chromosome preparations synchronized using methotrexate can be banded successfully simply by staining with Wright's stain.

5.3.2 *Synchronization with BrdU (39)*

The procedure is essentially the same as that using methotrexate (Section 5.3.1) with the substitution of BrdU (Sigma) for methotrexate. However, as well as synchronizing the cells, BrdU also has a useful tendency to inhibit chromosome condensation. Various standard banding methods can be used after this procedure, such as ASG (Section 4.2.1) and R-banding (Section 4.3); in addition, the fact that BrdU is incorporated into the

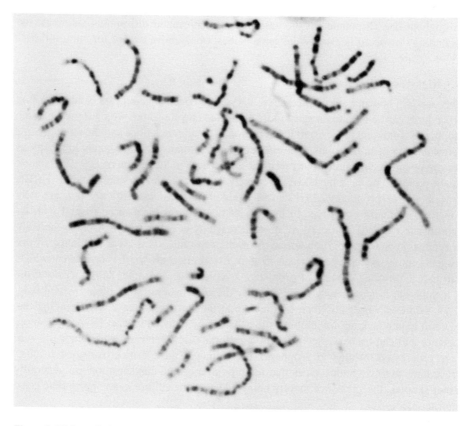

Figure 9. High-resolution banding of human chromosomes. Note the much larger number of bands compared with metaphase chromosomes (*Figure 3*). Micrograph kindly provided by Mr G.Spowart. Reprinted from Sumner,A.T. (1983), *Science Progress*, **68**, pp. 543–564.

DNA during this type of procedure means that the FPG method (35) can also be applied to obtain banding patterns (see also Section 5.2).

5.3.3 *Synchronization with FdU (13,40)*

Again FdU (Sigma) is used in place of methotrexate. It does, however, have the advantage that the block to DNA synthesis can simply be released by adding thymidine, which allows synthesis to proceed with the FdU still present. Standard banding methods can be used after this procedure. Alternatively, a mixture of BrdU and Hoechst 33258 can be used instead of thymidine to release the block to DNA synthesis, so that BrdU is incorporated into the chromosomal DNA and banding can be obtained with the FPG procedure (35). Both the BrdU and the Hoechst 33258 tend to inhibit condensation of the chromosomes.

5.3.4 *Inhibition of chromosome condensation using mercaptoethanol (41)*

Lymphocytes are cultured in the usual way, but at the end of culture, the cells are treated with hypotonic solution consisting of 0.075 M KCl and 0.075 M 2-mercaptoethanol at 37°C for 10 min. After this, colcemid is added for a further 10 min and chromosome preparations made in the usual way (Section 3). Chromosomes can be banded using trypsin–Giemsa (Section 4.2.2) or quinacrine (Section 4.4).

5.3.5 *Inhibition of chromosome condensation using ethidium bromide (42)*

Standard human lymphocyte cultures are exposed for 2 h before harvesting to 5–10 μg/ml of ethidium bromide (BDH) plus 0.02 g/ml of colcemid. Hypotonic treatment, fixation and making of chromosome spreads are all carried out by standard methods (Section 3), and G-banding can be obtained using the trypsin–Giemsa method (Section 4.2.2). Satisfactory Q-banding (Section 4.4) is also possible.

The results of HRB of human chromosomes are shown in *Figure 9*. This shows simultaneously the advantages and disadvantages of HRB. Although much more detail is visible compared with banding of metaphase chromosomes (*Figure 3*), the large amount of detail can be a problem. The patterns on the chromosomes are less clear, making identification more difficult, and because of the large number of bands it is quite likely that differences between homologues may occur purely due to technical factors. A large number of chromosome spreads may have to be examined to ensure that these are genuine differences between homologues. A further problem is that the greater length of the chromosomes makes it certain that there will be more overlapping chromosomes than would be found in a metaphase spread. Nevertheless, HRB has already proved to be valuable in identifying small deletions, particularly in neoplastic cells.

5.4 **Kinetochore labelling with CREST serum**

Methods for staining the kinetochores of chromosomes with Giemsa or with silver have been published but have not been found particularly successful. However, serum from patients with an autoimmune disease, the CREST variant of scleroderma, contains antibodies which react specifically with kinetochores (7). The immunocytochemical method given here is the method of choice for labelling kinetochores but cannot be regarded as routine for several reasons:

(i)   because the kinetochore antigen is very labile, normal chromosome preparations fixed with methanol–acetic acid cannot be used;
(ii)   autoimmune CREST serum used for this method is not widely available and is unobtainable commercially;
(iii)   different sera appear to be specific for different kinetochore antigens and may also label other, non-kinetochore antigens, making interpretation difficult;
(iv)   the quality of the metaphase spreads is generally low.

Nevertheless, the procedure is useful for studying kinetochores, particularly in dicentric chromosomes in which one of the centromeres may have active or 'inactive' (unlabelled) kinetochores.

(i)   Preparation of chromosome spreads: cells are cultured in the normal way, and treated with hypotonic solution to swell the cells (Section 3). Without fixation, the cells are then centrifuged on to glass slides using a Cytospin centrifuge (Shandon Southern Instruments). Remove the slides from the centrifuge and allow them to dry.

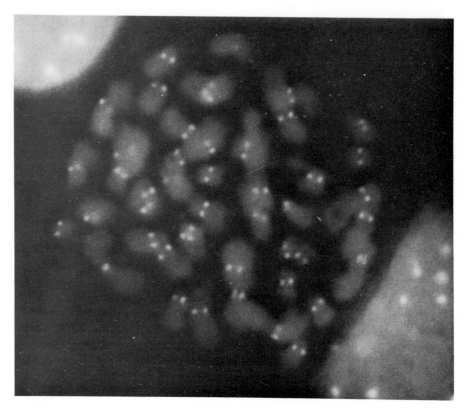

Figure 10. Immunofluorescent labelling of kinetochores of human chromosomes with CREST serum, followed by FITC-labelled anti-human IgG. Counterstained with ethidium bromide. Micrograph kindly provided by Mr G.Spowart.

(ii)   Fix the chromosome spreads in either pure acetone or pure methanol for 10 min at −20°C in a freezer and allow the slides to dry after fixation.

(iii) Incubate the chromosome preparations for 30 min at room temperature in CREST serum diluted with PBS (Dulbecco's solution A, Oxoid Ltd). It may be possible to obtain samples of CREST sera from hospital rheumatology departments. Although almost all such sera will label kinetochores, some also label other cellular components so the specificity must be tested for each sample. The optimal dilution for any serum must be found by experience but try 1:500 to start with. Place a drop of the diluted serum over the chromosome preparation and cover with a coverslip. The slide should then be placed in a moist chamber (a plastic container of suitable size with its bottom covered with moist filter paper).

(iv)   Rinse the slides with three lots of PBS and shake off excess fluid.

(v)   Incubate with a drop of FITC−conjugated anti-human-IgG, diluted 1:5 with PBS, for 30 min at room temperature in a moist chamber, as above (iii).

(vi)   Rinse the slides with three lots of PBS.

(vii) Mount the slides with Citifluor PBS mountant (AF3; Citifluor Ltd) and seal the coverslip with rubber solution (Section 4.4.1). This mountant is designed to inhibit the fading of fluorescein fluorescence.

It is sometimes possible to locate the chromosomes by their faint non-specific fluorescence, but it is better to counterstain them with a DNA-specific fluorochrome. Suitable counterstains are DAPI (Section 5.1), Hoechst 33258 (Section 5.2) and ethidium. Chromosomes may be stained with a solution of 10 mg ethidium bromide/ml of PBS for 5 min and rinsed in PBS before mounting. With DAPI or Hoechst 33258, DNA fluorescence is excited by UV light and the fluorescence of the chromosomes and kinetochores is not visible simultaneously. Both ethidium and FITC fluorescence are excited by blue light so that both the chromosomes and kinetochores are visible together (*Figure 10*). This is the most satisfactory arrangement but it is important not to overstain the chromosomes so that the ethidium fluorescence swamps that of the kinetochores. It may be necessary to dilute the ethidium bromide solution or reduce the staining time if this is a problem.

6. OBSERVATION AND RECORDING OF CHROMOSOME BANDS

Many of the features which can be demonstrated using chromosome banding techniques are at, or near, the limit of resolution of normal light microscopy. It follows, therefore, that a good quality microscope, equipped with high quality lenses and properly adjusted, is necessary for the observation of banded chromosomes. A high quality photographic attachment, or built-in camera, is also required. Not only are photographs required for publication but they are also generally used for the initial analysis of the karyotype. It is generally easier to compare and match up individual chromosomes cut out from a photographic print than to attempt such comparison by direct observation down the microscope.

Practical details of how to set up a light microscope to obtain optimal results have been given elsewhere (43; see also Chapter 1), and no attempt will be made to repeat this information here. Instead, this account will concentrate on aspects of microscopy specific to the banding methods described here. In general, the requirements for observations using absorbing dyes and for fluorescence are rather different but one point common to all observations should be made here. This is the desirability of using flat-field objectives for all observations of banded chromosomes. Not only is a flat-field objective necessary for photography (otherwise parts of the metaphase spread will be out of focus) but it is also helpful for direct observations, for the same reason. It is excessively tedious if the focus has to be adjusted continually to observe chromosomes in different parts of the field and, of course, comparison between different chromosomes becomes much more difficult if only one is in focus at a time.

6.1 Observations of banding with absorbing dyes

For the purpose of this chapter, the absorbing dyes are Giemsa and silver. Ideally, well-banded chromosome preparations should have just the right density of staining and contrast for comfortable observation and satisfactory observation, although in practice this is not always so. In some cases it may be possible to re-stain the slides, or in the case of Ag−NOR staining, to counterstain. Sometimes, however, this is impracticable or the re-staining does not produce the necessary improvement. It is therefore helpful to have available a quick method for assessing the quality of the preparation before it is mounted (so that it can be re-stained if necessary), and to have optical methods for improving the density and contrast of less than perfect specimens.

A low-power (×10) objective is necessary to locate metaphase spreads on the slide, where they may be quite rare compared with the interphase nuclei. Such a low power objective is, however, totally inadequate to see any banding. With a ×40 objective it is usually possible to see whether the chromosomes are banded or not but this magnification is inadequate for assessing the detailed quality of the preparations, let alone for analysing them. Even though certain plants and amphibia have large chromosomes that can be seen clearly at lower magnification it is still necessary to analyse them at the highest possible resolution to ensure that no details of banding are overlooked. Thus it is normally necessary to use a ×90 or ×100 oil-immersion lens to assess the quality of a preparation. This means that if the slide needs re-staining or any other further treatment (e.g. double staining with different banding techniques), the oil must be washed off the slide thoroughly first. This can be done by placing the slide in a Coplin jar of xylene until there is no visible trace of oil on the slide, washing in more xylene and then allowing the slide to dry in a fume cupboard. An alternative objective lens that has been recommended for cytogenetic work is the Zeiss Epiplan ×80 objective (32). This is a dry objective with a short working distance which must be used without a coverslip.

The contrast of banded chromosomes can be improved in two ways: by the use of phase contrast microscopy or by the use of appropriate filters. Differential interference (Normarski) contrast is not suitable since it only improves contrast in one direction at a time, whereas chromosomes lie in all directions. Phase contrast has been used for improving the visibility of R-bands which are sometimes rather pale. It can also be used with Ag−NOR staining, where the NORs themselves are black but the chromosome arms pale yellow and not easily seen. The visibility of the chromosomes can be greatly increased in this way; the Ag−NORs no longer appear black but as bright objects standing out against a darker background. It must be emphasized that phase contrast will only improve contrast if there is some staining, however faint, in the chromosome; since the specimens are mounted in a medium whose refractive index is intended to match that of the chromosomes, phase contrast will not produce an image with completely unstained specimens.

The use of filters to increase contrast is well established and choosing the correct filter is simply a matter of matching its transmission to the absorbance spectrum of the dye. In the case of Giemsa-stained chromosomes maximum absorbance is at about 550 nm and a green filter is required. For most purposes a broad band filter is adequate but for greater contrast a narrow band interference filter can be used. Use of a green filter helps to eliminate eye complaints in microscopists who have to examine chromosomes for prolonged periods (44). If the Giemsa staining is too dense the contrast can be reduced and the bands made more visible by using a red filter. This is, however, very tiring to work with for more than a few minutes and in such circumstances it is better to destain the slide and stain it again less strongly. With Ag−NOR preparations, the chromosome arms are stained yellow and the contrast can be improved by using a light blue filter. A deep blue filter may give greater contrast but, like a red filter, is very tiring to work with.

6.2 Observations of fluorescent banded chromosomes

Fluorescent banded chromosomes usually give a relatively low level of fluorescence and therefore a fluorescence microscope of the highest quality is required. The characteristics of such a system are described in detail elsewhere (45; Chapter 6). The important features are the greatest possible efficiency of illumination and of collection of the fluorescent light, which are achieved by using epi-illumination (incident illumination through the objectives) and objectives of the highest possible light transmission. Such objectives are available from the major microscope manufacturers specially for fluorescence work and should be used whenever possible. Unfortunately not everyone will have access to the best possible microscopes but it must be emphasized that with fluorescence the quality of the result is closely related to the quality of the microscope and some details will not be seen with inferior equipment.

Fluorescence microscopy requires the use of exciter filters, to ensure that only light of the required wavelengths reaches the specimen to excite fluorescence, and barrier filters, to cut out the exciting light and to transmit only the fluorescence. With epi-illumination these filters are combined with a dichroic mirror which supplements the effects of the filters. It has become apparent that for optimal results it is necessary to match the system of filters and dichroic mirror as closely as possible to the characteristics of the fluorochrome being used. Use of the wrong combination may result in excessive non-specific fluorescence and invisibility of the specific fluorescence that it is hoped to see. Nowadays the major microscope manufacturers incorporate the dichroic mirror and the exciter and barrier filters in a single module, and offer a wide variety specifically tailored to the characteristics of different fluorochromes. Once again, use of less sophisticated systems will mean that some features of the specimen will not be seen properly. The filter modules recommended by different manufactuers are listed in *Table 4*; unfortunately there is no standard system of nomenclature and the modules cannot be exchanged between different makes of microscopes.

A serious problem in all fluorescence microscopy is fading of the fluorescence during illumination. Fluorochromes used for chromosomes are subject to fading as much as any other fluorochromes. As already mentioned, fading of FITC fluorescence can be reduced by the use of an appropriate mountant (Section 5.6), and DAPI fluorescence becomes more stable if the slides are stored overnight before observation (Section 5.1). No method is known for inhibiting the fading of quinacrine fluorescence, which emphasizes the importance of using a microscope, and particularly objectives, with high

Table 4. Recommended filter modules for fluorescence with the fluorochromes described in this chapter.

Microscope manufacturer	Fluorochrome		
	DAPI	FITC	Quinacrine
Leitz	A	I2/3	E3
Nikon	UV−1A	B−2A,E or H	BV−1A
Olympus	A		B
Reichert	IU1	IB2	IV2
Zeiss (Oberkochen)	02	09	06

light transmission. This is particularly important for photography when it is vital to record the image before it fades.

6.3 Photography of banded chromosomes

Photography of banded chromosomes is required to provide a permanent record, for publication, and, particularly, for detailed analysis of banding patterns. In the case of fluorescent patterns analysis this is necessarily done on photographic prints since it is not practicable to do the analysis under the microscope before the fluorescence fades. A standard method of analysing karyotypes is to cut the chromosomes out from a print and match them up in pairs; this, of course, applies equally to fluorescent and non-fluorescent chromosomes.

General principles for photographing banded chromosomes have been described by Davidson (46). The first requirement is that the microscope should be set up correctly (43; see also Chapter 3). Whether a simple manual camera or a fully automatic camera is being used it is necessary to take a test strip both when using new equipment and when trying out a new type of film or developer. Different exposures are made by varying the exposure time with a manual camera or by varying the film speed setting with an automatic camera. A range of speeds both faster and slower than those recommended by the manufacturer should be tested. All data on lamp voltages, filters, exposure times and of course type of film and method of development must be recorded for future reference.

A suitable film for general use for photographing banded chromosomes of all types (G-, C-, R- and Q-banding) is Kodak Technical Pan developed in accordance with the manufacturer's instructions. For dim fluorescence, however, this film may be too slow and a faster film, such as Kodak Tri-X, should be used. Such films have a coarser grain and are therefore less satisfactory for recording fine detail. Nevertheless, this may be the only way to obtain an image of weakly fluorescent objects. Apart from the problem of fading of fluorescence, which necessitates the use of faster films, there is also the problem of reciprocity failure with dim objects requiring long exposures. In essence, this means that the longer the exposure, the slower the film speed. For very weak fluorescence, therefore, it may be necessary to rate the film at half or a quarter of the speed used for brighter objects. Again it is best to expose a test strip to determine the best exposure for this type of specimen.

Recently Kodak has introduced a film called T-max, which combines fine grain with high speed (400 ASA). This film evidently has great potential for photomicrography, and should be greatly superior to traditional films for photographing weak fluorescence.

6.4 Specialized procedures for investigating chromosome bands

So far this chapter has covered the preparation of banded chromosomes, their direct observation and photographic recording. In this section attention is drawn to some other methods, not in such widespread use, of obtaining further information from banded chromosomes. There is no standard procedure for any of these methods and therefore only general principles will be given, with reference to further information.

6.4.1 *Preparation of banding profiles*

A useful way of displaying banding patterns of chromosomes is by the preparation of banding profiles, that is, the graphical representation of the variations of staining density or fluorescence along the chromosomes. Such profiles are more convenient for making comparisons between chromosomes, permit the detection of small alterations in the banding patterns and are the basis for the quantitative and automated study of banding patterns.

Banding profiles may be obtained directly from the stained preparations, or, especially in the case of fluorescent chromosomes, from photographs. In the latter case, because of the characteristics of photographic emulsions, the heights of the peaks on the profiles will not be directly related to the intensity of fluorescence. The profiles can be obtained as a simple line scan long the axis of the chromosome, or the whole chromosome can be digitized as a 2-dimensional map for further processing. The former is simpler and often adequate but because it only uses part of the available information and is highly susceptible to minor deviations between the scan line and the axis of the chromosome, complete digitization is theoretically more satisfactory.

A wide variety of equipment has been used for preparing banding profiles. In the past the equipment has often been specially constructed or adapted, although standard instruments can be used. The important feature is that the sampling distance of the equipment should be sufficiently small; the distance between adjacent points measured may be as small as 0.05 μm (47).

Satisfactory banding profiles have been obtained in this laboratory using a basic machine, the Vickers M85 Microdensitometer. This machine can be connected to an $x-y$ plotter, the whole being set up as follows. The microdensitometer is set up to give the highest resolution, including the use of an achromatic oil-immersion condenser. The chosen chromosome is orientated so that the scanning spot can be scanned along its axis; either the x or the y scan direction can be chosen. The smallest scanning spot, a nominal 0.2 μm with the $\times 100$ objective, must be used for maximum resolution. The objective is a flat-field achromat ('Microplan'); since the measurements are made with monochromatic light, the extra colour correction afforded by an apochromatic objective is unnecessary. The spot is scanned along the chromosome slowly by hand. An output proportional to the position of the spot on the x or the y axis of the microdensitometer drives the x axis of the recorder, while the density output of the microdensitometer drives the y axis of the recorder. Typical results are illustrated in *Figure 11*. With a suitable adaptor placed in the plane of the field iris, the same instrument can be used to obtain profiles from photographic negatives (48).

Many of the modern image analysing systems now on the market are no doubt capable of producing banding profiles but, as with any system, the resolution is of vital importance. With image analysers this includes the television camera and the associated image processing electronics as well as the microscope itself. Before purchasing an image analysing system for preparing banding profiles, the intending user should obviously test its abilities for this purpose. The types of manipulations that can be done on the profiles once they have been obtained are obviously also important. It is particularly useful to be able to compare different chromosomes within cells and with stored data. It should be possible to normalize the lengths of homologous chromosomes from different cells, which will be contracted to different degrees. A particularly valuable feature is the ability to obtain profiles from bent chromosomes. Since few metaphases lack bent chromosomes, an inability to cope with them is a serious limitation. For an illustration of what can be done using sophisticated equipment see Piper (49).

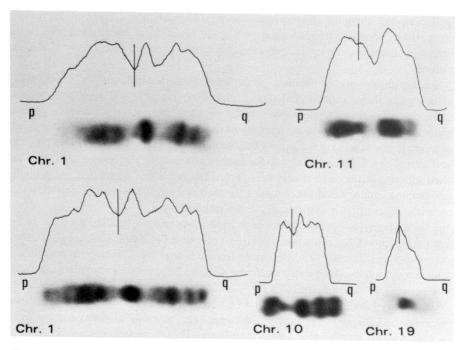

Figure 11. Optical density profiles of selected G-banded human chromosomes from a variety of cells. The vertical lines on the profiles indicate the position of the centromeres.

6.4.2 *Reflectance microscopy*

Pawlowitzki and his colleagues have described the use of oblique incidence reflectance microscopy for studying banded chromosomes. Essentially, the chromosomes are treated briefly with trypsin to remove the layer of protein covering them, and banded by an appropriate method. The chromosomes can then be covered with a thin layer of gold by vacuum evaporation, although this can be omitted if sufficient intensity of light is available. The preparations are examined unmounted, without a coverslip, using equipment based on the Leitz Opak system (50). The authors report that greater detail is visible by reflectance microscopy than by transmitted light.

6.4.3 *Measurement of chromosomal polymorphisms*

Polymorphisms or heteromorphisms (that is, differences between the homologues) are very common. They involve C-bands, where size polymorphisms appear to be universal throughout the animal and plant kingdoms; Ag−NOR bands, which also show extensive size polymorphisms; and heterochromatic Q-bands, which in human chromosomes show polymorphism of brightness.

Chromosome polymorphisms have been investigated in relation to mental retardation, reproductive wastage and cancer in man, although without any very clear correlations emerging. In mice, differences in C-bands are characteristic of different strains. The size of Ag−NOR bands is correlated to some extent with the number of rRNA genes. As practical tools, banding polymorphisms have applications as markers in paternity testing, tracing the origin of the extra chromosome or chromosome set in trisomy or triploidy, as markers in linkage studies and as markers of transplanted tissue.

Measurement of banding polymorphisms present particular problems. Firstly, there is the small size of the bands, so that the measurements have to be made near the limit of optical resolution and are therefore inherently subject to substantial errors. Secondly, the staining reactions are variable, so that a C-band may appear larger or smaller depending on its density of staining. This effect can be exacerbated in photographs. Thirdly, heterochromatic bands do not condense at the same rate as the euchromatic parts of the chromosomes, so that the relative sizes of the bands varies with the degree of contraction of the chromosomes.

Many workers do not attempt to measure the size of polymorphic C-bands, but make subjective estimates. The official recommendations for human chromosome polymorphisms (51) suggest classification into five size categories, although it is often difficult to recognize more than three (small, medium and large). In any case, in the absence of some standard, comparisons between individuals are difficult. The short arm of chromosome 16 (16p) has been proposed as a standard (52) and C-bands classified by their size in relation to 16p. Similar standards have not, apparently, been proposed for other species, although this could be done in many cases and is to be recommended if truly quantitative methods cannot be used.

It is possible to make length (53) or DNA (54) measurements of whole chromosomes containing a variable C-band and make comparison either with a homologue, or with some other chromosome not containing a significantly variable C-band [in the human karyotype, chromosome 2, or the F group (19−20)]. Although the variable band itself is not measured, methods of this type, particularly in conjunction with C-banding, are very useful.

Most authors have attempted to make direct measurements of C-band size in relation to the size of the whole chromosome. Such measurements may be obtained from simple profiles (55), or from integrated measurements over the whole chromosome (56). Recently, image analysers have been applied to the same problem (57). Whatever the method employed, there is a particular difficulty in addition to those described above, which is where to draw the boundary of the C-band. A comparison of various approaches to this problem was made by Mason *et al.* (56).

In some cases, C-bands are seen to be divided into subunits, which may be counted as a measure of C-band size (58). Wegner and Pawlowitzki (59) have proposed the use of reflectance microscopy to demonstrate these subdivisions. The method of counting subunits of C-bands is objective and does not have the same disadvantages as measurements. At the same time it is not clear that the subunits are all equivalent (58) and in most cases C-bands do not appear to be broken up into subunits.

The measurement of Ag−NORs differs from that of C-bands in two ways. An advantage is the high degree of contrast between the black NORs and the pale yellow surrounding area, so that problems due to variation in the level of staining are reduced. On the other hand, the NORs are extremely small, making measurement even more difficult. Successful measurements have been made with a high-resolution image analysis system, the Zeiss Micro-Videomat (Zeiss Oberkochen) (60). With this equipment, areas of silver-stained material were measured with an error of only 1%.

The problem of measuring quinacrine fluorescence polymorphisms of human chromosomes is different from that of measuring C-band or Ag−NOR polymorphisms, since the Q-band polymorphisms are differences in brightness, not in area. Subjective comparison with other, non-polymorphic chromosomal regions is recommended (51) but suffers from possible variations in the general level of fluorescence across the metaphase spread. If Q-band polymorphisms are being assessed from photographs it is important to ensure that information is not lost during photographic processing. Overton *et al.* (63) recommend printing the photographs at a series of different exposures so that Q-bands of all intensities can be properly recorded. Schnedl *et al.* (61) described the use of banding profiles for measuring the brightness of polymorphic regions. This is evidently a useful method, although the measured values will not be proportional to the actual fluorescence intensity unless special precautions are taken. For a full discussion of the quantification of fluorescence from photographic negatives, see van der Ploeg *et al.* (62).

It is clear that there is no universally acceptable method of measuring banding polymorphisms, although the importance of using quantitative data instead of subjective impressions means that measurement should be attempted if at all possible. High resolution image analysis systems, now becoming widely available, appear to have useful applications in this field. Nevertheless, the results depend upon the quality of the initial staining, the setting up of the microscope to give optimal results and, if a photographic process is involved, the faithfulness with which the image is reproduced. Understanding the processes involved and attention to detail to obtain the best results, are essential if reliable quantification is to be obtained.

7. ACKNOWLEDGEMENTS

I should like to thank Dr John Gosden for the method of preparing spermidine bis-acridine; Mr Norman Davison for advice on photographic techniques; and Mr George Spowart for general advice on banding techniques. Several microscope manufacturers kindly provided relevant information on their fluorescence microscopes. I am particularly grateful to Mrs Ann Kenmure for typing the manuscript and to the photographic department, MRC Human Genetics Unit, for preparing the illustrations.

8. REFERENCES

1. Sumner,A.T. (1982) *Cancer Genet. Cytogenet.*, **6**, 59.
2. Heitz,E. (1929) *Ber. Dtsch. Bot. Ges.*, **47**, 274.
3. John,B. and King,M. (1977) *Chromosoma*, **65**, 59.
4. Camacho,J.P.M., Viseras,E., Navas,J. and Cabrero,J. (1984) *Heredity*, **53**, 167.
5. Ochs,R.L. and Busch,H. (1984) *Exp. Cell Res.*, **152**, 260.
6. Markovic,V.D., Worton,R.G. and Berg,J.M. (1978) *Hum.Genet.*, **41**, 181.
7. Moroi,Y., Peebles,C., Fritzler,M.J., Steigerwald,J. and Tan,E.M. (1980) *Proc. Natl. Acad. Sci. USA*, **77**, 1627.
8. Pfeiffer,R.A. (1974) In Schwarzacher,H.G., Wolf,U. and Passarge,E. (eds), *Methods in Human Cytogenetics*. Springer-Verlag, Berlin, p. 1.
9. Zackai,E.H. and Mellman,W.J. (1974) In Yunis,J.J. (ed.), *Human Chromosome Methodology*. Academic Press, New York, 2nd edition, p. 95.
10. Macgregor,H.C. and Varley,J.M. (1983) *Working with Animal Chromosomes*. John Wiley and Sons, Chichester.
11. Watt,J.L. and Stephen,G.S. (1986) In Rooney,D.E. and Czepulkowski,B.H. (eds), *Human Cytogenetics: A Practical Approach*. IRL Press, Oxford, p. 39.

12. Worton,R.G. and Duff,C. (1979) In Jakoby,W.B. and Pastan,I.H. (eds), *Methods in Enzymology*. Academic Press, New York, Vol. 58, p. 322.
13. Harrison,C.J. (1986) In Rooney,D.E. and Czepulkowski,B.H. (eds), *Human Cytogenetics: A Practical Approach*. IRL Press, Oxford, p. 135.
14. Darlington,C.D. and La Cour,L.F. (1969) *The Handling of Chromosomes*. George Allen and Unwin, London, 5th edition, p. 43.
15. Schwarzacher,T., Ambros,P. and Schweizer,D. (1980) *Plant Syst. Evol.*, **134**, 293.
16. Evans,E.P., Breckon,G. and Ford,C.E. (1964) *Cytogenetics*, **3**, 289.
17. Breckon,G. (1982) *Stain Technol.*, **57**, 349.
18. Hungerford,D.A. (1971) *Cytogenetics*, **10**, 23.
19. Sumner,A.T. (1972) *Exp. Cell Res.*, **75**, 304.
20. Chandley,A.C. and Fletcher,J.M. (1973) *Humangenetik*, **18**, 247.
21. Sumner,A.T., Evans,H.J. and Buckland,R.A. (1971) *Nature New Biol.*, **232**, 31.
22. Seabright,M. (1972) *Chromosoma*, **36**, 204.
23. Gallimore,P.H. and Richardson,C.R. (1973) *Chromosoma*, **41**, 259.
24. Sehested,J. (1974) *Humangenetik*, **21**, 55.
25. Buckton,K.E. (1976) *Int. J. Radiat. Biol.*, **29**, 475.
26. van de Sande,J.H., Lin,C.C. and Deugau,K.V. (1979) *Exp. Cell Res.*, **120**, 439.
27. Howell,W.M. and Black,D.A. (1980) *Experientia*, **36**, 1014.
28. Tantravahi,R., Miller,D.A. and Miller,O.J. (1977) *Cytogenet. Cell Genet.*, **18**, 364.
29. Howell,W.M. and Black,D.A. (1978) *Human Genet.*, **43**, 53.
30. Schweizer,D., Ambros,P. and Andrle,M. (1978) *Exp. Cell Res.*, **111**, 327.
31. Donlon,T.A. and Magenis,R.E. (1983) *Human Genet.*, **65**, 144.
32. Benn,P.A. and Perle,M.A. (1986) In Rooney,D.E. and Czepulkowski,B.H. (eds), *Human Cytogenetics: A Practical Approach*. IRL Press, Oxford, p. 57.
33. Schmidt,M., Stolzmann,W.M. and Baranovskaya,L.I. (1982) *Chromosoma*, **85**, 405.
34. Willard,H.F. (1977) *Chromosoma*, **61**, 61.
35. Perry,P. and Wolff,S. (1974) *Nature*, **251**, 156.
36. Wolff,S. (1977) *Annu. Rev. Genet.*, **11**, 183.
37. Limon,J. and Gibas,Z. (1985) In Sandberg,A.A. (ed.), *The Y Chromosome, Part A: Basic Characteristics of the Y Chromosome*. Alan R.Liss, New York, p. 317.
38. Yunis,J.J. (1976) *Science*, **191**, 1268.
39. Dutrillaux,B. and Viegas-Pequignot,E. (1981) *Human Genet.*, **57**, 93.
40. de Braekeleer,M., Keushnig,M. and Lin,C.C. (1985) *Can. J. Genet., Cytol.*, **27**, 622.
41. Kao,Y.S., Whang-Peng,J. and Lee,E. (1983) *Am. J. Clin. Pathol.*, **79**, 481.
42. Ikeuchi,T. (1984) *Cytogenet. Cell Genet.*, **38**, 56.
43. Bradbury,S. (1984) *An Introduction to the Optical Microscope*. Oxford University Press, Oxford.
44. Wulf,H.C. (1983) *Mikroskopie*, **40**, 1.
45. Ploem,J.S. and Tanke,H.J. (1987) *Introduction to Fluorescence Microscopy*. Oxford University Press, Oxford.
46. Davidson,N.R. (1973) *J. Med. Genet.*, **10**, 122.
47. Wayne,A.W. and Sharp,J.C. (1981) *J. Microsc.*, **124**, 163.
48. Sumner,A.T. and Finlayson,D. (1978) *J. Microsc.*, **114**, 85.
49. Piper,J. (1982) *Anal. Quant. Cytol.*, **4**, 233.
50. Enk,D. and Pawlowitzki,I.H. (1987) *J. Microsc.*, **143**, 301.
51. ISCN (1978) *Cytogenet. Cell Genet.*, **21**, 309.
52. Patil,S.R. and Lubs,H.A. (1977) *Human Genet.*, **38**, 35.
53. Cohen,M.M., Shaw,M.W. and MacCluer,J.W. (1966) *Cytogenetics*, **5**, 34.
54. Sumner,A.T. (1977) *Cytogenet. Cell Genet.*, **19**, 250.
55. Cavalli,I.J., Mattevi,M.S., Erdtmann,B., Sbalqueiro,J.J. and Maia,N,A. (1985) *Human Hered.*, **35**, 379.
56. Mason,D., Lauder,I., Rutovitz,D. and Spowart,G. (1975) *Comput. Biol. Med.*, **5**, 179.
57. Lopetegui,P.H. (1980) *Japan J. Hum. Genet.*, **25**, 29.
58. Drets,M.E. and Seuanez,H. (1974) In Coutinho,E.M. and Fuchs,F. (eds), *Physiology and Genetics of Reproduction*, Part A. Plenum Publishing Corporation, New York, p. 29.
59. Wegner,H. and Pawlowitzki,I.H. (1981) *Human Genet.*, **58**, 302.
60. Schmid,M., Löser,C., Schmidtke,J. and Engel,W. (1982) *Chromosoma*, **86**, 149.
61. Schnedl,W., Roscher,U. and Czaker,R. (1977) *Human Genet.*, **35**, 185.
62. van der Ploeg,M., Vossepoel,A.M., Bosman,F.T. and van Duijn,P. (1977) *Histochemistry*, **51**, 269.
63. Overton,K.M., Magenis,R.E., Brady,T., Chamberlin,J. and Parks,M. (1976) *Am. J. Human Genet.*, **28**, 417.

9. FURTHER READING

MacGregor,H.C. and Varley,J.M. (1983) *Working with Animal Chromosomes*. John Wiley and Sons, Chichester.

Rooney,D.E. and Czepulkowski,B.H. (1986) *Human Cytogenetics: A Practical Approach*. IRL Press, Oxford.

APPENDIX

Suppliers of specialist items

AMS, Shirehill, Saffron Walden, Essex CB11 3AQ, UK
AVT GmbH, D-7080 Aalen, FRG
Agfa-Gavaert, Brentford, Middlesex, UK
Aldrich Chemical Co. Ltd, The Old Brick Yard, New Road, Gillingham, Dorset SP8 4JL, UK
Amersham International plc, Little Chalfont, Amersham, Bucks HP7 9NA, UK
Arnold and Richter Cine Technik GmbH, D-8000 München, FRG
Azlon Products Ltd, 172 Brownhill Road, London SE6 2DL, UK
BDH Chemicals Ltd, Broom Road, Poole, Dorset BH12 4NN, UK
BRL-Gibco Ltd, PO Box 35, Renfrew Road, Paisley PA3 4EF, UK
Bio-Rad Lasersharp Ltd, 7 Suffolk Way, Drayton Road, Abingdon, Oxford OX14 5JX, UK
Biovision (Perceptics Corp.), Knoxville, TN, USA
Brian Reece Scientific Instruments, Newbury, Berks RG14 5HG, UK
Buehler UK Ltd, Coventry CV4 9XJ, UK
C Z Scientific Instruments Ltd, Borehamwood, Herts WD6 1NH, UK
C.Zeiss Jena—See C Z Scientific Instruments Ltd
Cambridge Instruments Ltd, Cambridge CB3 8EL, UK
Carl Zeiss (Oberkochen) Ltd, PO Box 78 Woodfield Road, Welwyn Garden City, Herts, UK
Cinema Products Corporation, Los Angeles, CA 90025, USA
Citifluor Ltd, Connaught Building, Northampton Square, London EC1V 0HB, UK
Clay Adams Co., Parsippany, NJ, USA
Cohu Inc., San Diego, CA 92138-5623, USA
Colorado Video Inc., Boulder, CO 80306, USA
Dage-MTI Inc., Michigan City, IN 46360, USA
Dakopatts Ltd, High Wycombe, Bucks HP1 12HT, UK
Data Translation Inc., Marlboro, MA 01752, USA
Datacube Inc., Peabody, MA 01960, USA
Difco Laboratories Ltd, PO Box 14B, Central Avenue, East Molesey, Surrey KT8 0SE, UK
Diffraction Gratings & Optics Ltd, Chobham, Surrey, UK
E.Leitz (Instruments) Ltd, Luton, Beds LU1 3HP, UK
EOS Electronics AV Ltd, Barry, South Glamorgan, UK
Eastman Kodak Co., Rochester, NY 14650, USA
Emscope Laboratories Ltd, Ashford, Kent TN23 2LN, UK
Enzo Biochem Inc., 325 Hudson Street, New York, NY 10013, USA
Flow Laboratories, Woodcock Hill, Harefield Road, Rickmansworth, Herts WD3 1PQ, UK
For-A Company Ltd, Tokyo 160, Japan
Fritz Hesselbein Chemische Fabrik, D-2000 Norderstedt, FRG
Gallenkamp Group Service Organisation, PO Box 290, Technico House, Christopher St, London EC2P 2ER, UK
Gibco Ltd, PO Box 35, Renfrew Road, Paisley PA3 4EF, UK

Goodfellow Metals Ltd, Cambridge CB4 4DJ, UK
Graticules Ltd, Morley Road, Tonbridge, Kent, UK
Gurr—Available from BDH Ltd
Hamamatsu Photonics K.K., Hamamatsu City, 431–32, Japan—Also marketed by PMI
Heidelberg Instruments GmbH (marketed by Leitz), D-6900 Heidelberg, FRG
Holdtite Stick on Soles Ltd, Petersfield, Hants, UK
Hughes Aircraft Corp. (marketed through Dage-MTI), Carlsbad, CA 92008, USA
ICN Biomedicals Ltd, Lincoln Road, Cressex Industrial Estate, High Wycombe, Bucks HP12 3XJ, UK
ISI Europe, Buxton, SK17 9JB, UK
Ilford Ltd, Mobberley, Cheshire, UK
Imaging Technology Inc, Woburn, MA 01801, USA
Imaging Transform Inc., 17 Old Church Lane, London NW9 / Hollywood, CA, USA
Institute of Ophthalmology, Judd Street, London, UK
Janssen—See ICN Biomedicals Ltd
Joyce-Loebl, Gateshead NE11 0P2, UK—Also available from Nikon Inc.
Kenda Electronic Systems, Southampton SO4 3NB, UK
Kodak Ltd, Harrow, Middlesex, UK
Matrox Electronic Systems Ltd, Montreal, Quebec H4T 1H4, Canada
McCrone Research Assoc. Ltd, 2 McCrone Mews, Belsize Lane, London NW3 5BG, UK
Micro Instruments (Oxford) Ltd, 7 Little Clarendon Street, Oxford OX1 2HP, UK
Minolta Camera Co. Ltd, Osaka 541, Japan
Minolta (UK) Ltd, 235 Regent St, London W1, UK
Mitec GmbH, D-8011 Hofolding, FRG
Modulation Optics Inc., Greenvale, NY 11548, USA
Molecular Probes Inc., Eugene, OR 94702, USA
Motion Analysis Inc., Santa Rosa, CA 95401, USA
National Diagnositics Ltd, Aylesbury, Bucks, UK
Nikon UK Ltd, Telford, Salop TF7 4EW, UK
Nikon Inc., Garden City, NY 11530, USA
Nyegaard AS, Postbox 4220, Torshov, Oslo 4, Norway
O.Kindler GmbH, D-7800 Freiburg, FRG
Olympus Optical Co. (UK) Ltd, London EC1Y 0TX, UK
Oxoid Ltd, Wade Road, Basingstoke, Hants RG24 0PW, UK
PMI Photonic Microscopy Inc., Oak Brook, IL 60521, USA
Pang (UK) Ltd, Newmarket, Suffolk CB8 7A4, UK / Dunfermline, Fife, Scotland, UK
Pentax UK Ltd, South Hill Ave, Harrow, Middx, UK
Perceptics Corporation, Knoxville, TN 37933-0991, USA
Pharmacia Ltd, PO Box 175, S-75104 Uppsala-1, Sweden
Polaroid (UK) Ltd, St Albans, Herts AL1 5PR, UK
Polaroid Inc., Cambridge, MA 02139, USA
Polaron Ltd, 52 Greenhill Crescent, Watford Industrial Estate, Watford WD1 8QS, UK
Polysciences Ltd, Northampton NN3 1HY, UK
P.Z.O.—Marketed by Aico International, Newbury, Berks RG13 2AD, UK

Quantel, Newbury, Berks RG13 2NE, UK

Quantex Corporation, Sunnyvale, CA 94086, USA

Reichert/Cambridge Instruments and **Reichert-Jung**—Marketed by Cambridge Instruments

Rolyn Optics Co., Covina, CA 91006, USA

Scan Systems, Stockport, Cheshire, UK

Seescan Ltd, Cambridge CB1 2LG, UK

Shandon Southern Products Ltd, Chadwick Road, Astmoor, Runcorn, Cheshire WA7 1PR, UK

Sight Systems, Newbury, Berks RG13 3HD, UK

Sigma Chemical Co. Ltd, Fancy Road, Poole, Dorset BH17 7NH, UK / PO Box 14508, St Louis, MO 63178, USA

Synoptics Ltd, Cambridge CB4 4BH, UK

Tetenal GmbH, D-2000 Norderstedt, FRG

Tracor Europa, Milton Keynes MK8 0AJ, UK

Tracor Northern, Middletown, WI 53562, USA

Triton Instruments Ltd, Market Place, Westbury, Wilts BA13 4Q1, UK

Universal Imaging Corporation, Media, PA 19063, USA

Vickers Instruments, Haxby Road, York YO3 7SD, UK

Whatman Lab. Sales Ltd, Unit 1, Coldred Road, Parkwood, Maidstone, Kent ME15 9XN, UK

INDEX